Algorithms in Structural Molecular Biology

Computational Molecular Biology

Sorin Istrail, Pavel Pevzner, and Michael Waterman, editors

A complete list of the books published in the Computational Molecular Biology series appears at the back of this book.

Algorithms in Structural Molecular Biology

Bruce R. Donald

The MIT Press
Cambridge, Massachusetts
London, England

This book was set in Syntax and Times Roman by Westchester Book Composition.

Library of Congress Cataloging-in-Publication Data

Donald, Bruce R.
Algorithms in structural molecular biology / Bruce R. Donald.
    p.   ;   cm. — (Computational molecular biology)
Includes bibliographical references and index.
ISBN 978-0-262-01559-2 (hardcover : alk. paper)
ISBN 978-0-262-54879-3 (paperback)   1. Biophysics—Mathematical models.
I. Title. II. Series: Computational molecular biology.
[DNLM: 1. Computational Biology—methods.   2. Algorithms.   3. Magnetic Resonance Spectroscopy—methods.
4. Proteins. QU 26.5]
QH505.D66   2011
572′.33—dc22

                                                                    2010040934

Chapter 5 is a revised version of a tutorial article by Anthony Yan and Bruce Donald.

Chapters 15–18 are a revised version of a review article published in an earlier form in *Progress in NMR Spectroscopy* (2009) by Bruce Donald and Jeff Martin.

Some of the figures in this book were adapted from a free use Web image repository called Wikimedia Commons. My thanks to all the contributors to this repository.

For Aleta, Maia, and Jenni

Brief Contents

Contents

Preface

Some of the most challenging and influential opportunities for computational scientists arise in developing and applying information technology to understand the molecular machinery of the cell. The past decade has shown that many novel algorithms may be developed and fruitfully applied to the challenges of computational molecular biology. This research has led to computer systems and algorithms that are useful in structural molecular biology, proteomics, and rational drug design.

Concomitantly, a wealth of interesting computational problems arise in proposed methods for discovering new pharmaceuticals. These include algorithms for interpreting X-ray crystallography and nuclear magnetic resonance (NMR) data, disease classification using mass spectrometry of human serum, and protein design. For example, this book presents chapters on provable algorithms that have recently been used to reveal the enzymatic architecture of organisms high on the CDC bioterrorism watch list (chapter 47), for probabilistic cancer classification from human peripheral blood (chapter 19), and to redesign an antibiotic-producing enzyme to adenylate a novel substrate (chapter 12).

In the postgenomic era, key problems in molecular biology center on the determination and exploitation of three-dimensional protein structure and function. For example, modern drug design techniques use protein structure to understand how a drug can bind to an enzyme and inhibit its function. Structural proteomics will require high-throughput experimental techniques, coupled with sophisticated computer algorithms for data analysis and experiment planning.

Novel computational methods are enabling high-throughput structural and functional studies of proteins. One recent subfield is structural genomics, whose goal is (in the broadest terms) to determine the three-dimensional structures of all proteins in nature, through a combination of direct experiments and theoretical analysis. By determining the structures of proteins, we are better able to understand how each protein functions normally and how faulty protein structures can cause disease. Scientists can use the structures of disease-related proteins to help develop new therapeutics and diagnostic techniques.

At the molecular level, many genes provide the blueprint for proteins, and it remains very expensive and time-consuming to determine what these proteins do, and how they do it. Modern automated techniques are revolutionizing many aspects of biology, for example, supporting extremely fast gene sequencing and massively parallel gene expression testing. Protein structure determination, however, remains a long, hard, and expensive task. High-throughput, automated, algorithmic methods are required to apply modern techniques such as computer-aided drug design on a much larger scale. New algorithms in structural biology and molecular biophysics are advancing our long-range goal of understanding biopolymer interactions in systems of significant biochemical as well as pharmacological interest. For example, to analyze noncrystallographic symmetry in X-ray diffraction data of biopolymers, one must "recognize" a finite subgroup of $SO(3)$ (the Lie group of 3-dimensional rotations) out of a large set of molecular orientations. The problem may be reduced to clustering in $SO(3)$ modulo a finite group, and solved efficiently by "factoring" into a clustering on the unit circle followed by clustering on the 2-sphere S^2, plus some group-theoretic calculations. This yields a polynomial-time algorithm that is efficient in practice, and which recently enabled biological crystallographers to reveal the architecture of a parasite's enzyme (chapter 47), which will help researchers reduce the threat of certain diseases among those with weak immune systems.

The preceding example illustrates an algorithm for *analyzing* biological data and biological problems. However, another family of algorithms can be applied to *synthetic* problems such as protein engineering. Recently developed ensemble-based scoring and search algorithms for protein design have been applied to modify the substrate specificity of an antibiotic-producing enzyme in the nonribosomal peptide synthetase (NRPS) pathway (chapter 12). Realization of novel molecular function requires the ability to alter molecular complex formation. Enzymatic function can be altered by changing enzyme-substrate interactions via modification of an enzyme's active site. A redesigned enzyme may perform either a novel reaction on its native substrates or its native reaction on novel substrates. A number of computational approaches have been developed to address the combinatorial nature of the protein redesign problem. These approaches typically search for the global minimum energy conformation among an exponential number of protein conformations (chapter 11). New algorithms for protein redesign combine a statistical mechanics-derived ensemble-based approach to computing the binding constant with the speed and completeness of a branch-and-bound pruning algorithm. In addition, efficient deterministic approximation algorithms have been obtained, capable of approximating biophysical scoring functions to arbitrary precision. These techniques include provable ε-approximation algorithms for estimating partition functions to model binding and protein flexibility (chapter 12).

These two examples illustrate specific studies where advanced algorithms have provided leverage in biological macromolecular structure determination and in the design of enzymes. This book explains the algorithmic foundations and computational approaches underlying these, and other technical subfields in structural biology, including NMR structural biology; X-ray crystallography; design of proteins, peptides, and small molecules; analysis of protein:protein interactions; modeling of protein flexibility; protein:ligand binding; protein loops; and intrinsically disordered proteins (see table P.1).

This book is based on lectures from a course I teach at Duke University, called "Algorithms in Structural Molecular Biology and Molecular Biophysics." While some undergraduate and

Table P.1
Ten biological and biophysical topics for which algorithms are developed in this book

Topic	Chapters
NMR structural biology	1–8, 13–18, 39, 29, 31, 33–35, 43–45
X-ray crystallography	
Crystallography methodology	40, 47, 48
Analysis and use of X-ray structures	9–12, 14, 20–30, 40, 41, 49, 50.
Design of proteins, peptides, and small molecules	9–12, 25, 46, 45, 27, 28, 30, 41, 26, 49
Protein:protein interactions	9, 10, 23, 27, 28.
Protein flexibility	5, 9–12, 20–26, 30, 41, 42, 25
Protein:ligand binding	9, 10, 12, 25–29, 40, 41, 46
Protein loops	20, 22, 24
Intrinsically disordered proteins	7, 39, 42
Protein kinematics	1, 2, 9, 15–18, 13, 20–24, 30
Modular enzymes	10, 12, 27, 28, 46

graduate students may have had a course in sequence-based computational biology (for example, genomics or even systems biology), in general, the algorithmic content of such a course will not prepare them for the challenges of computational structural biology or algorithms for molecular biophysics. This book is intended to fill that gap, but is a standalone course that does not presume a formal background in computational biology.

I hope the exposition in this book, and the algorithms I describe, will be generally useful both to computer scientists and to the structural biology community, not only for the earlier examples but also in other areas, including studies of protein:ligand binding and protein redesign that are experimentally driven. For example, provable approximation algorithms encode quite general techniques for computer-assisted drug design, and for docking flexible ligands to flexible active sites. I believe there are broad potential applications of these algorithmic techniques for determining structures, modeling protein flexibility, designing proteins, and modeling the biophysical processes of binding and catalysis in protein biochemistry. In each of these topics, computational techniques are central, and the applications present intriguing problems to computer scientists who design algorithms and implement systems. The next generation of computational structural biologists will need training in geometric algorithms, provably good approximation algorithms, scientific computation, and an array of algorithmic techniques for handling noise and uncertainty in combinatorial geometry and computational biophysics. This book is designed to speed young scientists on their way to research success in this exciting endeavor.

How to Use This Book

This textbook is intended for first- or second-year graduate students, or advanced undergraduates (juniors or seniors, in the American system). It should also serve as a reference book to more advanced students and researchers, and as a resource on algorithms important to computational structural biology and molecular biophysics. The book is organized into chapters called "lectures," each of which covers concepts essential and important to the field. Many of the primary research papers on these subjects cover difficult algorithmic problems and are dense, so my goal was to have short chapters that, when carefully read and studied, could be pondered like koans by a student, and provide the foundation for launching her into research.

Each "lecture" focuses on a key topic and is typically much shorter than a review article, covering the computational process and biophysical highlights from the original, primary papers in a compact and (I hope) thought-provoking style. An exception to the brief chapter format is provided in the "short course" in lectures 15–18, covering algorithmic and computational issues related to NMR structural biology. The "short course" allows the student to build up sufficient momentum to master the fascinating yet thorny computational problems that require a sustained attack and uninterrupted exposition. These four lectures could also provide a standalone short course for a laboratory, departmental, or center retreat, or they could be a subtheme of a longer course.

This book interleaves important themes in structural computational biology, including algorithms, NMR, design of proteins and other molecules, and macromolecular flexibility. For difficult topics, a didactic style of increasingly deeper repetition is employed (for example, for Dead-end Elimination in protein design, and residual dipolar couplings in NMR), in which a biophysical theme or algorithm is gently introduced in a brief form, and then periodically revisited in a cyclic and iteratively deeper fashion throughout the book.

A key theme of this textbook is understanding the interplay between biophysical experiments and computational algorithms. I try to emphasize the mathematical foundations of computational structural biology while balancing between algorithms and a nuanced understanding of experimental data. The book concentrates on the information content of the experiments, and the methods for design and analysis of protein structure, together with techniques for macromolecular structure determination, providing an emphasis, where possible, on provable algorithms with guarantees of soundness, completeness, and complexity bounds. In particular, rather than describing a competition between computer programs, this book tries to evaluate the strengths and weaknesses of the underlying ideas (algorithms). There are several reasons. First, I believe no one will be using the same programs in 10 years (and if we are, that would reflect poorly on the field). However, the underlying mathematical relationships between the data and the biological structures should prove enduring, warranting a characterization of the completeness, soundness, and complexity of algorithms in structural molecular biology and molecular biophysics.

This book describes the development of algorithms for a broad array of biological, biophysical, and biochemical topics in structural biology. A list of ten important topics is given in table P.1. I used several guiding principles in designing this book. First, several excellent textbooks already exist on sequence-based computational biology. There are fewer books on computational aspects of structural biology, and fewer still that focus on the interplay between advanced algorithms and raw experimental data. This book emphasizes computational biology from a structural and biophysical point of view, and tries to place a greater emphasis on algorithms, especially provable algorithms, and the underlying mathematics. Three emerging areas are stressed because they are particularly fertile ground for research students: NMR methodology, design of proteins and other molecules, and the modeling of protein flexibility. Where possible, the focus is on combinatorially precise algorithms with provable properties. In particular, I have tried to convey the rich geometric and algebraic structure of the underlying algorithms.

NMR and X-ray Crystallography   At the same time, the topics in this book are not exhaustive. They have been selected with the following intent. First, where possible, the algorithms should be nontrivial and therefore have the possibility to inspire computer scientists to work on algorithms for structural molecular biology. The exposition, however, should be at such a level that biochemistry students will be moved to learn about useful structural algorithms for their work. In order for the course to be feasible to complete within one year, I have made some choices about what to cover. Even though the majority of the protein and DNA structures in the protein data bank (PDB) have been determined by X-ray crystallography, this book gives more emphasis to algorithms for solution-state NMR. One reason is that, while both techniques are undergoing rapid technology development, NMR is currently less automated than crystallography and therefore there are probably more algorithmic opportunities for computational scientists to make their mark in developing new algorithms to process, analyze, and make inferences from the new biophysical experiments being developed, on a seemingly weekly basis, by NMR spin gymnasts. This having been said, some fascinating computational problems arising in experimental X-ray crystallography of proteins are discussed in chapters 40, 47, and 48, and, of course, the analysis and use (for example, in design, docking, and the modeling of protein flexibility) of these crystal structures are covered in chapters 9–12, 14, 20–30, 40, 41, 49, and 50.

Design of Proteins, Peptides, and Small Molecules   The design of proteins in general and enzymes in particular represents one of the most important problems in biochemistry and synthetic biology. An example of the computational and experimental challenges in protein design is given at the beginning of this Preface. Similarly, design, identification, and discovery of small molecule inhibitors are of central biochemical and pharmacological importance. Therefore, many chapters of this book are devoted to the design of proteins, peptides, and small molecules: chapters 9–12, 25, 46, 45, 27, 28, 30, 41, 26, and 49.

Protein Flexibility   In the early days of structural biology, it was an experimental challenge even to obtain a single structural model of a biological macromolecule such as a protein. Furthermore, computational techniques were limited both in terms of algorithmic sophistication and processor speed, and therefore could only deal with a single three-dimensional model. More recently, experimental techniques have yielded models of proteins that encompass some of the dynamics and mobility of proteins and nucleic acids in solution-state physiological conditions. These range from ensemble representations (sets of conformations) to more direct measurements of mobility such as motion tensors (from NMR relaxation experiments or residual dipolar couplings). Moreover, computer processor speeds have increased dramatically, and massively parallel distributed cluster or cloud computations have become routine. Algorithms have improved in scope and sophistication, and now are beginning to handle some of the challenges of modeling macromolecular flexibility. Therefore, algorithms for modeling and computing with representations of protein flexibility are emphasized in this book, since they will be of increasing importance in the future of structural biology: chapters 5, 9–12, 20–26, 30, 41, 42, and 25.

Protein:Ligand Binding Biological macromolecules such as proteins do not exist in isolation. They move, flex, and dynamically bind to partner ligands. Enzymes must selectively bind to substrates and cofactors in order to perform their reaction chemistry. Protein:protein interactions are ubiquitous for cell signaling, regulation, transcription, and translation. Algorithms to predict the functional interactions of proteins with other molecules must be able to model the affinity of protein:ligand binding. These algorithms are developed and discussed in chapters 9, 10, 12, 23, 25–29, 40, 41, and 46.

Stochastic and Heuristic Techniques I do not explicitly cover in separate chapters the stochastic techniques of Monte Carlo, simulated annealing, Metropolis algorithms, or genetic algorithms. These techniques are well-treated in both other books and excellent reviews, and therefore they are discussed as an ingredient of certain lectures (e.g., 6, 9, 10, 11, 20, 22–26, 30, 31, 41, 43, 44, 45, and 49) rather than as the main course. This also fits well with our emphasis on provable algorithms, since these stochastic techniques are heuristic, typically admitting no provable guarantees. In previous reviews and textbooks, provable algorithms are less generously covered, and therefore they are given more attention in this book. Again, this choice is intentional, and is motivated to lure young computer scientists into the field.

Protein Kinematics On the other hand, there are some rigorous and well-developed areas that are mentioned only briefly in this book, because excellent textbooks and reviews of these techniques already exist. These areas include molecular dynamics simulations, geometric hashing, protein threading, and the Vereshchagin-Featherstone linear-time recursive Lagrangian dynamics algorithm. These four topics have also been successfully integrated into many best-practices algorithms. With regard to algorithms for kinematics, I take a middle way, and provide material on the interplay between geometry, algebra, and experimental restraints in defining protein kinematics (lectures 1, 2, 9, 15–18, 13, 20–24, and 30). This having been said, my book is not a general text on kinematics, a subject that is also well-represented by other authors. Finally, it must be mentioned that for many of the topics in this book there is a large literature that we do not cover. For example, an entire course could be taught on any one of the following subjects: normal mode analysis, protein loops, enzyme kinetics, Markov random fields, energy minimization, computational topology, protein design, singular value decomposition and its applications, and the structural bioinformatics of rotamer and small molecule libraries. Though each of these is discussed briefly, to fit within fifty chapters, I had to choose the material I thought was most accessible for my student audience. Consequently, a lot of thoughtful and significant papers are not mentioned. I hope that engaged students will pursue their interests further by doing a detailed search on Medline for the most recent papers complementing the brief chapters in this book.

Background and Prerequisites   Students taking a course based on this book should have a background in algorithms (for example, an advanced undergraduate or early graduate course on algorithms) plus some exposure to the notions of NP-completeness. The latter may be waived for open-minded students with a mathematics or physics background, and what is sometimes called "mathematical sophistication." It is also helpful to have a background in basic biochemistry or to be taking a structural biochemistry course concurrently. For students in the life sciences, this textbook and a course based thereupon should be valuable if they have some background in computer science and are willing to read more. The book is organized as a year-long lecture course, to be taught in the order I provide. As a two-semester or two-quarter course, it allows time for presentations of student projects during the last two weeks of the second semester. It is recommended that student projects should consist of (a) choosing an algorithm covered in one of the lectures, (b) implementing it from scratch, and (c) trying it on real data obtainable from one of the data or structure repositories (e.g., PDB, BMRB, etc.). When I use this book in teaching, I have computer scientists, computational biologists, and biochemists in the class. Therefore, being somewhat relaxed about prerequisites is recommended to include different backgrounds and have a more lively discussion. For projects and assignments, it is useful to pair students with a "dry" (computational) background with those from a wet lab (experimental) background.

Courses of Different Lengths

There are alternative paths through this book for a shorter, one-semester or one-quarter class. An excellent one-semester class would cover only lectures 1–5, 8, 9, 11, 12, 15–18, 20–23, 25, 31, 32, 37, 38, 45, 47, and 50. As mentioned earlier, a 3- to 4-day short course on NMR methodology could be taught stand alone based on lectures 15–18. Alternative paths through the book could emphasize different themes. One path for one semester would emphasize algorithmic and computational approaches in NMR structural biology and cover lectures 1–8, 13–18, 39, 29, 31, 33–35, and 43–45. Another route would emphasize design (especially protein design) covering lectures 9, 10, 11, 12, 25, 46, 45, 27, 28, 30, 41, 26, and 49, in that order.

A third path would emphasize the general theme of protein flexibility, covering lectures 5, 9–12, 20–26, 30, 41, 42, and 25. Last, a running example is given, of particular biological systems, from the family of enzymes known as nonribosomal peptide synthetases (NRPS). The reason for this is fourfold. First, it is useful to have an example where particular residues, active sites, and substrates can be named and visualized. Second, although we envision all these algorithms as quite general, in practice most algorithms still have some interplay or dependence on the kind of systems they have been developed and demonstrated on. We have not achieved a complete decoupling or genericity of the algorithms from the underlying data and structures. Therefore, describing particular biological systems mirrors the reality that they are usually interleaved with the algorithms when we read about them in the literature. Third, vacationing from mathematical algorithms into the countryside of the biochemistry can be a comforting and inclusive excursion for students with a life science background. Fourth, and perhaps most important, this system of modular enzymes presents fascinating opportunities for computational design and therefore may inspire a new generation of students to tackle them in their research. At any rate, a short subtheme or path covering the NRPS (and related) enzymes is given in the sequence of lectures 10, 12, 27, 28, 46. If my recommended sequence for a year-long or semester-long course does not fit the instructor's needs, the NMR short course and the four paths above (namely, NMR, design, protein flexibility, and NRPS) can be mixed and matched to create a course that is tailored for a particular occasion or purpose.

The preceding discussion about alternative paths and subthemes may give the mistaken impression that these are the only topics this book covers in depth. To the contrary, as can be seen from table P. 1 or by browsing the table of contents, we cover a wide range of topics in computational structural biology. For example, I hope instructors will note the two chapters on computational mass spectometry (19, 43), an increasingly valuable technique in biophysics and proteomics. However, the paths and subthemes of NMR, design, protein flexibility, and NRPS are distinguished in that the later lectures in each path presume some knowledge of the earlier lectures in that track. However, most of the lectures outside these tracks can be read independently. Therefore, they can also serve as reference chapters for researchers in the field, or be put together in an order that suits an individual instructor for a particular course. In particular, a nice section of a course on protein:ligand binding could be based on chapters 9, 10, 12, 25–29, 40, 41, and 46, read in that order.

Finally, the chapter on belief propagation (25) could serve as a general introduction to Markov Random Fields and factor graphs; the chapters on distance geometry (1, 31, 32, 36) or computational topology (50) would be at home in an algorithms or computational geometry course; and any of these could be followed by the chapters on graph cuts (37–38). That being said, these lectures are all motivated by the underlying biophysics and hence more trenchant when nestled in the context of macromolecular structural biochemistry within the logical sequence of this book.

Acknowledgments

I'm grateful to my students and colleagues over many years, who gave me a great deal of assistance in writing this book. Their contributions sharpened the writing, and I am indebted to them for whatever clarity and insight this text manifests. However, I alone am responsible for any remaining errors or flaws.

I would like to thank Xin Guo, Jeff Martin, Lirong Xia, and Michael Zeng, teaching assistants and scribes for my course at Duke, for transcribing and formatting these lectures. Thanks also to the students and postdoctoral fellows in the Donald, Bailey-Kellogg, and Hartemink labs, and to all students in this class, who contributed to the lectures and the notes. I particularly thank Serkan Apaydin, Chittu Tripathy, Ivelin Georgiev, Raluca Gordan, and Pablo Gainza for their help and contributions. I thank Celeste Hodges for assistance in typesettting, proofreading, and figures.

I thank my colleagues for collaborations and advice, without which this book would not have been possible: Chris Bailey-Kellogg, Jeffrey Boyles, James Chou, Lauren Cowles, Jeff Hoch, Dan Keedy, Shobha Potluri, David and Jane Richardson, Gerhard Wagner, Tony Yan, Anna Yershova, and Pei Zhou. I thank the following colleagues for helpful discussions, which greatly improved the chapters on the interplay of algorithms and biophysics: Marcelo Berardi, Vincent Chen, David Cowburn, Brian Hare, Janet Huang, Terry Oas, Tomás Lozano-Pérez, Guy Montelione, Art Palmer, Len Spicer, Bruce Tidor, Ron Venters, and Peter Wright. This book was improved by the suggestions of two anonymous readers. Thanks to Janet Huang, Lydia Kavraki, George I. Makhatadze, James Masse, and Guy Montelione for providing figures. I thank all members of the Donald laboratory, past and present, for helpful discussions and comments. I thank Chittu Tripathy, Pablo Gainza, Swati Jain, Michael Zeng, and Celeste Hodges for proofreading.

I thank my editors and the staff at MIT Press for invaluable assistance in developing this book.

This book was made possible by Grants to B.R.D. (Numbers R01 GM-65982, R01 GM-78031, and T32 GM-71340) from the National Institute of General Medical Sciences, National Institutes of Health. Its contents are solely the responsibility of the author, and do not necessarily reflect the views of the U.S. Department of Health and Human Services.

A sabbatical leave from Duke allowed me to complete this book while visiting Chapel Hill and Issy-les-Moulineaux, Paris. I would like to thank the University of North Carolina and Jack Snoeyink, Ming Lin, and Dinesh Manocha for hosting me in Chapel Hill.

1 Introduction to Protein Structure and NMR

This chapter presents the basics of protein structure, and then discusses some principles in protein structure determination by nuclear magnetic resonance (NMR). We will use these basic facts about proteins and the NMR technique in developing several algorithms and computational frameworks in this book.

This first chapter introduces and ties together the concepts of protein kinematics, experimental data for geometric restraints, and sophisticated algorithms to determine structures given restraint data. We begin with simple examples, to illustrate the general principles that will be investigated in depth in this book.

1.1 Protein Structure

Proteins are linear polymers of amino acids, and different sequences of amino acids generally have different three-dimensional protein structures. Figure 1.1 shows the cartoon structure of a prototypical amino acid, where R stands for any of 20 natural amino acid side-chains.

There are two important exceptions. The glycine side-chain is simply a single hydrogen atom bonded to the C^α, and so effectively glycine has two H^α atoms. Thus, glycine has exceptional conformational flexibility compared to the other amino acids. Proline is the only cyclic amino acid.

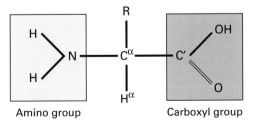

Figure 1.1
The structure of a typical amino acid.

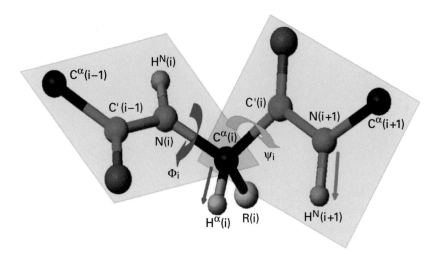

Figure 1.2 (plate 1)
The ϕ and ψ torsion angles of peptide backbones.

The cyclic structure of proline's side-chain restricts its ϕ backbone dihedral angle to approximately -75 degrees, giving proline more conformational rigidity than other amino acids.

Sometimes C^α will be written as C_α. The carbonyl carbon is often written as C', and the protons bonded to C^α and N, respectively, are written as H^α and H^N.

Through a reaction of the corresponding amino and carboxyl groups, amino acids form peptide bonds. We refer to the N, C^α, O, and C' atoms in the peptide as the *peptide backbone*. The characteristics of the peptide bond is shown in figure 1.2 (plate 1). The peptide bond is ideally, although not in reality, planar and rigid, and the rotation is only allowed by the ϕ ($C^\alpha - N$) and ψ ($C^\alpha - C'$) torsion angles. Given a sequence of n amino acids, we are interested in finding all torsion angles ($\phi_1, \psi_1, \phi_2, \psi_2, \ldots, \phi_n, \psi_n$) for all peptide units to determine the three-dimensional structure of the protein backbone. In some models the C^β atom is considered with the backbone (chapter 11), which requires handling glycine and proline as special cases.

1.2 Structure Determination of Proteins with NMR Spectroscopy

At a high level, some NMR spectra can give us some biophysical (frequency) signals, which encode geometric information about the protein structure, such as the distance and angle information. Determination of the three-dimensional structure (ϕ and ψ angles) of the protein backbone can be regarded as a graph realization or distance geometry problem, using measured distance or angle constraints. This problem can be solved easily if the *exact* distance constraints between *all* pairs of atoms are known. However, the NMR data unavoidably is uncertain and may have missing data. In fact, the noise in NMR spectra increases the complexity of our computational task. Finally, in general, distances can only be measured between pairs of protons less than 6 Å apart.

Nuclear Overhauser Enhancement (NOE)   The NOE is a consequence of dipole-dipole coupling and cross-relaxation between different nuclear spins [1], and is used to measure the distance between atoms (usually protons) in proteins. The relation between the observed NOE intensity I and the distance r is ideally $I \propto 1/r^6$.

NMR Assignment Problem   The NMR resonance assignment problem is to find the mapping from each observed frequency (resonance) in NMR spectra to each NMR-active nucleus in the protein. More precisely, let S denote the set of atoms, and Ω denote the set of observed resonances. Physics defines a function (which we cannot write down precisely), $f : S \rightarrow \Omega$. We have the inverse problem: to compute f^{-1}, where f^{-1} is called the *assignment*. We will discuss the resonance assignment problem in later lectures.

Distance Geometry Problem   Let $d_{ij} = ||x_i - x_j||$ where $x_1, x_2, \ldots, x_n \in \mathbb{R}^3$ denote the (3D) coordinates for atoms $a_1, a_2, \ldots, a_n$ in the protein structure. We can represent all d_{ij} by a matrix, called the *distance matrix*, where each item intercrossed at the ith row and the jth column corresponds to d_{ij}. Now, the distance geometry problem is to find coordinates[1] $x_1, x_2, \ldots, x_n$ such that the distance constraints in the given distance matrix are all satisfied. We classify the distance geometry problem into several cases according to different kinds of distance matrices:

1. *Problem with All Exact Distances*   If the exact distances between all pairs of atoms are known, the problem can be solved in $O(n^3)$ time by using the Singular Value Decomposition method, where n denotes the number of atoms. More recently, Dong and Wu [2] have shown that this problem can be solved in linear time by using a "Geometric Build-Up" algorithm.

2. *Problem with Sparse Sets of Distances*   Given only a subset of the distances, the distance geometry problem becomes difficult to solve. Saxe [3] has shown that this problem is (strongly) NP-hard.

3. *Problem with All Distances But with Some Errors*   This problem has been examined by Berger, Kleinberg, and Leighton [5]. They mainly considered the approximation solution of this problem.

4. *Problem with Interval Bounds of Distances*   Because of the noise from experimental data, we may only get some lower and upper bounds on the distances. It has been demonstrated that such a problem is still hard to solve. More and Wu [4] considered the ε-approximation[2] of this problem. They showed that if ε is smaller than a certain value, finding an ε-optimal solution to the distance geometry problem is still NP-hard.

---

[1] Using kinematics, the coordinates are related to the ϕ and ψ angles in the conformation of the protein backbone.

[2] The solution is called ε-*optimal* if distance errors in the solution are within a tolerance of ε.

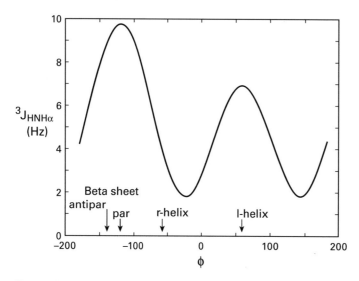

Figure 1.3
The Karplus relationship between the three-bond coupling constant $^3J_{HNH\alpha}$ and the torsional angle ϕ. Credit:
http://www.cryst.bbk.ac.uk/PPS.

Later chapters in this book will describe upper and lower bounds for distance geometry problems.

Scalar (Spin-Spin) Coupling Constant $^3J_{HNH\alpha}$ The 3-bond scalar coupling constant $^3J_{HNH\alpha}$ provides a measurement of torsion angle ϕ information, which is different in different kinds of regular secondary structures, such as α helix and β sheet. The correspondence between $^3J_{HNH\alpha}$ and ϕ can be found from the empirically derived Karplus curve, as shown in figure 1.3. More specifically, the following values of $^3J_{HNH\alpha}$ in some regular secondary structures are obtained [1]:

1. Right-handed alpha helix, $\phi = -57$, $^3J_{HNH\alpha} = 3.9$ Hz
2. Right handed 3.10 helix, $\phi = -60$, $^3J_{HNH\alpha} = 4.2$ Hz
3. Antiparallel beta sheet, $\phi = -139$, $^3J_{HNH\alpha} = 8.9$ Hz
4. Parallel beta sheet, $\phi = -119$, $^3J_{HNH\alpha} = 9.7$ Hz
5. Left-handed alpha helix, $\phi = 57$, $^3J_{HNH\alpha} = 6.9$ Hz

The preceding data provide a method for identifying regular secondary structures based on the measurements of $^3J_{HNH\alpha}$.

The reader may wonder, if our goal is to compute ϕ and ψ angles, why not measure them *directly* using scalar couplings? There are four reasons. (1) The relationship between $^3J_{HNH\alpha}$ and ϕ is multivalued (figure 1.3), necessitating a logical "OR" approach and a tree-search. (2) The data are uncertain and noisy, and generally only provide loose bounds on dihedral orientations. (3) Higher-resolution structures are possible by combining different NMR data. Nevertheless, scalar couplings provide valuable geometric measurements when they can be recorded. Finally, (4) these experiments are much less sensitive for large proteins, compared to other NMR measurements.

References

[1] Kurt Wuthrich. *NMR of Proteins and Nucleic Acids*. New York: John Wiley & Sons, 1986.

[2] Q. Dong and Z. Wu. A linear-time algorithm for solving the molecular distance geometry problem with exact inter-atomic distances. *Journal of Global Optimization,* 22 2002: 365–375.

[3] J. B. Saxe. Embeddability of weighted graphs in k-space is strongly NP-hard. *Proceedings of the 17th Allerton Conference on Communications, Control, and Computing,* 1979, 480–489.

[4] Jorge More and Zhijun Wu. ε-Optimal solutions to distance geometry problems via global continuation, in P. M. Pardalos, D. Shalloway, and G. Xue (eds.,). *Global Minimization of Nonconvex Energy Functions: Molecular Conformation and Protein Folding.* pp. 151–168, American Mathematical Society, 1996.

[5] Bonnie Berger, Jon Kleinberg, and Tom Leighton. Reconstructing a three-dimensional model with arbitrary errors. *Journal of the ACM (JACM),* 46(2) 1999: 212–235.

2 Basic Principles of NMR

In this chapter we give a brief introduction to the principles of nuclear magnetic resonance (NMR), and a survey of using NMR for studies of structures and dynamics of proteins and nucleic acids.

While some of the basic concepts were introduced in chapter 1, we must now develop a deeper understanding for the physics and geometry of the experimental measurements. The foundation we build in chapters 2 and 3 will be expanded by the algorithms in later chapters.

2.1 Overview of NMR

NMR is a physical experiment that measures the electronic properties of each nucleus in a magnetic field by exploiting the spin properties of an atom's nucleus. All nuclei containing an odd number of neutrons and some that contain an even number of neutrons but odd charge have an intrinsic magnetic moment. NMR works by applying a strong magnetic field (the stronger the field, the higher the resolution) to a sample and then measuring how the system responds to radio-frequency waves. The radio-frequency wave is electromagnetic radiation with an appropriate frequency that can be absorbed by a nucleus and flip its spin. Currently, NMR spectroscopy and X-ray crystallography are the only techniques capable of determining the three-dimensional structures of macromolecules at atomic resolution. In contrast to X-ray crystallography, the NMR approach is complementary in different ways [1]:

1. NMR directly probes the nuclei of protons (^{1}H), ^{13}C, and ^{15}N.

2. NMR can be applied to molecules for which no crystals are available. Thus, new structures can be obtained that are not available from X-ray studies.

3. NMR spectroscopy is a powerful technique for investigating intramolecular dynamics on many timescales, for example, picoseconds to seconds. In addition, for a characterization of internal dynamics of biomacromolecular structures, NMR provides direct, quantitative measurements of the frequencies of certain high activation energy motional processes and at least semiquantitative information on additional high-frequency processes. In comparison, X-ray structure determinations may include an outline of the conformation space covered by high-frequency structural fluctuations.

2.2 The Physical Basis of NMR Spectroscopy

Preliminaries: Where Does the Magnetization Come From? ^{1}H nuclei (or spin $I = 1/2$ nuclei) possess a spin $1/2$ that is associated with an angular momentum and a magnetic dipole moment. Each ^{1}H nucleus thus behaves like a little magnet that spins around its dipole axis. A single spin $1/2$ has the quantum mechanical property that, in an external magnetic field B_0 (measured in tesla, abbreviated T), only two orientations (called "parallel" and "antiparallel" or "α" and "β"; see figure 2.1) with respect to B_0 have a defined energy. The energy difference between these two states, $\triangle E$, is given by

$$\triangle E = h\nu = -\frac{h\gamma B_0}{2\pi}$$

where h is Planck's constant, and γ is the gyromagnetic ratio (2.67×10^8 radians per second per Tesla for ^{1}H). Because of the angular momentum associated with spins, spins don't align with the magnetic field B_0, but precess around an axis parallel to B_0 with a frequency ν_0 (the so-called Larmor frequency, $\omega_0 = 2\pi\nu_0 = -\gamma B_0$; see figure 2.2 and [4]).

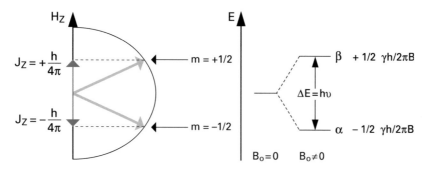

Figure 2.1
Energy levels of the $\alpha-$ and $\beta-$ states of ^{1}H (and $I = 1/2$ nuclei).

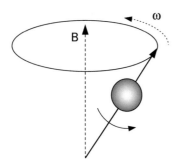

Figure 2.2
Rotation of nuclear momentum about its own axis and about the magnetic field axis.

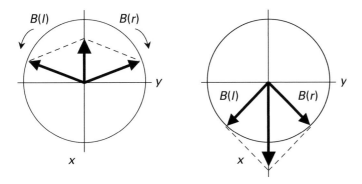

Figure 2.3
Representation of a linear alternating field (max $2B$) as the sum of two rotating fields, $B(r)$ (clockwise) and $B(l)$ (anticlockwise).

In an NMR experiment, many spins are present. Since the energy difference between the states α and β is small compared to kT (k Boltzmann constant, T absolute temperature), there is, at room temperature, only a small excess of spins in the energetically lower ground state, even in the highest available magnetic fields. In thermal equilibrium, the average magnetization from all molecules constitutes a macroscopic magnetization M parallel to the magnetic field.

The NMR Experiment in a Nutshell In a high-resolution NMR experiment, a glass tube containing a water solution with the molecule of interest is placed in a *static magnetic field* B_0 and then subjected to irradiation by one (or several) radio-frequency (RF) fields. Typically, the solution is quite concentrated. Suppose an RF pulse B' is applied along the direction of the x-axis. B' can be represented by two vectors with the same magnitude, B_1, rotating in the x,y-plane with the same frequency, ν_L. One of the vectors, $B(r)$, rotates clockwise, and the other, $B(l)$, rotates anticlockwise (figure 2.3).

Only the component precessing around B_0 with the same frequency and the same direction as the precessing nuclear dipoles can interact with M (the macroscopic magnetization), denoted as B_1. If instead of the fixed coordinate system x, y, z, we use a rotating coordinate system x', y', z, which rotates with the same frequency as B_1, the effect of B_1 is to turn the vector M about the x'-axis, that is, in the y', z-plane. The angle θ through which M is tipped increases with the amplitude B_1 and the duration of the pulse (figure 2.4a). A $90° x$ pulse is an RF pulse with the precise duration that takes M to become transverse (i.e., perpendicular to B_0) on its way to the southern hemisphere (figure 2.4b).

If we place a detector (i.e., a coil) along the y-axis, after a $90° x$ pulse, the system will induce a current in the detector, which is called *free induction decay (FID)* (figure 2.5). The FID decays exponentially with a decay constant T. The decay constant T (also called "relaxation time") is generated by the magnetization vector M gradually returning to equilibrium state when the RF pulse is switched off. Two different time constants describe this behavior:

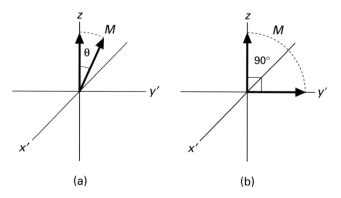

Figure 2.4
Direction of the macroscopic magnetization vector M in the rotating coordinate system: (a) after a pulse of arbitrary angle $\theta x'$; (b) after a $90°x'$ pulse.

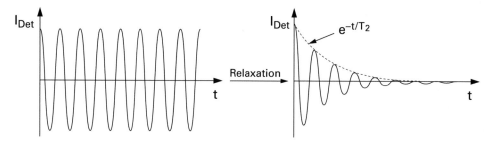

Figure 2.5
Left: signal in absence of transverse relation; right: real FID.

1. The reestablishment of the equilibrium α/β state distribution (T_1);
2. Dephasing of the transverse component (destruction of the coherent state, T_2).

The NMR spectrum is obtained from Fourier transformation of the FID by plotting the amplitude of the sine and cosine functions versus their frequencies (figure 2.6). The Fourier transformation is a mathematical procedure to represent any signal as a sum of sine and cosine functions with different frequencies.

2.3   Chemical Shifts

As mentioned in section 2.2, an atomic nucleus can have a magnetic moment (nuclear spin), which gives rise to different energy levels and resonance frequencies in a magnetic field. The total magnetic field experienced by a nucleus includes local magnetic fields induced by currents of electrons in the molecular orbitals. The electron distribution of the same type of nucleus (e.g., ^{1}H,

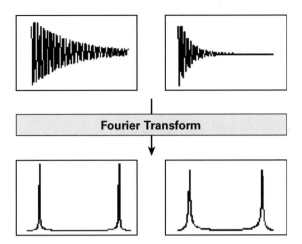

Figure 2.6
Slowly decaying FIDs lead to narrow lines (left), rapidly decaying ones to broad lines (right).

^{13}C, ^{15}N) usually varies according to the atom's local electronic environment (including electrons from nearby atoms, i.e., covalent bonds, binding partners, bond lengths, angles between bonds, etc.). Therefore, the electronic environment of an atom shields or deshields it from the external magnetic field and the local magnetic field at each nucleus (different atoms with different local electronic environments "feel" a different external field and therefore resonate at different frequencies). This variation of NMR frequencies of the same kind of nucleus, due to variations in the electron distribution, is called the *chemical shift*. The size of the chemical shift is given with respect to a reference frequency or reference sample, usually a molecule with a minimally distorted electron distribution. Since an atom's chemical shift is very sensitive to the local electronic environment (and thus structure), the chemical shifts can be considered as a sort of fingerprint for a molecule, and play an important role in NMR spectroscopy to identify the atoms and reveal the aspects of their local geometry. (See chapter 2 of [3] for details.)

Note　Chemical shift is usually expressed in parts per million (*ppm*) by frequency, because it is calculated from

$$\delta = \frac{\omega_{\text{reference}} - \omega_{\text{observed}}}{\omega_{\text{reference}}} \times 10^6.$$

2.4　Introduction to NMR Experiments

NMR Spectrometer　NMR spectrometers have a certain operating frequency, which depends on the strength of the applied magnetic field. Generally speaking, the higher the operating frequency, the better the resolution of the spectrum. For example, a 500-MHz spectrometer has a much stronger magnet than a 60-MHz spectrometer. In structural biology, 500, 600, 700, 800, and even 900 MHz instruments are commonly used.

Protein NMR   Protein NMR is performed on aqueous samples of highly purified protein. The source of the protein can be either natural or produced in an expression system using recombinant DNA techniques through genetic engineering. Recombinantly expressed proteins are usually easier to produce in sufficient quantity, and make isotopic labeling possible.

The most abundant isotopes of carbon and oxygen, ^{12}C and ^{16}O, have no net nuclear spin, and thus NMR spectroscopy cannot be directly exploited to identify them. And the most abundant isotope of nitrogen, ^{14}N, although it has a net nuclear spin, also has a large quadrupolar moment, a property of the atomic nuclei that prevents high-resolution information to be obtained from this isotope. In fact, NMR of proteins from natural sources is sometimes restricted to utilizing NMR based solely on protons. Nevertheless, the less common isotopes, ^{13}C and ^{15}N, are ideal for NMR experiments, and therefore labeling the proteins with these nuclei opens up possibilities for doing more advanced experiments that detect or use these nuclei. Isotopic labeling is generally done by growing the expression host in a growth media enriched with the desired isotopes.

Note   There are many different kinds of NMR experiments, and each type reveals some information about the structure or dynamics of the proteins. The different experiments also select for some specific kinds of atoms, for example, all protons or all protons that are part of a methyl group. However, it is not generally possible to select for a single residue.

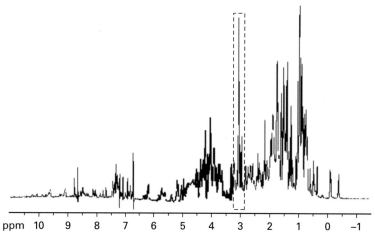

Figure 2.7
In a 1D spectrum assignment is very difficult because of the crowding/overlap.

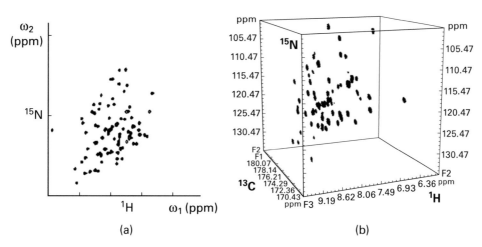

Figure 2.8
(a) Two-dimensional ^{1}H-^{15}N correlation spectrum; (b) three-dimensional ^{1}H-^{13}C-^{15}N correlation spectrum.

Multidimensional Spectra Although one-dimensional NMR spectral analysis, such as the ^{1}H-NMR spectrum, is useful for elucidating the structures of small organic molecules, for more complex samples arising from biopolymers, the resulting spectrum is far too complicated for interpretation as most of the signals overlap heavily (figure 2.7). Therefore, we often require two- and three-dimensional analysis for these samples using ^{13}C and ^{15}N nuclei.

Two-dimensional (2D) analysis basically measures the spectra of two different nuclei in a sample and plots them against each other. Figure 2.8a is an example of a 2D ^{1}H-^{15}N correlation spectrum called *heteronuclear single-quantum coherence* (HSQC). These ^{1}H-^{15}N correlation experiments exploit the relatively strong shielding effects (J coupling) between the amide hydrogens and their corresponding amide nitrogens in the protein samples. However, complex protein samples may still have some degree of peak overlap with 2D-NMR. Therefore higher dimensional experiments have been devised to deal with this problem. Currently, 3D and 4D spectra are routinely done, and 5D and 6D experiments are possible in advanced laboratories.

References

[1] Kurt Wuthrich. *NMR of Proteins and Nucleic Acids*. New York: John Wiley & Sons, 1996.

[2] P. J. Hore. *Nuclear Magnetic Resonance*. Oxford: Oxford University Press, 1995.

[3] Horst Friebolin. *Basic One- and Two-Dimensional NMR Spectroscopy*. Weinheim: VCH Pub., 1990.

[4] John Cavanagh, Wayne J. Fairbrother, Arthur G. Palmer III, Nicholas J. Skelton, and Mark Rance. *Protein NMR Spectroscopy: Principles and Practice*. New York: Academic Press, 2006.

3 Proteins and NMR Structural Biology

As mentioned in the previous chapter, one-dimensional NMR techniques, which yield extremely useful information in small molecules, are of limited applicability to the complex, highly overlapped spectra of biological macromolecules. To effectively use the information available from NMR spectroscopy of biological macromolecules, we introduced multidimensional NMR spectroscopy. It combines data from several kinds of spectra to establish the mappings for resonance assignments. In general, there are two types of multidimensional spectra, *homonuclear* spectra, in which each axis represents the same type of atom (usually protons), and *heteronuclear* spectra, in which each axis represents a different type of atom. In this lecture, we briefly introduce several essential multidimensional NMR experiments and techniques.

3.1 COSY

Homonuclear correlation spectroscopy (COSY) is two-dimensional, and shows the frequencies for a single isotope (usually protons) along both axes of the spectrum. As shown in figure 3.1, the COSY spectrum consists of a series of peaks on the diagonal, and the peaks that appear off of the diagonal are called *cross-peaks*. These cross-peaks are symmetrical (above and below the diagonal) and indicate which protons are spin-spin coupled with each other. Therefore, by matching the center of a cross-peak with the center of each of two corresponding diagonal peaks, one can determine which atoms are connected to one another by chemical bonds. The peaks on the diagonal and the matched cross-peaks are coupled to each other.

3.2 $^3J_{HNH\alpha}$

The three-dimensional HNH$^\alpha$ ($^3J_{HNH\alpha}$) experiment is designed to accurately determine three-bond H^N-H$^\alpha$ *J-coupling constants* (also called *scalar couplings*). While the NOE provides information on the distance between two nuclei, this J-coupling constant is dependent on the dihedral angle in the structure. In particular, this coupling constant provides information about the dihedral angles subtended by the coupled atoms. For example, we can derive the dihedral angle ϕ by mapping J-coupling constants in the Karplus curve as shown in figure 3.2.

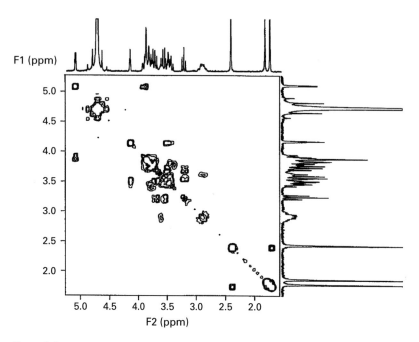

Figure 3.1
COSY spectrum of proton-proton scalar couplings.

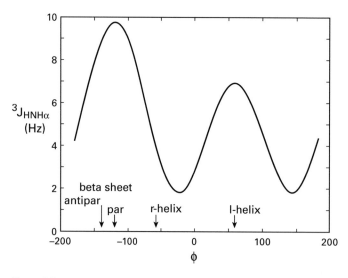

Figure 3.2
The Karplus relationship between the three-bond coupling constant $^3J_{HNH\alpha}$ and the torsional angle ϕ. The Karplus curve is calculated as $^3J(\phi) = A\cos^2(\phi - 60) + B\cos(\phi - 60) + C$, where A, B, and C are empirically derived parameters. Credit: http://www.cryst.bbk.ac.uk/PPS.

3.3 H^N-^{15}N HSQC

Heteronuclear Single Quantum Correlation (HSQC) is one of the most important two-dimensional NMR experiments. It correlates the nitrogen atom of an NH moiety with the directly attached proton, and each signal in an HSQC spectrum represents a proton that is bound to a nitrogen atom. Since every residue except proline has (ideally) a unique H^N-^{15}N pair on the protein backbone and ideally has distinct frequency signals, the HSQC spectrum can serve as the identification ("hash code") of each residue, and reference interaction for all other spectra. The HSQC spectrum also contains signals from the NH groups of the side-chains, such as Asn and Gln and the aromatic H^N protons of Trp and His. Unlike homonuclear spectra, the HSQC spectrum has no diagonal because different nuclei are observed during the relaxation periods T_1 and T_2 (figure 3.3).

3.4 ^{15}N TOCSY

3D TOtal Correlation SpectroscopY (TOCSY) is a three-dimensional (3D) heteronuclear NMR experiment. In contrast to the H^N-^{15}N HSQC, TOCSY experiments record the side-chain protons as the third axis in the spectrum. Figure 3.4 shows a TOCSY spectrum by fixing the ^{15}N atoms from 3D data, in which the peaks are from side-chain protons and H$^\alpha$, and the *strips* group atoms

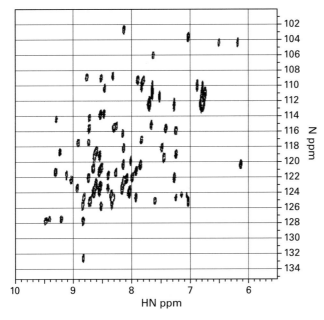

Figure 3.3
The NMR spectrum of H^N-^{15}N HSQC.

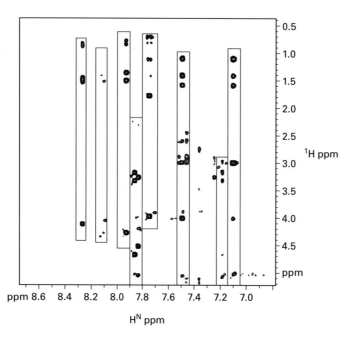

Figure 3.4
The NMR spectrum of ^{15}N TOCSY. Each strip indicates atoms grouped by the spin system.

by the spin system. In this spectrum, the x-axis is the H^N shift; the y-axis shows the H^α and the side-chain shifts, and the z-axis (projected out) gives the ^{15}N shift. Given the fact that the chemical shifts of protons for different amino acids are characteristically different, we can regard these *strips* (number and position of peaks) as the "fingerprints" of the protons to identify probable amino acid types.

Having constrained the amino acid type for each peak in HSQC from TOCSY experiments, in order to obtain actual assignments, we still need to establish the sequential connectivities between spin systems. Generally, we combine information from different NMR experiments by cross-referencing peaks to reveal the sequential connectivities, for example, HN(CO)CA and HNCA. These correlate the H^N, N, and C^α nuclei in a residue, and in the previous residue in the protein. Note that prolines will not generate peaks in any spectrum that involves the backbone amide, for example, HSQC, NH(CO)CA, HNCA, ^{15}N TOCSY.

3.5 NOESY

The Nuclear Overhauser Effect (NOE) is caused by dipolar coupling between nuclei, that is, the local field at one nucleus is affected by the presence of another nucleus. The intensity of the interactions is (ideally) a function of the distance between the nuclei according to the following equation

$$I = A(1/r^6),$$

where I is the intensity that is measured, A is a scaling constant, and r is the distance between nuclei. In other words, NOE provides a link between an experimentally measurable quantity, I, and internuclear distance. NOEs are measured with NOE spectroscopy (NOESY) experiments. For example, 3D ^{15}N-HSQC NOESY, a 3D heteronuclear NMR experiment, measures the through-space NOE between an amide proton H^N and a neighboring ^{1}H, when they are within a distance less than 6 Å. In a NOESY experiment, the height and volume of the peak gives the intensity, and is related to the distance between the two atoms: Bigger peaks indicate close distance, and weak peaks indicate that the atoms are further apart and/or moving.

3.6 RDC

Residual dipolar couplings (RDCs) [2], which were first proposed in the 1980s, have emerged as an important tool in NMR to study macromolecular structure and function in a solution environment. In typical NMR experiments, the protein is in water (or water and deuterium) and it tumbles isotropically. However, if the molecules in solution exhibit a partial alignment, for example, after adding an alignment medium to the solution, this leads to incomplete averaging of spatially anisotropic dipolar couplings, and RDCs between two spins will be measured.

RDCs are complementary to NOEs; they provide *global orientational information* as opposed to *local distance information* (NOEs). RDCs also contain distance information as well as measuring the angles that are formed by a vector connecting the two atoms within a tensor axis system. In addition, both NOEs and RDCs can be used as restraints by algorithms for protein structure determination.

RDCs are defined in terms of a coordinate system called the *principal order frame (POF)*. The general formula to extract bond orientations using RDCs for two spins a and b, as described in figure 3.5, is as follows:

$$D = \frac{\mu_0 \gamma_a \gamma_b \hbar}{4\pi^2 \langle r_{a,b}^3 \rangle} \left\langle \frac{3\cos^2 \theta - 1}{2} \right\rangle, \tag{3.1}$$

where D is the residual dipolar coupling, measured in units of Hertz, $\hbar$ is Planck's constant, γ is the gyromagnetic ratio, r is the interspin distance, θ is the angle between the interspin vector and the external magnetic field, and a and b are the spins. Then, we can simplify the equation as (see chapter 5, or [3] for details)

$$D = D_{\max} \mathbf{v}^T \mathbf{S} \mathbf{v}, \tag{3.2}$$

where $D_{\max}$ is a physical constant, v is the orientation of the bond vector in the POF, and S is a 3×3 symmetric and traceless matrix, which contains 5 degrees of freedom, called the *alignment tensor* (or *Saupe matrix*). In practice, the dipolar coupling D is measured directly, and the matrix $\mathbf{S}$ and the vector $\mathbf{v}$ are initially unknown. However, given a putative model, we can solve for $\mathbf{S}$ (via singular value decomposition) and back-compute RDCs. The difference between the computed

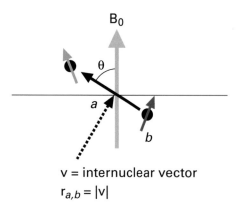

Figure 3.5
Extracting bond orientations using RDCs: Spins a and b are in the magnetic field B_0; v and $r_{a,b}$ indicate the orientation and the distance between the spins a and b, respectively.

and observed RDCs ($\mathrm{RDC}_o - \mathrm{RDC}_c$) can be used to refine the model. $\langle \cdot \rangle$ denotes the ensemble average. When a and b are covalently bonded, we typically know this distance from empirical crystal structures and replace $\langle r_{a,b}^3 \rangle$ by a scalar distance $r_{a,b}^3$.

3.7 Summary

There are many NMR experiments. Let us summarize the ones we have discussed so far.

Multidimensional NMR Experiments

$$
\begin{array}{rcl}
[\text{HN(CO)CA+HNCA}] & \Rightarrow & \text{Sequential connectivity via } C^\alpha \text{ and other backbone} \\
 & & \text{chemical shifts} \\
\text{HSQC, [HN(CO)CA+HNCA]} & \Rightarrow & \text{Sequential connectivity of HSQC peaks} \\
\text{HSQC, TOCSY} & \Rightarrow & \text{Amino acid types} \\
{}^3J_{\text{HNH}\alpha} & \Rightarrow & \text{J-coupling } (\phi \text{ angles}) \\
{}^{15}\text{N NOESY} & \Rightarrow & \text{H}^N\text{-}{}^1\text{H distance } (\leq 6\text{Å})
\end{array}
$$

Data Type and Information Content

$$
\begin{array}{rcl}
\text{Chemical shifts} & \Rightarrow & \text{Resonance frequencies (chemical shifts)} \\
\text{Scalar couplings} & \Rightarrow & \text{Dihedral angular constraints (local)} \\
\text{NOEs} & \Rightarrow & \text{Distance contraints} \\
\text{Residual dipolar couplings} & \Rightarrow & \text{Global orientation constraints}
\end{array}
$$

NOEs and scalar couplings are local restraints. RDCs are global restraints.

More Applications of NMR   Solvent accessibility, dynamics, protein folding assays, protein interaction assay, SAR by NMR, and others.

References

[1] John Cavanagh, et al.: *Protein NMR Spectroscopy : Principles and Practice* 2nd ed. New York: Academic Press, 2007.

[2] R. S. Lipsitz and N. Tjandra. Residual dipolar couplings in NMR structure analysis. *Annual Review Biophysics Biomolecular Structure* 2004. 33: 387–413.

[3] S. Grzesiek. EMBO short course on RDCs. Accessed: 2006. http://www.embl-heidelberg.de/nmr/sattler/embo/handouts/grzesiek_RDC.pdf

[4] Kurt Wuthrich. *NMR of Proteins and Nucleic Acids*. New York: John Wiley & Sons, 1996.

[5] P. J. Hore. *Nuclear Magnetic Resonance*. New York: Oxford University Press, 1995.

[6] Horst Friebolin. *Basic One- and Two-Dimensional NMR Spectroscopy*. Weinheim: VCH Pub., 1998.

4 MBM, SVD, PCA, and RDCs

In this chapter, we review the basic ideas of maximum bipartite matching (MBM), singular value decomposition (SVD), principal component analysis (PCA), and residual dipolar coupling (RDC) geometry. These algorithms are useful in computational biology in general, and structural biology and molecular biophysics in particular, and will be used in several applications in this book.

Each of these topics could be covered in some depth and breadth, even requiring several lectures. It is useful to have the basic primer on these techniques in one short chapter here. Later, we will explore details and applications of these fundamental concepts.

4.1 MBM

The input of an MBM problem is a complete bipartite graph $G = (V, E)$ with a weight function $w : E \to \mathbb{R}^+$, such that $V = L \cup R$, $L \cap R = \emptyset$, and $E = R \times L \cup L \times R$. More generally, $E \subseteq R \times L \cup L \times R$, but we can add edges with weight 0 to the graph to make it complete bipartite. Here, L can be visualized as the set of vertices on the left, and R as the set of vertices on the right.

A *matching* is a set of nonintersecting edges. The output of MBM is a matching E_m between L and R with the highest weight. That is,

$$E_m = \operatorname*{argmax}_{E' \subseteq E, s.t. \forall e_1, e_2 \in E', e_1 \cap e_2 = \emptyset} \sum_{e \in E'} w(e).$$

Example 4.1 *In the assignment problem, we want to match the residues in the model with the peaks observed in HSQC. This problem is modeled as an MBM problem as follows:*

1. Let $L = \{$All the residues in the model$\}$, $R = \{$Peaks in HSQC$\}$.

2. For any $i \in L$, $j \in R$, let

$$w(i, j) = |D_i - v_j S v'_j|.$$

Here D_j is the actual RDC data obtained from experiments, and $v_j S v'_j$ is the simulated RDC data back-calculated from the model.

The **Hopcroft–Karp algorithm** finds the maximum bipartite matching in $O(\sqrt{|V|}|E|)$ time [1, 2].

An MBM problem can be very sensitive to noise, as shown in the following example.

Example 4.2 *Let $L = \{l_1, l_2\}$, $R = \{r_1, r_2\}$, the ground truth weight is $w(l_1, r_1) = 10$, $w(l_2, r_2) = 10$, $w(l_1, r_2) = 10$, $w(l_2, r_1) = 9.9$. Then, the maximum matching is $E_m = \{(l_1, r_1), (l_2, r_2)\}$. However, if there is a noise on the weight of the edge (l_1, r_2) such that the weight we observed is 11, then the maximum matching will become $\{(l_1, r_2), (l_2, r_1)\}$, which does not share any edge with the ground truth.*

4.2 SVD

4.2.1 Definition

SVD is a technique to decompose a matrix into the product of three matrices. For any m-by-n matrix X, there exists a decomposition of X to the product of three matrices $X = USV^T$ such that

1. U is an $m \times m$ unitary matrix, that is, $M \times M^T = I_m$ where I_m is the identity matrix.

2. S is a diagonal matrix, which means that all the entries outside the main diagonal of S are zero. Thus, $S = \text{diag}(s_1, \dots, s_k, 0, \dots, 0)$, $s_1, \dots, s_k > 0$, $k = \text{rank}(S) = \text{rank}(X)$.

3. V^T is the transpose of a unitary n-by-n matrix V.

4.2.2 Properties

One of the most important properties of SVD is that it can be used to calculate the best solution to a linear system $Ax = b$ that minimizes the *least-square norm*. The problem is defined as follows:

Problem 4.1 *Let A be an $m \times n$ matrix, and $\mathbf{b}$ be an $m \times 1$ vector. Find an $n \times 1$ vector $\hat{\mathbf{x}}$ that minimizes $\|A\mathbf{x} - \mathbf{b}\|$, where $\|A\mathbf{x} - \mathbf{b}\| = \sqrt{\sum_{i=1}^{m}((A\mathbf{x})_i - \mathbf{b}_i)^2}$, $(A\mathbf{x})_i$ is the ith component of $A\mathbf{x}$, etc. That is*

$$\hat{\mathbf{x}} = \arg \min_{\mathbf{x}} \|A\mathbf{x} - \mathbf{b}\|.$$

$\hat{\mathbf{x}}$ can be calculated as follows by applying SVD to A.

- Calculate the SVD of A. Let $A = USV^T$, where $S = \text{diag}(s_1, \dots, s_k, 0, \dots, 0)$. Let $S^\dagger = \text{diag}(\frac{1}{s_1}, \dots, \frac{1}{s_r}, 0, \dots, 0)$ be an $n \times m$ diagonal matrix.
- Let $A^\dagger = VS^\dagger U^T$ be the *pseudoinverse* of A.
- Let $\hat{\mathbf{x}} = A^\dagger \mathbf{b}$.

It can be proved that $\hat{\mathbf{x}}$ is the minimal-norm least-square solution to $A\mathbf{x} = \mathbf{b}$.

When A is *symmetric* $(A = A^T)$, and *positive semidefinite* (that is, for any vector $\mathbf{x}$, $\mathbf{x}^T A \mathbf{x} \geq 0$), then $U = V$ and the $s_1, \dots, s_r$ are the *eigenvalues* of A.

4.3 PCA

PCA is a technique to reduce the dimension of high-dimensional data, in order to simplify the analysis. PCA is an orthogonal linear transformation that transforms the data set to another coordinate system such that the distance variance among data is mainly in the first several dimensions. PCA is useful because in many real-life instances, we do not know the structure of the problem, so we have to model the data in a very high-dimensional space. However, the data set itself has some underlying structure such that the variance of data can be approximately characterized in a space of much lower dimension. For example, imagine a case where, in $\mathbb{R}^{100}$, all of the data are generated approximately in a line (1-dimensional space). Then PCA can find a new coordinate system such that this line is the first coordinate (also known as the first *principal component*). To analyze the distance between points in the data set, we do not need to consider the remaining 99 dimensions.

4.3.1 Calculating PCA by SVD

Let $X = (p_1, \ldots, p_n)$ be an $m \times n$ matrix, in which each column p_i is a data vector in $\mathbb{R}^m$. Without loss of generality, we assume that X has zero empirical mean, that is, $\sum_{i=1}^{n} p_i = 0$.

To find the subspace that best represents the variance of X, let $C = XX^T$ be the *covariance*. Then, compute the SVD for C. Notice that C is symmetric and positive semidefinite, so the SVD of C must be $C = VSV^T$, where the diagonal of S consists of all the eigenvalues of C, and the column vectors of V are the normalized eigenvectors (recall that v_i is the normalized eigenvector of an eigenvalue λ_i of C, if $\|v_i\| = 1$ and $Cv_i = \lambda_i v_i$). Then, we choose the largest eigenvalue eigenvectors as the axes to project X on. Without loss of generality, suppose $\lambda_1, \ldots, \lambda_k$ are the k largest eigenvalues of C, and let $v_1, \ldots, v_k$ be their eigenvectors, respectively. Then, $V' = (v_1^T, \ldots, v_k^T)$ is the transformation matrix, and $V'X$ is the projection of the original data vector X to a k-dimensional subspace that best represents the variance of the data in X.

4.4 RDCs

NOEs provides interproton distance information for a protein. However, such information is only available when the distance between two nuclei is short (under 6 Å). RDCs, on the other hand, can provide global structural information between spins even when the distance between them is very large.

Definition 4.1 *The* residual dipolar coupling *between two spins a and b, is defined as follows:*

$$D_{IS} = D = \frac{\mu_0 \gamma_a \gamma_b \hbar}{4\pi^2 \langle r_{a,b}^3 \rangle} \left\langle \frac{3\cos^2\theta - 1}{2} \right\rangle, \tag{4.1}$$

where $\hbar$ is the Planck's constant, γ is the gyromagnetic ratio, $r_{a,b}$ is the interspin distance, θ is the angle between the interspin vector and the external magnetic field **B** *(see figure 3.5). In NMR, spins a and b are sometimes called I and S (see figure 4.1).*

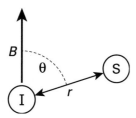

Figure 4.1

For example, in NVR, RDC is used to measure the orientation of the NH bond vector for each residue (chapter 14). Therefore, the distance r in (eq. 4.1) is known. In this case the RDC measures the global orientation of the NH bond.

Exercise   How can SVD be used to analyze RDC data? What is the relationship between this use of SVD, and the PCA method?

References

[1] John E. Hopcroft and Richard M. Karp. An $n^{5/2}$ Algorithm for Maximum Matchings in Bipartite Graphs. *SIAM Journal of Computing* 2(4) (1973): 225–231.

[2] Thomas H. Cormen, Charles E. Leiserson, Ronald L. Rivest, and Clifford Stein. *Introduction to Algorithms*, 2nd ed. Cambridge, MA: MIT Press, pp. 696–697.

5 Principal Components Analysis, Residual Dipolar Couplings, and Their Relationship in NMR Structural Biology

Anthony K. Yan and Bruce R. Donald

In biochemistry, nuclear magnetic resonance (NMR) is a powerful tool for analyzing the structure and dynamics of proteins in solution. Recently, research has taken advantage of a type of NMR data known as *residual dipolar couplings* (RDC). The physics of RDC's tells us they are formed from a time and ensemble average. This ensemble average is mathematically similar to the data analysis technique known as *principal components analysis* (PCA). In this chapter, we consider the formal relationship between the two, and consider the information content of the RDC formalism with respect to protein flexibility. After mastering this chapter, you will understand the fundamentals of RDCs, which are important in structural biology, and know the foundations of PCA, one of the most important algorithms in computational science and applied statistics.

In classical physics (or molecular dynamics, MD), ensemble averages are over continuous degrees of freedom. To keep the discussion simple for our broad audience, in this chapter we pretend that the ensembles are discrete. By way of intuition, in quantum mechanics the ensemble is discretized, so our derivation is analogous to averaging over quantum states.

At any rate, the reader who is familiar with continuous probability distributions and Jacobians will find it easy to generalize all the math in this chapter to continuous ensembles. Basically, the key ideas in the lecture are very simple, and we didn't want to have intimidating integrals everywhere: This is why we feel that discrete averages make for a cleaner presentation.

5.1 Introduction

Residual diploar couplings (RDCs) are a type of data from NMR that has been useful for solving protein structures and solving the assignment problem. According to the physics of RDCs, they are based on the relationship between internuclear vectors and a set of parameters known as the Saupe alignment tensor.

In this chapter, we consider the relationship between the Saupe alignment tensor and a data analysis technique known as *principal components analysis* (PCA). It turns out that the Saupe matrix is mathematically similar to the covariance matrix in PCA. The Saupe matrix is the result of an ensemble average of internuclear vectors, and contains statistical information about the ensemble that is very similar to the statistical information contained in PCA.

Intuitively, PCA gives us the spread or variance of a set of points along the principal axes of the ellipsoid that best fits the point set. One might have hoped that the Saupe matrix would contain exactly the same information as PCA. However, this is not the case. Instead, the Saupe matrix contains an upper bound of the variances of its ensemble. It will be shown that the Saupe matrix along with the ensemble average contains information precisely equal to PCA. However, the ensemble average is not contained within the RDC data; consequently, we are able to compute only an upper bound on the ensemble's variances. If the average of the ensemble were estimated by alternative means, we could use the Saupe matrix to perform PCA on the ensemble.

Below, we derive PCA, and explain its properties. Although this may be a review to some readers, it is useful for the exposition here as we will also derive the Saupe matrix formalism and explain its relationship to the derivation of PCA.

The structure of this chapter is as follows. First we derive PCA in section 2. In section 3, we derive the Saupe alignment tensor. In section 4, we compare the Saupe matrix to PCA, and explain their relationship. Finally, in section 5, we discuss future work and possible applications. In the appendix, we explain how to efficiently perform PCA on data sets with very large dimension and a relatively small number of data points.

5.2 Introduction to PCA

We motivate PCA by considering the statistics of discrete sets of $\mathbb{R}$. Let $A = \{a_i\}$, where $a_i \in \mathbb{R}$ and $i \in \{1, 2, 3, \ldots, N\}$. The standard statistical characterizations of A are the mean and variance (standard deviation):

$$\langle a_i \rangle = \frac{1}{N} \sum_{i=1}^{N} a_i \tag{5.1}$$

$$\sigma^2(A) = \frac{1}{N} \sum_{i=1}^{N} (a_i - \langle a_i \rangle)^2. \tag{5.2}$$

Very often, our data is not in $\mathbb{R}$, but instead is in $\mathbb{R}^m$. We would like to define the direct analogs of mean and standard deviation in higher dimensions. Consider a point set in $\mathbb{R}^m$. Let $P = \{\mathbf{p}_i\}$, where $\mathbf{p}_i \in \mathbb{R}^m$ and $i \in \{1, 2, 3, \ldots, N\}$. The mean is easy to define:

$$\langle \mathbf{p}_i \rangle = \frac{1}{N} \sum_{i=1}^{N} \mathbf{p}_i. \tag{5.3}$$

The analog of the standard deviation is less easy to define in $\mathbb{R}^m$. However, we can simplify things by projecting our data points P onto a one-dimensional line. We will then compute the *directional variance* along the line.

To simplify our presentation, let us assume that our data P has already been zero-meaned:

$$\langle \mathbf{p}_i \rangle = \frac{1}{N} \sum_{i=1}^{N} \mathbf{p}_i = 0. \tag{5.4}$$

We project P onto a line through the origin, which is specified by the unit vector $\mathbf{u}$. Let $\mathbf{q}_i(\mathbf{u})$ be the projection of $\mathbf{p}_i$ onto the line specified by $\mathbf{u}$:

$$\mathbf{q}_i = (\mathbf{u} \cdot \mathbf{p}_i)\mathbf{u} = l_i \mathbf{u} \tag{5.5}$$

where $l_i = (\mathbf{u} \cdot \mathbf{p}_i)$ is the (signed) length of $\mathbf{q}_i$ along the line specified by $\mathbf{u}$. We can now compute the average and variance of the l_i. The mean $\langle l_i \rangle$ is zero:

$$\langle l_i \rangle = \frac{1}{N} \sum_i l_i = \frac{1}{N} \sum_i (\mathbf{u} \cdot \mathbf{p}_i) = \mathbf{u} \cdot \left(\frac{1}{N} \sum_i \mathbf{p}_i \right) = \mathbf{u} \cdot \langle \mathbf{p}_i \rangle = 0 \tag{5.6}$$

since we assume the $\mathbf{p}_i$ are zero-meaned.

Now let's compute the *directional variance*, which we define as follows:

Definition 5.1 *If* $\langle \mathbf{p}_i \rangle = 0$, *then the* directional variance *along unit vector* $\mathbf{u}$ *is*

$$\sigma_{\mathbf{u}}^2(P) \equiv \left(\frac{1}{N} \sum_i l_i^2 \right). \tag{5.7}$$

Playing with algebra, we have:

$$\sigma_{\mathbf{u}}^2(P) = \frac{1}{N} \sum_i l_i^2 = \frac{1}{N} \sum_i (\mathbf{u} \cdot \mathbf{p}_i)^2 = \frac{1}{N} \sum_i (\mathbf{u} \cdot \mathbf{p}_i)(\mathbf{p}_i \cdot \mathbf{u}) \tag{5.8}$$

$$= \frac{1}{N} \sum_i (\mathbf{u}^T \mathbf{p}_i)(\mathbf{p}_i^T \mathbf{u}) = \frac{1}{N} \sum_i \mathbf{u}^T (\mathbf{p}_i \mathbf{p}_i^T)\mathbf{u} = \mathbf{u}^T \left(\frac{1}{N} \sum_i \mathbf{p}_i \mathbf{p}_i^T \right) \mathbf{u}. \tag{5.9}$$

$$\tag{5.10}$$

So we have,

$$\sigma_{\mathbf{u}}^2(P) = \mathbf{u}^T (M_P) \mathbf{u} \tag{5.11}$$

$$M_P \equiv \frac{1}{N} \sum_i \mathbf{p}_i \mathbf{p}_i^T. \tag{5.12}$$

The term $\mathbf{p}_i \mathbf{p}_i^T$ is the *outer product* of $\mathbf{p}_i$ with itself and is an $M \times M$ matrix. It is not to be confused with $\mathbf{p}_i^T \mathbf{p}_i$, which is the inner product of $\mathbf{p}_i$ with itself, which is a scalar. If P is zero-meaned, then M_P is known as the *covariance matrix* of P. For this chapter, we shall say that if P is arbitrary (not zero-meaned), then M_P is the *pseudo covariance matrix* of P. The pseudo covariance matrix equals the covariance matrix when P is zero-meaned. Later on, we shall be forced to consider the relationship between pseudo covariance matrices and true covariance matrices.

Let's pause now to consider where we are. Given P, we can compute the covariance matrix M_P. Equation (5.11) tells us that given the covariance matrix, we know the directional variance for *all* directions $\mathbf{u}$. Intuitively, M_P is a complete description of the variance of P in $\mathbb{R}^m$.

Now that we have a notion of standard deviation in $\mathbb{R}^m$, we should study its geometric properties. First notice that M_P is symmetric by construction. This means M_P has real eigenvalues and orthogonal eigenvectors. Furthermore, we know that M_P is positive semidefinite because $\mathbf{u}^T(M_P)\mathbf{u} = \sigma_{\mathbf{u}}^2(P) \geq 0$ for all unit vectors $\mathbf{u}$. This is because each term l_i^2 of the sum in equation (5.8) is non-negative. Consequently, M_P has non-negative eigenvalues.

Let $\mathbf{v}_i$ be eigenvectors of M_P with eigenvalues λ_i:

$$(M_P)\mathbf{v}_i = \lambda_i \mathbf{v}_i \tag{5.13}$$

$$\mathbf{v}_i \cdot \mathbf{v}_i = \delta_{ij} \tag{5.14}$$

$$\lambda_i \geq 0 \tag{5.15}$$

where δ_{ij} is the Kronecker delta. In short, M_P is our favorite kind of matrix (orthogonal eigenvectors, non-negative eigenvalues).

We now consider the geometry associated with M_P. We'll consider two aspects of M_P. First, we'll make a plot of the *directional standard deviation* and second, we'll consider the shape of the associated Gaussian distribution in $\mathbb{R}^m$.

Definition 5.2 *If* $\langle \mathbf{p}_i \rangle = 0$, *then the* directional standard deviation *along unit vector* $\mathbf{u}$ *is just the square root of the directional variance:*

$$l_i = \mathbf{u} \cdot \mathbf{p}_i \tag{5.16}$$

$$\sigma_{\mathbf{u}}(P) \equiv \sqrt{\sigma_{\mathbf{u}}^2(P)} = \sqrt{\frac{1}{N}\sum_i l_i^2}. \tag{5.17}$$

One can imagine plotting the directional standard deviation for all directions. One would do this by plotting all the points $(\sigma_{\mathbf{u}}(P))\mathbf{u}$ for all unit vectors $\mathbf{u}$. Intuitively, this is a polar plot where, instead of using angles to specify a direction, one uses a unit vector. The radius of the plotted point is the directional standard deviation. For an illustration of this, see figure 5.1, which is a plot in $\mathbb{R}^2$. If the eigenvalues of the covariance matrix are all identically equal, then the directional standard deviation is isotropic, hence our plot is circular. On the other hand, if the eigenvalues are not equal to each other, then the plot has a lobes and a dimpled structure. Perhaps surprisingly, the plot is not an ellipsoid, but might more aptly be called a "dimpleoid."

Next, we consider the Gaussian distribution, which is associated with the covariance matrix. We would like to know the shape of a Gaussian distribution which has the same directional standard

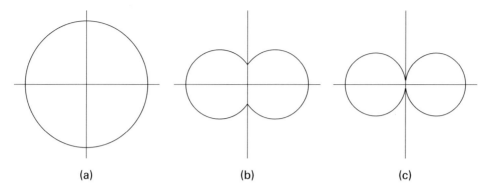

Figure 5.1
Examples of directional standard deviations in $\mathbb{R}^2$. In these three examples, we make polar plots of the directional standard deviation for three different covariance matrices. All three plots are in the Principal Order Frame (POF) of the covariance matrix, so that the eigenvectors are along the x and y axes. In (a) the eigenvalues are equal. In (b) the eigenvalues are not equal and nonzero, with the y eigenvalue being smaller than the x eigenvalue. In (c) the y eigenvalue is zero. Credit: Bruce Donald and Anthony Yan.

deviations as our covariance matrix. Let M_P be our covariance matrix. Let $\mathcal{N}(0, M_P)$ be a zero-meaned Gaussian with covariance matrix M_P. It can be shown that the directional standard deviations of $\mathcal{N}(0, M_P)$ are equal to the directional standard deviations of M_P. The directional standard deviation of a probability distribution $\mathcal{P}$ in $\mathbb{R}^m$ is computed from its covariance matrix, which is the expected value (average) of the outer product:

$$M_{\mathcal{P}} = \int_{\mathbb{R}^m} \mathbf{x}\mathbf{x}^T \mathcal{P}(\mathbf{x}) \, d\mathbf{x} \tag{5.18}$$

$$\sigma_{\mathbf{u}}^2(\mathcal{P}) = \mathbf{u}^T (M_{\mathcal{P}})\mathbf{u}, \tag{5.19}$$

where the integral is over $\mathbb{R}^m$, $d\mathbf{x}$ is a volume element of $\mathbb{R}^m$, and $\mathcal{P}(\mathbf{x})$ is a probability distribution over $\mathbb{R}^m$. We have abused notation in equation (5.19) since $\mathcal{P}$ is a probability distribution, not a set of points.

To make things simple, let's work in the POF of the covariance matrix. In the POF, M_P is diagonal, and so is its inverse M_P^{-1}. It is now easy to write down $\mathcal{N}(0, M_P)$ explicitly:

$$M_P = \begin{bmatrix} \lambda_1 & & & & \\ & \lambda_2 & & & \\ & & \lambda_3 & & \\ & & & \ddots & \\ & & & & \lambda_m \end{bmatrix} \tag{5.20}$$

$$M_P^{-1} = \begin{bmatrix} \lambda_1^{-1} & & & & \\ & \lambda_2^{-1} & & & \\ & & \lambda_3^{-1} & & \\ & & & \ddots & \\ & & & & \lambda_m^{-1} \end{bmatrix}$$
(5.21)

$$\mathcal{N}(0, M_P)|_{\mathbf{x}} = A \exp\left(\mathbf{x}^T M_P^{-1} \mathbf{x}\right).$$
(5.22)

The probability distribution $\mathcal{N}(0, M_P)$ can be visualized by its level sets. In particular, let's consider the level set that occurs when the exponent is equal to unity. In this case, we have the set of all $\mathbf{x} \in \mathbb{R}^m$ that satisfies $\mathbf{x}^T M_P^{-1} \mathbf{x} = 1$. If we write this explicitly in the POF, we get

$$\mathbf{x}^T M_P^{-1} \mathbf{x} = \sum_i \left(\frac{x_i^2}{\lambda_i}\right) = 1,$$
(5.23)

where x_i is the ith component of $\mathbf{x}$ in the POF; λ_i is the eigenvalue associated with the eigenvector pointing along the ith coordinate axis (and is associated with the component x_i). Recall that in the POF, the eigenvectors are along the coordinate axes.

The level set represented by equation (5.23) is easily recognized as the equation for a hyperellipsoid with semiaxes $\sqrt{\lambda_i}$. The length of the semiaxes are equal to the directional standard deviation along the semiaxes, and are also the standard deviations of the Gaussian distribution $\mathcal{N}(0, M_P)$ (see figure 5.2).

However, for directions *not* along an eigenvector, the ellipsoid's radius is not equal to the directional standard deviation. To illustrate this point, see figure 5.3, which overlays the level set of the Gaussian with its directional standard deviation.

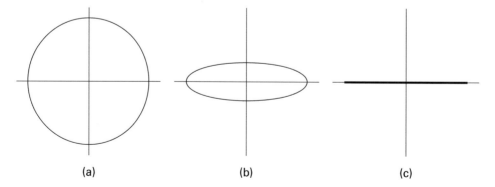

(a)   (b)   (c)

Figure 5.2
Examples of Gaussian ellipsoids in $\mathbb{R}^2$. In these three examples, we make polar plots of the level sets for three Gaussians associated with three different covariance matrices. All three plots are in the Principal Order Frame (POF) of the covariance matrix, so that the eigenvectors are along the x and y axes. In (a) the eigenvalues are equal. In (b) the eigenvalues are not equal and nonzero, with the y eigenvalue being smaller than the x eigenvalue. In (c) the y eigenvalue is zero. Credit: Bruce Donald and Anthony Yan.

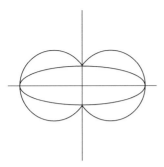

Figure 5.3
Overlay of directional standard deviation and Gaussian ellipsoid (level set of the Gaussian) in $\mathbb{R}^2$. In this example, we overlay a polar plot of the directional standard deviation and a plot of the associated Gaussian ellipsoid (level set). Our plot is in the Principal Order Frame (POF) so that the eigenvectors of the covariance matrix are along the coordinate axes. Notice that along the eigenvectors, the semiaxes of the ellipse are equal to the directional standard deviation. However in directions not along any eigenvectors, the radius of the ellipse is not equal to the directional standard deviation. Credit: Bruce Donald and Anthony Yan.

Finally, we discuss the largest and smallest directional variances. Let $\mathbf{u}$ be a unit vector in $\mathbb{R}^m$, and let P be an arbitrary point set:

$$\sigma_{\mathbf{u}}^2(P) = \mathbf{u}^T (M_P)\mathbf{u} = \lambda_1 u_1^2 + \lambda_2 u_2^2 + \lambda_3 u_3^2 + \cdots + \lambda_m u_m^2 \tag{5.24}$$

$$\mathbf{u}^T \mathbf{u} = u_1^2 + u_2^2 + u_3^2 + \ldots u_m^2 = 1 \tag{5.25}$$

$$u_i^2 \geq 0. \tag{5.26}$$

From equations (5.24)–(5.26) we see that in general, $\sigma_{\mathbf{u}}^2(P)$ is a *convex sum* of the λ_i's with coefficients u_i^2. That is, $\sigma_{\mathbf{u}}^2$ is a weighted average of the λ_i's. As is well known, the convex sum (convex combination) of m points lies within the convex hull of the m points. Here, we apply this fact trivially, because our points λ_i live in only one dimension, $\mathbb{R}$. So our "convex hull" is simply the largest and smallest values of λ_i:

$$\lambda_{\min} \equiv \min_i \lambda_i \tag{5.27}$$

$$\lambda_{\max} \equiv \max_i \lambda_i \tag{5.28}$$

$$\lambda_{\min} \leq \sigma_{\mathbf{u}}^2(P) \leq \lambda_{\max}. \tag{5.29}$$

PCA is the procedure of computing the eigenvalues λ_i and the corresponding eigenvectors $\mathbf{v}_i$, and is used to characterize the directional variances of the data P. It is beyond the scope of this book to give a full exposition of PCA. For additional details and applications, we refer the reader to any number of books on statistics, dimensionality reduction, and clustering. It is hoped, though, that this brief introduction will give the reader a solid intuition for PCA. (See the appendix for how to efficiently compute PCA in high dimensions with relatively few data points.)

5.3   Residual Dipolar Couplings in Structural Biochemistry

In NMR, we have the interaction of nuclear spins. Intuitively, each spin can be thought of as a magnetic dipole (a tiny magnet) that is spinning like a gyroscope. If there is a very strong external magnetic field, then the dipoles will align themselves with the external field.

In this case, the dipole-dipole interaction of spins can be written as

$$D = \frac{\mu_0 \gamma_a \gamma_b \hbar}{4\pi^2 r_{ab}^3} \left\langle \frac{3\cos^2\theta - 1}{2} \right\rangle \tag{5.30}$$

$$D = K \left\langle \frac{3\cos^2\theta - 1}{2} \right\rangle \tag{5.31}$$

where γ_a and γ_b are gyromagnetic ratios of the two spins, $\hbar$ is Planck's constant, r_{ab} is the distance between the two spins, and θ is the angle between the external field $\mathbf{B}$ and the internuclear vector $\mathbf{v}$ (figure 5.4). K is a fixed constant equal to $\mu_0 \gamma_a \gamma_b \hbar / 4\pi^2 r_{ab}^3$. The angle brackets represent an average over time and over the ensemble of proteins in solution, where the scale of the time average is much shorter than can be measured experimentally. For simplicity, we assume here that the nuclei a and b are covalently bonded, so the distance r_{ab} is not an ensemble average but known from crystal structures.

We now play a series of algebra games, which are the standard manipulations for deriving the Saupe matrix formalism. This is basically the same algebra we used to derive the covariance matrix, where a dot product $(\mathbf{B} \cdot \mathbf{v})^2$ or a $\cos^2\theta$ term is expanded as an outer product. To simplify the derivation, let's normalize $\mathbf{B}$ and $\mathbf{v}$:

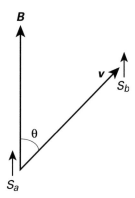

Figure 5.4
Dipole-dipole interaction in a strong external field. If an external magnetic field $\mathbf{B}$ is very strong, it will align the two spins S_a and S_b; $\mathbf{v}$ is the internuclear vector that points from the location of S_a to S_b. θ is the angle between $\mathbf{B}$ and $\mathbf{v}$. Credit: Bruce Donald and Anthony Yan.

$$\|\mathbf{B}\| = 1 \tag{5.32}$$

$$\|\mathbf{v}\| = 1 \tag{5.33}$$

$$D = (K/2) \langle 3\cos^2\theta - 1 \rangle \tag{5.34}$$

$$= (K/2) \langle 3(\mathbf{B} \cdot \mathbf{v})^2 - 1 \rangle \tag{5.35}$$

$$= (K/2) \langle 3(\mathbf{v} \cdot \mathbf{B})(\mathbf{B} \cdot \mathbf{v}) - 1 \rangle \tag{5.36}$$

$$= (K/2) \langle 3(\mathbf{v}^T \mathbf{B})(\mathbf{B}^T \mathbf{v}) - 1 \rangle \tag{5.37}$$

$$= (K/2) \langle 3\mathbf{v}^T (\mathbf{B}\mathbf{B}^T)\mathbf{v} - 1 \rangle \tag{5.38}$$

$$= (K/2) \langle 3\mathbf{v}^T (\mathbf{B}\mathbf{B}^T)\mathbf{v} - \mathbf{v}^T I \mathbf{v} \rangle, \tag{5.39}$$

where I is the identity matrix.

We now make a physical approximating assumption. Let's consider our protein to be perfectly rigid. With a rigid protein, we can define a *molecular coordinate frame* where $\mathbf{v}$ is constant, and $\mathbf{B}$ varies because the entire ensemble of proteins are tumbling in solution. Since $\mathbf{v}$ is constant, we can factor it out:

$$D = (K/2)\mathbf{v}^T \langle 3(\mathbf{B}\mathbf{B}^T) - I \rangle \mathbf{v} \tag{5.40}$$

$$D = (K/2)\mathbf{v}^T S \mathbf{v} \tag{5.41}$$

$$S \equiv \langle 3(\mathbf{B}\mathbf{B}^T) - I \rangle. \tag{5.42}$$

S is known as the *Saupe matrix* or *alignment tensor*. S is a symmetric, real-valued, and traceless matrix. To show that S is traceless, one only needs to show that $\text{Tr}(\mathbf{B}\mathbf{B}^T) = 1$. Let $\{\mathbf{B}_i\}$ represent the direction of the magnetic field in the molecular frames of the ensemble of proteins in solution, where $i \in \{1, 2, 3, \ldots, N\}$. Then,

$$\langle \mathbf{B}\mathbf{B}^T \rangle = \frac{1}{N} \sum_{i=1}^N \mathbf{B}_i \mathbf{B}_i^T. \tag{5.43}$$

Consider the trace of each term:

$$\text{Tr}(\mathbf{B}_i \mathbf{B}_i^T) = \text{Tr} \begin{bmatrix} B_x^2 & B_x B_y & B_x B_z \\ B_y B_x & B_y^2 & B_y B_z \\ B_z B_x & B_z B_y & B_z^2 \end{bmatrix} = B_x^2 + B_y^2 + B_z^2 = 1 \tag{5.44}$$

$$\text{Tr} \langle \mathbf{B}_i \mathbf{B}_i^T \rangle = \frac{1}{N} \sum_i^N 1 = 1 \tag{5.45}$$

$$\text{Tr} \langle 3\mathbf{B}_i \mathbf{B}_i^T - I \rangle = 3 - 3 = 0. \tag{5.46}$$

In a homogeneous solution, the distribution of **B** will be spherically uniform since the protein will be tumbling isotropically in all directions. A consequence of this isotropic tumbling is that S will become the zero matrix, and the measured RDC effect will be zero. To perform RDC experiments, aligning media are introduced into the solution. The aligning media make the solution anisotropic and cause the proteins to tumble anisotropically. Examples of aligning media include bicelles, phage, and stretched polyacrimide gels. Bicelles are very large flat planar walls of molecules. The wall-like bicelles are packed densely enough that they are roughly parallel, and form a liquid crystal. If the protein is not spherical, then steric collisions with the wall tend to favor some orientations of the protein and disfavor other orientations. As a result, the protein will not tumble isotropically, and the Saupe matrix S will not be zero. Other aligning media are believed to work in similar ways, although other aligning effects may be due to charge interactions between surface charges in the media and charges on the protein.

Notice that equations (5.39), (5.42), and (5.43) are similar to a covariance matrix. Because the $\{\mathbf{B}_i\}$ are not zero-meaned, (5.42) is a pseudo covariance matrix with the identity subtracted off.

While the preceding is the most common formalism presented for the Saupe matrix, we next consider a variation in which we do not assume the protein is perfectly rigid. In fact, we are more interested in the structure and dynamics of the protein than in the Saupe matrix S itself. But the same formalism that describes S and $\{\mathbf{B}_i\}$ can be used to characterize the distribution of internuclear bond vectors $\mathbf{v}_i$ due to protein flexibility. More precisely, there is an alignment tensor for $\mathbf{v}_i$ that has exactly the same form as S. Below, we derive this (well-known) formalism.

First, we need a special kind of matrix multiplication, which is essentially the dot product between two matrices.

Definition 5.3   *Given two matrices A and B, which are both $N \times M$, the* Frobenius inner product *of A and B is*

$$A \odot B \equiv \sum_{i=1}^{N} \sum_{j=1}^{M} A_{ij} B_{ij} \tag{5.47}$$

where A_{ij} and B_{ij} are the ij^{th} component of the respective matrix.

Notice that $\odot$ is commutative and distributive just like the vector dot product. If A and B are square, this can also be written as $A \odot B = \text{Tr}(A(B^T))$ where $\text{Tr}(\cdot)$ is the trace of a matrix. If B is symmetric, then this reduces to $A \odot B = \text{Tr}(AB)$.

Now let's return to equation (5.39), which precedes our assumption that the protein is rigid. Notice we have a term of the form $\mathbf{v}^T \mathbf{B}\mathbf{B}^T \mathbf{v}$. We have

Let $G = \mathbf{v}\mathbf{v}^T$ \hfill (5.48)

Let $H = \mathbf{B}\mathbf{B}^T$ \hfill (5.49)

$\mathbf{v}^T \mathbf{B}\mathbf{B}^T \mathbf{v} = \mathbf{B}^T \mathbf{v}\mathbf{v}^T \mathbf{B} = G \odot H.$ \hfill (5.50)

Furthermore, note that

$$(\mathbf{v}\mathbf{v}^T) \odot I = \begin{bmatrix} v_x^2 & v_x v_y & v_x v_z \\ v_y v_x & v_y^2 & v_y v_z \\ v_z v_x & v_z v_y & v_z^2 \end{bmatrix} \odot \begin{bmatrix} 1 & & \\ & 1 & \\ & & 1 \end{bmatrix} = \mathrm{Tr}\,(\mathbf{v}\mathbf{v}^T) = v_x^2 + v_y^2 + v_z^2 = 1 \qquad (5.51)$$

because $\|\mathbf{v}\| = 1$.

Using equations (5.50) and (5.51), one can factor equation (5.39) into two parts:

$$D = (K/2) \left\langle 3\mathbf{v}_i^T (\mathbf{B}_i \mathbf{B}_i^T) \mathbf{v}_i - 1 \right\rangle \qquad (5.52)$$

$$= (K/6) \left\langle 9\mathbf{v}_i^T (\mathbf{B}_i \mathbf{B}_i^T) \mathbf{v}_i - 3 \right\rangle \qquad (5.53)$$

$$= (K/6) \left\langle 9(\mathbf{B}_i \mathbf{B}_i^T) \odot (\mathbf{v}_i \mathbf{v}_i^T) - 3 \right\rangle \qquad (5.54)$$

$$= (K/6) \left\langle 9(\mathbf{B}_i \mathbf{B}_i^T) \odot (\mathbf{v}_i \mathbf{v}_i^T) - 3 - 3 + 3 \right\rangle \qquad (5.55)$$

$$= (K/6) \left\langle 9(\mathbf{B}_i \mathbf{B}_i^T) \odot (\mathbf{v}_i \mathbf{v}_i^T) - 3(\mathbf{B}_i \mathbf{B}_i^T) \odot I - 3I \odot (\mathbf{v}_i \mathbf{v}_i^T) + 3 \right\rangle \qquad (5.56)$$

$$= (K/6) \left\langle (3\mathbf{B}_i \mathbf{B}_i^T - I) \odot (3\mathbf{v}_i \mathbf{v}_i^T - I) \right\rangle \qquad (5.57)$$

$$D = (K/6) \left\langle E \odot F \right\rangle \qquad (5.58)$$

where

$$E = 3\mathbf{B}_i \mathbf{B}_i^T - I \qquad (5.59)$$

$$F = 3\mathbf{v}_i \mathbf{v}_i^T - I. \qquad (5.60)$$

At this point, we assume that the overall tumbling of the protein (which determines E) is uncorrelated with the flexing of the protein (which determines F in the molecular frame). If two variables are uncorrelated, then the expectation of their product is the product of their expectations:

$$D = (K/6) \left\langle E \right\rangle \odot \left\langle F \right\rangle = (K/6) S \odot T. \qquad (5.61)$$

We define the Saupe matrix $S \equiv \langle E \rangle = \left\langle 3\mathbf{B}_i \mathbf{B}_i^T - I \right\rangle$.

We define the alignment tensor for $\mathbf{v}_i$ to be $T \equiv \langle F \rangle = \left\langle 3\mathbf{v}_i \mathbf{v}_i^T - I \right\rangle$.

The literature often defines $S \equiv (K/2) \langle E \rangle$. Different constants may be found in the definition of T according to different primary researchers.

If the $\mathbf{v}_i$ are identical (i.e., the protein is rigid) in the molecular frame, then mathematically equation (5.61) reduces to equation (5.41).

Both the Saupe matrix S and $\mathbf{v}_i$'s alignment tensor have the same form. For protein flexibility (dynamics), we would like to know the distribution of $\{\mathbf{v}_i\}$ where the $\mathbf{v}_i$ are expressed in the molecular frame. (Because the protein is flexing, a molecular frame may not be well defined; however, it may be reasonably defined.)

It turns out that finding T is often an underconstrained problem. However, using a procedure described by Tolman [1], it seems possible to estimate T. Tolman's procedure requires RDCs in five aligning media which is considered to be a high number of media by current experimental standards. Also, Tolman's problem is underconstrained, so he is required to make some reasonable assumptions, which can be rationalized. Tolman's procedure can make reasonable (or at least plausible) estimates of T.

Once we have T, we would like to characterize the distribution $\{\mathbf{v}_i\}$ in the molecular frame. This, in turn, will help us understand protein flexibility.

Now that we have presented the alignment tensor formalism, and motivated our interest in it, we proceed to the next section where we describe in detail the relationship between alignment tensors and covariance matrices.

5.4 RDCs and PCA

In this section we relate RDCs to PCA by showing the connection between the alignment tensor and covariance matrices. To begin, let's compare the two, side by side.

Let $\{\mathbf{v}_i\}$ be an arbitrary set of unit vectors.

$$\mathbf{v}_i \in \mathbb{R}^m, i \in \{1, 2, 3, \ldots, m\} \text{ and } \|\mathbf{v}_i\| = 1 \tag{5.62}$$

$$\langle \mathbf{v}_i \rangle = \frac{1}{N} \sum_{i=1}^{N} \mathbf{v}_i \tag{5.63}$$

$$\text{Covariance Matrix: } M_P = \frac{1}{N} \sum_{i=1}^{N} (\mathbf{v}_i - \langle \mathbf{v}_i \rangle)(\mathbf{v}_i - \langle \mathbf{v}_i \rangle)^T \tag{5.64}$$

$$\text{Alignment Tensor: } T = (K/6) \frac{1}{N} \sum_{i=1}^{N} \left(3\mathbf{v}\mathbf{v}^T - I\right) = (K/6)(3Q - I) \tag{5.65}$$

$$\text{where } Q \equiv \frac{1}{N} \sum_{i=1}^{N} \mathbf{v}\mathbf{v}^T \text{ is a pseudo covariance matrix.} \tag{5.66}$$

Our goal is to get as much information about the distribution $\{\mathbf{v}_i\}$ as possible. In particular, we would like to perform PCA on $\{\mathbf{v}_i\}$, or if we can't, we would like to have as much statistics on $\{\mathbf{v}_i\}$ as possible.

The two differences between M_P and T are that the outer products in T are not from zero-meaned vectors, and the final result has the identity subtracted off (The identity can be pulled out of the summation because it is constant, so its average is itself.) As we shall see, subtracting the identity is essentially no problem, but the lack of zero-meaning of the data is a big problem.

First, let us address the subtraction of the identity. It is trivial to see that T has the same eigenvectors as Q, and the eigenvalues of T have a simple relationship. This is because, if $\mathbf{x}$ is an eigenvector of Q, then multiplication by either Q or the identity I, does not change the direction of $\mathbf{x}$:

$$Q\mathbf{x} = (\lambda_Q)\mathbf{x} \tag{5.67}$$

$$\Longleftrightarrow T\mathbf{x} = (K/6)\,(3Q - I)\,\mathbf{x} = (K/6)\,(3Q\mathbf{x} - \mathbf{x}) \tag{5.68}$$

$$\Longleftrightarrow T\mathbf{x} = (K/6)\left(3\lambda_Q\mathbf{x} - \mathbf{x}\right) \tag{5.69}$$

$$\Longleftrightarrow T\mathbf{x} = (K/6)\left(3\lambda_Q - 1\right)\mathbf{x} \tag{5.70}$$

$$\Longleftrightarrow T\mathbf{x} = (\lambda_T)\mathbf{x} \text{ where } \lambda_T = (K/6)\left(3\lambda_Q - 1\right). \tag{5.71}$$

So if we know T and K, we can solve for the pseudo covariance matrix Q. In practice, we know T only up to a scale factor, in which case we can solve for Q up to a scale factor.

Given T, we know the pseudo covariance matrix for $\{\mathbf{v}_i\}$. What is the relationship between the pseudo covariance matrices and true covariance matrices? As we will show later, the directional variance of a pseudo covariance matrix is an upper bound on the eigenvalues of the true covariance matrix (directional variances of the true covariance matrix). To prove this, we first show that the pseudo covariance matrix Q of $\{\mathbf{v}_i\}$ is actually the true covariance matrix of a different point set. Then, we will show that the directional variance of that different point set is always greater than or equal to the directional variance of the original point set $\{\mathbf{v}_i\}$. Finally, we will show that if the mean of $\{\mathbf{v}_i\}$ were known, then we could convert the pseudo covariance matrix Q into the true covariance matrix for $\{\mathbf{v}_i\}$:

$$\text{Let } V^+ = \{\mathbf{v}_i\} \text{ where } \mathbf{v}_i \in \mathbb{R}^m, i \in \{1, 2, 3, \ldots, m\} \text{ and } \|\mathbf{v}_i\| = 1 \tag{5.72}$$

$$\text{Let } V^- = \{-\mathbf{v}_i \mid \mathbf{v}_i \in V^+\} \tag{5.73}$$

$$\text{Let } W = V^+ \cup V^- \tag{5.74}$$

$$\text{Let } Q = \frac{1}{N} \sum_i \mathbf{v}_i \mathbf{v}_i{}^T \text{ pseudo covariance matrix.} \tag{5.75}$$

We now claim that Q is the true covariance matrix of $W = V^+ \cup V^-$. First, we show that W is zero-meaned:

$$\frac{1}{2N} \sum_{\mathbf{w} \in W} \mathbf{w} = \frac{1}{N} \sum_{i=1}^{N} (\mathbf{v}_i - \mathbf{v}_i) = 0. \tag{5.76}$$

Next, we compute the true covariance matrix of W and show it is equal to Q:

$$M_W = \frac{1}{2N} \sum_{\mathbf{w} \in W} \mathbf{w}\mathbf{w}^T = \frac{1}{2} \left[\frac{1}{N} \sum \mathbf{v}_i \mathbf{v}_i^T + \frac{1}{N} \sum (-\mathbf{v}_i)(-\mathbf{v}_i^T) \right] \qquad (5.77)$$

$$= \frac{1}{2} \left[\frac{1}{N} \sum \mathbf{v}_i \mathbf{v}_i^T + \frac{1}{N} \sum \mathbf{v}_i \mathbf{v}_i^T \right] \qquad (5.78)$$

$$= \frac{1}{2} [2Q] \qquad (5.79)$$

$$= Q. \qquad (5.80)$$

Intuitively, we have performed PCA not on the original data $V^+ = \{\mathbf{v}_i\}$, but on "the original data plus its reflection," namely, PCA on $W = V^+ \cup V^-$. See figure 5.5 for an example in $\mathbb{R}^2$.

Next, we show that the directional variance of $V^+ \cup V^-$ is always greater than or equal to the directional variance of V^+. To do this, we start with zero-meaned data and show that if the data

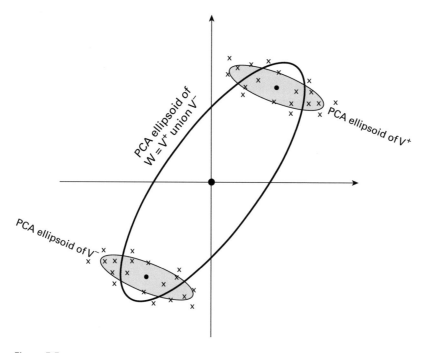

Figure 5.5
PCA of $W = V^+ \cup V^-$ in $\mathbb{R}^2$. The x's mark the location of data points $\mathbf{v}_i \in V^+$. The shaded ellipse represents PCA on V^+ and is the level set of the associated Gaussian for V^+. The large unshaded ellipse represents PCA on $W = V^+ \cup V^-$ and is the level set of the associated Gaussian for W. The means of the three sets W, V^+, and V^- are marked by dark filled circles. Credit: Bruce Donald and Anthony Yan.

is translated in any direction, the directional standard deviation will remain the same or increase. In fact, the directional variance along the direction of translation will always increase:

Let $X_0 = \{\mathbf{x}_i\}$ where

$$\mathbf{x}_i \in \mathbb{R}^m \text{ and } i \in \{1, 2, 3, \ldots, N\} \tag{5.81}$$

By assumption, X_0 is zero-meaned:

$$\langle \mathbf{x}_i \rangle = 0 \tag{5.82}$$

Let $\mathbf{h} \in \mathbb{R}^m$ be an arbitrary offset. $\tag{5.83}$

Let $X_\mathbf{h}^+ = \{\mathbf{y}_i \mid \mathbf{y}_i = \mathbf{x}_i + \mathbf{h}\}$ $\tag{5.84}$

where $\mathbf{x}_i \in X_0$

Let $X_\mathbf{h}^- = \{-\mathbf{y}_i \mid \mathbf{y}_i \in X_\mathbf{h}^+\}$. $\tag{5.85}$

Let $W_\mathbf{h} = X_\mathbf{h}^+ \cup X_\mathbf{h}^-$. $\tag{5.86}$

We claim that for any direction $\mathbf{u}$, the directional variance of $W_\mathbf{h} = X_\mathbf{h}^+ \cup X_\mathbf{h}^-$ is greater than or equal to the directional variance of X_0. First, let's consider the covariance matrix of X_0 and $W_\mathbf{h}$:

$$M_{X_0} = \frac{1}{N} \sum_{i=1}^{N} \mathbf{x}_i \mathbf{x}_i^T \tag{5.87}$$

$$M_{W_\mathbf{h}} = \frac{1}{2N} \left[\sum_{i=1}^{N} (\mathbf{x}_i + \mathbf{h})(\mathbf{x}_i + \mathbf{h})^T + \sum_{i=1}^{N} [-(\mathbf{x}_i + \mathbf{h})][-(\mathbf{x}_i + \mathbf{h})]^T \right] \tag{5.88}$$

$$= \frac{2}{2N} \left[\sum_{i=1}^{N} (\mathbf{x}_i + \mathbf{h})(\mathbf{x}_i + \mathbf{h})^T \right] \tag{5.89}$$

$$= \frac{1}{N} \sum_{i=1}^{N} \left(\mathbf{x}_i \mathbf{x}_i^T + \mathbf{x}_i \mathbf{h}^T + \mathbf{h} \mathbf{x}_i^T + \mathbf{h} \mathbf{h}^T \right) \tag{5.90}$$

$$= \frac{1}{N} \sum_{i=1}^{N} \left(\mathbf{x}_i \mathbf{x}_i^T + \mathbf{h} \mathbf{h}^T \right) + \frac{1}{N} \sum_{i=1}^{N} (\mathbf{x}_i \cdot \mathbf{h} + \mathbf{h} \cdot \mathbf{x}_i) \tag{5.91}$$

$$= \frac{1}{N} \sum_{i=1}^{N} \left(\mathbf{x}_i \mathbf{x}_i^T + \mathbf{h} \mathbf{h}^T \right) + 2\mathbf{h} \cdot \left[\frac{1}{N} \sum_{i=1}^{N} \mathbf{x}_i \right] \tag{5.92}$$

$$= \frac{1}{N} \sum_{i=1}^{N} \left(\mathbf{x}_i \mathbf{x}_i^T + \mathbf{h} \mathbf{h}^T \right) + 2\mathbf{h} \cdot \langle \mathbf{x}_i \rangle \tag{5.93}$$

$$= \frac{1}{N} \sum_{i=1}^{N} \left(\mathbf{x}_i \mathbf{x}_i^T + \mathbf{h}\mathbf{h}^T \right) \tag{5.94}$$

$$= \frac{1}{N} \sum_{i=1}^{N} \mathbf{x}_i \mathbf{x}_i^T + \frac{1}{N} \sum_{i=1}^{N} \mathbf{h}\mathbf{h}^T \tag{5.95}$$

$$= M_{X_0} + \frac{N}{N} \mathbf{h}\mathbf{h}^T \tag{5.96}$$

$$M_{W_{\mathbf{h}}} = M_{X_0} + \mathbf{h}\mathbf{h}^T. \tag{5.97}$$

Given the covariance matrices for X_0 and $X_{\mathbf{h}}^+ \cup X_{\mathbf{h}}^-$, we can now compare their directional variances:

$$\sigma_{\mathbf{u}}^2(X_0) = \mathbf{u}^T (M_{X_0}) \mathbf{u} \tag{5.98}$$

$$\sigma_{\mathbf{u}}^2(W_{\mathbf{h}}) = \mathbf{u}^T (M_{X_0} + \mathbf{h}\mathbf{h}^T) \mathbf{u} \tag{5.99}$$

$$\sigma_{\mathbf{u}}^2(W_{\mathbf{h}}) = \mathbf{u}^T (M_{X_0}) \mathbf{u} + \mathbf{u}^T \mathbf{h}\mathbf{h}^T \mathbf{u} \tag{5.100}$$

$$\sigma_{\mathbf{u}}^2(W_{\mathbf{h}}) = \sigma_{\mathbf{u}}^2(X_0) + (\mathbf{u} \cdot \mathbf{h})(\mathbf{h} \cdot \mathbf{u}) \tag{5.101}$$

$$\sigma_{\mathbf{u}}^2(W_{\mathbf{h}}) = \sigma_{\mathbf{u}}^2(X_0) + (\mathbf{u} \cdot \mathbf{h})^2. \tag{5.102}$$

Both $W_{\mathbf{h}}$ and X_0 are positive semidefinite, and $(\mathbf{u} \cdot \mathbf{h})^2$ is also always greater than or equal to zero. So all the terms of equation (5.102) are non-negative. As a result, we can conclude that

$$\sigma_{\mathbf{u}}^2(W_{\mathbf{h}}) = \sigma_{\mathbf{u}}^2(X_{\mathbf{h}}^+ \cup X_{\mathbf{h}}^-) \geq \sigma_{\mathbf{u}}^2(X_0). \tag{5.103}$$

Intuitively, the nonzero meaned pseudo covariance matrix always has a larger direction variance than the zero-meaned covariance matrix because the nonzero-meaned case is "stretched apart" by its shadow. (see figure 5.5).

So given $M_{W_{\mathbf{h}}}$ (which is equal to $M_{X_{\mathbf{h}}^+}$), we have upper bounds on the directional variances of X_0. In the event that we know $\mathbf{h}$ (which is the mean of $X_{\mathbf{h}}^+$), we can use equation (5.97) to compute the original covariance matrix M_{X_0} from the pseudo covariance matrix $M_{X_{\mathbf{h}}^+}$ because $M_{W_{\mathbf{h}}} = M_{X_{\mathbf{h}}^+}$. This would allow us to perform PCA on X_0 even though we were only given the pseudo covariance matrix.

One might wonder if equations (5.97) and (5.102) can be used to solve for M_{X_0} by minimizing $\left[\sigma_{\mathbf{u}}^2(W_{\mathbf{h}}) - (\mathbf{u} \cdot \mathbf{h})^2 \right]$. Intuitively, the idea is to find the translation $\mathbf{h}$ that minimizes the directional variances as much as possible. Unfortunately, this is not the case. The intuition is reasonable, however, we cannot perform all the required steps.

Suppose we are given a pseudo covariance matrix M_A of some point set $A = \{\mathbf{x}_i\}$, which is arbitrary (not zero-meaned). Now we wish to compute the pseudo covariance matrix of $A_{\mathbf{h}}$, which is the translation of A:

$$A_{\mathbf{h}} = \{(\mathbf{x}_i + \mathbf{h}) \mid \mathbf{x}_i \in A\}. \tag{5.104}$$

We are given only M_A and $\mathbf{h}$. Can we compute $M_{A_\mathbf{h}}$? The answer is no:

$$M_{A_\mathbf{h}} = \frac{1}{N}\sum_i (\mathbf{x}_i + \mathbf{h})(\mathbf{x}_i + \mathbf{h})^T \tag{5.105}$$

$$M_{A_\mathbf{h}} = \frac{1}{N}\sum_i (\mathbf{x}_i\mathbf{x}_i^T + \mathbf{h}\mathbf{h}^T) + \frac{1}{N}\sum_i [(\mathbf{x}_i \cdot \mathbf{h}) + (\mathbf{h} \cdot \mathbf{x}_i)] \tag{5.106}$$

$$M_{A_\mathbf{h}} = \frac{1}{N}\sum_i (\mathbf{x}_i\mathbf{x}_i^T + \mathbf{h}\mathbf{h}^T) + 2\frac{1}{N}\sum_i (\mathbf{h} \cdot \mathbf{x}_i) \tag{5.107}$$

$$M_{A_\mathbf{h}} = \frac{1}{N}\sum_i (\mathbf{x}_i\mathbf{x}_i^T + \mathbf{h}\mathbf{h}^T) + 2\mathbf{h} \cdot \left(\frac{1}{N}\sum_i \mathbf{x}_i\right) \tag{5.108}$$

$$M_{A_\mathbf{h}} = \frac{1}{N}\sum_i (\mathbf{x}_i\mathbf{x}_i^T + \mathbf{h}\mathbf{h}^T) + 2\mathbf{h} \cdot \langle \mathbf{x}_i \rangle . \tag{5.109}$$

So we can see that we need to know the average of $\{\mathbf{x}_i\}$ to solve for $M_{A_\mathbf{h}}$. In case that the data is already zero-meaned, then $\langle \mathbf{x}_i \rangle = 0$ and the above reduces to equation (5.97).

5.5   Conclusions and Future Work

Though we don't know the average $\langle \mathbf{x}_i \rangle$ from the RDC data, there may be reasonable ways to estimate it based on the area of application. For example, if the protein's X-ray or NMR structure is known, then the internuclear vectors of the known structure might be reasonable estimates of the average $\langle \mathbf{x}_i \rangle$. Given such an estimate, it becomes possible to estimate M_{X_0} and perform PCA on $\{\mathbf{x}_i\}$.

In some cases, the vectors $\{\mathbf{x}_i\}$ may be closely clustered on the unit sphere. If their directional variances are much less than unity, then they can be considered to be tightly clustered. In this case, the principle eigenvector of the pseudo covariance matrix may be a reasonable approximation for the average $\langle \mathbf{x}_i \rangle$. However, normally in RDC data, the Saupe matrix is nearly zero because the vast majority of proteins are tumbling isotropically in solution. Only a tiny fraction of proteins interact with the aligning media and tumble anisotropically. As a result, we know that the directional variance will be almost the same in all directions, and in fact, the mean will be close to zero. In a sense, we are fortunate that the Saupe matrix subtracts the identity from the pseudo covariance, because if this were not the case, then the eigenvalues of the Saupe matrix would be nearly identical. That, in turn, would make computation of the eigenvectors of the Saupe matrix less numerically stable. If two eigenvalues are equal, their directions must lie in the plane spanned by both of them, but their directions are otherwise unrestricted in that plane. Thus, if the two eigenvalues are nearly equal, their directions are nearly unconstrained in the plane that they span.

Finally, it may be possible to add information or assumptions that result in a lower bound on the directional variances.

Appendix

In this appendix we consider how to compute PCA efficiently when $M \gg N \gg 1$, where M is the dimension of our data, and N is the number of data points. We are given $\mathbf{x}_i \in \mathbb{R}^M$ with $i \in \{1, 2, 3, \ldots, N\}$. We would like to compute PCA on $\{\mathbf{x}_i\}$ while not being overwhelmed by M.

First, construct a data matrix where the columns are the data vectors $\mathbf{x}_i$:

$$A \equiv \left[\; \mathbf{x}_1 \; \middle| \; \mathbf{x}_2 \; \middle| \; \mathbf{x}_3 \; \middle| \cdots \middle| \; \mathbf{x}_M \; \right]. \tag{5.110}$$

We can compute the covariance matrix $M_A = AA^T$. But, M_A is now $M \times M$, which is huge because $M \gg N \gg 1$. Working with M_A is computationally infeasible. Notice, though, that N points in any dimension live in a subspace whose dimension is at most $N - 1$, which is much less that M. (For example, three points determine a two-dimensional plane, no matter what dimension the points live in themselves.) So it seems natural to suspect that we can work with a matrix that is $N \times N$ rather than $M \times M$.

The matrix $A^T A$ is $N \times N$. What is its relationship to AA^T? Notice that both $A^T A$ and AA^T are symmetric real matrices. So both will have orthogonal eigenvectors. Suppose that we have an eigenvector of AA^T:

$$AA^T \mathbf{v} = \lambda \mathbf{v}. \tag{5.111}$$

Then we can multiply both sides by A^T:

$$A^T AA^T \mathbf{v} = \lambda A^T \mathbf{v} \tag{5.112}$$

$$A^T A \left(A^T \mathbf{v} \right) = \lambda \left(A^T \mathbf{v} \right). \tag{5.113}$$

So $A^T \mathbf{v}$ is an eigenvector of $A^T A$ with the same eigenvalue λ.

Similarly, suppose we have an eigenvector of $A^T A$:

$$A^T A \mathbf{u} = \gamma \mathbf{u} \tag{5.114}$$

$$AA^T A \mathbf{u} = \gamma A \mathbf{u} \tag{5.115}$$

$$AA^T \left(A\mathbf{u} \right) = \gamma \left(A\mathbf{u} \right). \tag{5.116}$$

$$\tag{5.117}$$

So $A\mathbf{u}$ is also an eigenvector of AA^T with the same eigenvalue γ.

As a result, we can solve the eigen-problem for the smaller matrix $A^T A$ and then transform the eigenvalues and eigenvectors into solutions of the eigen-problem of the larger matrix $A A^T$. This can save a tremendous amount of computation, and allows PCA to be feasible in practice.

Reference

[1] J. R. Tolman. A novel approach to the retrieval of structural and dynamic information from residual dipolar couplings using several oriented media in biomolecular NMR spectroscopy. *Journal American Chemical Society,* 124 (2002): 12020–12030.

6 Orientational Sampling of Interatomic Vectors

6.1 Introduction

Sampling of tensorial properties is one of the fundamental concepts in computational structural biology. We introduce the basic concepts in this chapter, while describing RDCs in more detail. Second-rank tensors, such as the Saupe alignment tensor, can be determined using a set of experimental data and interatomic bond vectors. We describe a framework from Cowburn and co-workers [1] that quantifies how well a distribution of bond vectors samples the components of a tensor, and therefore provides a measure of tensor accuracy and quality. In particular, the framework measures (1) how well orientation space is sampled by a distribution of bond vectors, (2) how well this distribution samples the components of a second-rank tensor, and (3) how well the distribution completely characterizes the tensor.

   The sampling properties of a set of 1736 proteins, representing all experimentally determined protein folds, were analyzed using the framework [1]. The results show that NH and carbonyl C′O bond vectors are the least uniformly distributed, reflecting secondary structure properties and hydrogen bonding in proteins. We also describe applications of the framework to experimental optimization, resulting in the determination of more accurate tensor properties.

6.2 Theory

The accuracy of a second-rank tensor D, determined using a set of interatomic bond vectors, can be computed as follows. We describe (1) the sampling tensor (Ω) and (2) the generalized sampling parameter (Ξ), which mathematically characterize the uniformity of the bond vector distribution. We also calculate (3) the average constant (D_{av}) and (4) the generalized quality factor (Λ), which quantify the accuracy of the tensor as sampled by the distribution.

6.2.1 Sampling Tensor

The sampling tensor (Ω), represented by a 3×3 matrix, characterizes the sampling of interatomic bond vector orientations along three axes of an arbitrary reference frame. The elements of the sampling tensor are calculated by

$$\Omega_{ij} = \frac{3\langle \mathbf{r}_i \mathbf{r}_j \rangle - \delta_{ij}}{2},$$ (6.1)

where i, j are axes of an arbitrary reference frame, $\mathbf{r}_i$ is the projection of a unit bond vector on the axis i, and δ_{ij} is the Kronecker delta.

The sampling tensor can be diagonalized to yield the principal axis frame oriented in the direction of best sampling, and the principal values Ω_i of Ω. The principal values Ω_i can be used to calculate the sampling fractions f_i,

$$f_i = \langle \mathbf{r}_i^2 \rangle = \frac{2\Omega_i + 1}{3},$$ (6.2)

which represent the fraction of bond vectors oriented along each axis i, with $f_z + f_y + f_x = 1$. Since the tensor is diagonalized in the direction of best sampling, $\Omega_z \geq \Omega_y \geq \Omega_x$, and $f_z \geq f_y \geq f_x$. If the distribution of vectors is uniform, $f_z = f_y = f_x = \frac{1}{3}$, and $\Omega_i = 0$. If the distribution of vectors is perfectly nonuniform, $f_z = 1$, $f_y = f_x = 0$, $\Omega_z = 1$, and $\Omega_y = \Omega_x = -\frac{1}{2}$. The triangle spanned by (f_x, f_y, f_z) is called a *simplex* (see figure 6.1). The properties of sets of simplices will be discussed further in chapter 50.

6.2.2 Generalized Sampling Parameter
The generalized sampling parameter, a scalar, quantifies the uniformity of the distribution of bond vectors. The sampling parameter is calculated by

$$\Xi = \frac{1}{4}[(3f_z - 1)^2 + 3(f_y - f_x)^2],$$ (6.3)

where f_i are the sampling fractions. If the distribution of vector orientations is uniform, $\Xi = 0$, and if the distribution of vector orientations is perfectly nonuniform, which means that all of the vectors are in the same direction, $\Xi = 1$.

6.2.3 Average Constant
While Ω and Ξ characterize the orientational uniformity of a set of bond vectors, D_{av} quantifies how well the tensor of interest, D, is sampled by that set. D_{av} can be calculated by

$$D_{av} = \frac{1}{3}\text{Tr}[D] + \frac{2}{3}\sum_{i,j=x',y',z'} \Omega_{ij}D_{ij},$$ (6.4)

where Ω_{ij} are elements of the sampling tensor Ω and D_{ij} are elements of D. If all parts of the tensor are sampled equally well, $D_{av} = \frac{1}{3}\text{Tr}[D] = D_{iso}$.

6.2.4 Generalized Quality Factor
The generalized quality factor, a scalar similar to the generalized sampling parameter, measures how well a set of bond vectors samples all elements of D, the tensor of interest. The quality factor is calculated by

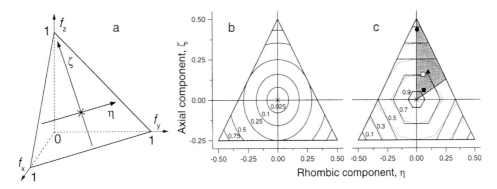

Figure 6.1
Geometric representation of sampling fractions [1]. Reprinted with permission. Copyright 2000 American Chemical Society.

$$\Lambda = 1 - \left| \frac{D_{av} - D_{iso}}{D_z - D_{iso}} \right|, \tag{6.5}$$

where D_{av} is the average constant, D_{iso} is the optimal average constant, and D_z is the z principal component of D.

6.2.5 Geometric Representation

The sampling fractions for a distribution of bond vectors can be represented by a point $\mathbf{f} = (f_x, f_y, f_z)$ on a plane in $\{f_x, f_y, f_z\}$-space. This plane (figure 6.1a) can be parametrized (figure 6.1b and c) in terms of a rhombic component (η) and an axial component (ξ), calculated by

$$\eta = \frac{1}{2}(f_y - f_x), \tag{6.6}$$

$$\xi = \frac{1}{4}(3f_z - 1). \tag{6.7}$$

The rhombic and axial components represent the degree of directional asymmetry of the sampling tensor. Thus, points lying close to the parametrized origin ($\eta = \xi = 0$) represent a higher degree of uniformity, whereas points lying away from the parametrized origin represent a lower degree of uniformity. Contour lines represent levels of the generalized sampling parameter, Ξ (figure 6.1b), and the generalized quality factor, Λ (figure 6.1c). Because the ordering $f_z \geq f_y \geq f_x$ is imposed, points may only lie in the shaded area in figure 6.1c.

6.3 Results

To determine which types of bond vectors (NH, NC$^\alpha$, C$^\alpha$H$^\alpha$, C$^\alpha$C$'$, C$'$O, C$_i'$N$_{i+1}$) sample orientation space optimally, the sampling properties of 1,736 protein structures (879 single proteins, 857 multisubunit proteins) were measured using the framework of Cowburn and co-workers [1].

The results are graphically illustrated in figure 6.2. The triangles shown correspond to the shaded area in figure 6.1c. Each point represents the sampling properties, (η, ξ), for a particular set of bond vectors in a particular structure.

NH bond vectors represent the least uniformly distributed sets. C'O bond vectors also demonstrate nonuniform distributions. This result has a biophysical, structural basis. NH bond vectors and C'O bond vectors lie antiparallel in the same peptide plane, and thus sample orientation space similarly. The result also reflects secondary structure properties in proteins. N-H$\cdots$O=C hydrogen bonding is important to forming α-helix and β-sheet secondary structures.

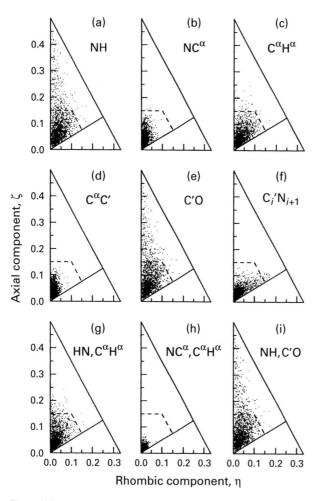

Figure 6.2
Sampling distributions for interatomic bond vectors [1]. Reprinted with permission. Copyright 2000 American Chemical Society.

Generalized sampling parameter (Ξ) values were calculated for ideal α-helices and β-sheets. In α-helices, $\Xi = 0.84$ for NH bond vectors, indicating a nonuniform distribution. NH bond vectors are all oriented in approximately the same direction, donating a hydrogen bond to the carbonyl group of the amino acid four residues earlier. $C^{\alpha}H^{\alpha}$ bond vectors sample orientation space more optimally, with $\Xi = 0.03$. $C^{\alpha}H^{\alpha}$ bond vectors are oriented more uniformly, in different directions around the outside of the helix. NH and $C^{\alpha}H^{\alpha}$ bond vectors together sample orientation space better than NH bond vectors alone, with $\Xi = 0.15$.

In β-sheets, both NH and $C^{\alpha}H^{\alpha}$ bond vectors are distributed nonuniformly, with $\Xi = 0.80$ for NH bond vectors and $\Xi = 0.95$ for $C^{\alpha}H^{\alpha}$ bond vectors. To sample orientation space more uniformly, another set of bond vectors, such as $C^{\alpha}C'$, must be used.

6.4 Applications

This framework presents opportunities for experimental optimization. When determining a second-rank tensor for a particular structure or substructure, orientation space may not be optimally sampled, and tensor accuracy may be low, if a particular set of bond vectors, a particular set of residues, or a particular aligning medium is used. By quantifying tensor accuracy and the uniformity of the bond vector distribution using the framework, the experiment can be optimized such that a more uniform distribution and a better sampled, more accurate tensor is obtained.

Fushman et al. [1] determined a rotational diffusion tensor for a protein, using NH bond vectors and ^{15}N relaxation data. For all residues, the bond vector distribution was uniform, $\Xi = 0.0232$, and the tensor was well sampled, $\Lambda = 0.9256$. However, using only α-helical residues, $\Xi = 0.7160$. We can conclude that using NH bond vectors, data from the α-helical residues alone is insufficient to optimally sample orientation space and accurately determine the rotational diffusion tensor. To characterize the rotational diffusion tensor from α-helical residues, additional sets of vectors, such as $C^{\alpha}H^{\alpha}$ or $C^{\alpha}C'$, should be used.

The authors also determined a Saupe alignment tensor for a protein by recording NH residual dipolar couplings in a liquid-crystalline medium. While the bond vectors sampled orientation space relatively well, $\Xi = 0.1084$, tensor quality was not optimal, $\Lambda = 0.7724$. The quality factor (Λ) changes if the alignment tensor frame changes relative to the sampling tensor frame. By changing the aligning medium, we can change the relative angle, and improve Λ. This can be accomplished by doping the medium with ions, or using a different aligning medium. Changing from a liquid-crystalline medium to a purple membrane medium changes the angle of the alignment tensor, raising the value of the quality factor and resulting in a more accurate, better sampled alignment tensor, $\Lambda = 0.96$.

Reference

[1] D. Fushman, R. Ghose, and D. Cowburn. The effect of finite sampling on the determination of orientational properties: A theoretical treatment with application to interatomic vectors in proteins. *Journal of the Amercian Chemical Society* 122(2000): 10640–10649.

7 Solution Structures of Native and Denatured Proteins Using RDCs

In the previous chapters, we introduced the principles of the Residual dipolar coupling (RDC), which can provide global orientational restrains on internuclear vectors. In this lecture, we introduce some RDC applications, that is, using RDCs to determine the solution structures of native [1, 2] and denatured [3] proteins.

7.1 Determining Native Protein Structure

We first introduce a quartic equation and two simple trigonometric equations that can compute, *exactly* and *in constant time*, the backbone dihedral angles for a residue from RDCs in two media on any single backbone vector type. Furthermore, based on these exact solutions we introduce a systematic search algorithm for determining protein backbone substructure consisting of both α-helices and β-sheets [1]. "Exact" solutions is a mathematical term, meaning that the polynomials can be solved in closed form (see chapters 15,16). It should not be confused with a guarantee of "biological exactness."

7.1.1 Theoretical Background

Based on the equations of NH RDCs measured in two media, we can derive a quartic equation that allows us to compute the vector orientation. For instance, let $\mathbf{v} = (x \; y \; z)^T$ be the NH bond vector in the POF of medium 1. With respect to x, we have

$$f_4 u^4 + f_3 u^3 + f_2 u^2 + f_1 u + f_0 = 0,$$

$$u = 1 - 2\left(\frac{x}{a}\right)^2, \tag{7.1}$$

where the coefficients $f_0, f_1, f_2, f_3, f_4, a$, as well as y and z can be computed from the equations of NH RDCs (appendix A in [1]). As shown in figure 7.1, the number of real solutions of equation (7.1) is at most 8, or in other words, given NH RDCs measured in two media, we can obtain up to 8 possible vector orientations, one of which is the correct orientation.

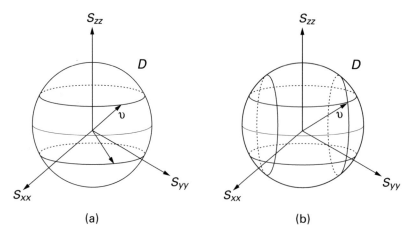

Figure 7.1
An example of extracting NH unit vector using RDCs in two media: (a) unit vector determined by one RDC; (b) unit vector determined by two RDCs. Consequently, 0, 2, 4, 6, or 8 solutions for the internuclear vector orientation can be obtained from two RDCs. Credit: Bruce Donald and Jeff Martin.

In order to compute the dihedral angles for a single residue, we can prove that if the directions of any two vectors $\mathbf{v}_i$ and $\mathbf{v}_{i+1}$ in consecutive peptide planes i and $i+1$ are known, together with the orientation of peptide plane i, then the intervening backbone angles (ϕ_i, ψ_i) can be computed from the following two trigonometric equations:

$$\sin(\phi_i + a_1) = b_1, \qquad \sin(\psi_i + a_2) = b_2, \tag{7.2}$$

where a_1, b_1, a_2, and b_2 are computed from the six bond angles between two consecutive residues, $\mathbf{v}_i$, and $\mathbf{v}_{i+1}$. The full expressions for all these four coefficients are computed from backbone kinematics (appendix B in [1]). Note that these two equations can be solved exactly for $\sin \phi_i$ and $\sin \psi_i$, and therefore at most two possible (ϕ_i, ψ_i) solutions exist for each orientation of the pair (v_i, v_{i+1}) of bond vectors.

Furthermore, if the first peptide plane is given, we can in principle compute all the possible (ϕ, ψ) angles of a fragment based on the following two observations: (1) A peptide plane i with respect to a principal order frame (POF) for medium 1 can be determined by its NC^α vector and its NH vector; (2) a unique NC^α vector for the peptide plane $i + 1$ can be computed from the NC^α vector of the peptide plane i and (ϕ_i, ψ_i). Therefore, we can consecutively compute all the possible discrete (ϕ, ψ) solutions for all the residues of the fragment by using a depth-first search (DFS) strategy. The optimal solution, under a scoring function, may be computed by DFS plus backtracking.

7.1.2 The Algorithm

The algorithm for determining a 3-dimensional backbone substructure containing α-helices and β-sheets employs the following inputs: (1) assigned backbone NH RDCs in two media,

(2) identified α-helices and β-sheets with known H-bonds between paired strands, and (3) very sparse NOE distance restraints. The algorithm is divided into three stages:

1. *Computation of alignment tensors*   Using an ideal helix model built with the backbone, the Saupe matrices for both media can be computed from the model by SVD (chapters 4, 5). This is essentially a bootstrapping method to obtain initial estimates of the alignment tensors. The tensors are refined by refitting with SVD either during or after step 2 below. Then, an optimal first NC^α vector is computed using an m-residue fragment built with the average dihedral angles from either an α-helix or a β-strand, and standard bond lengths and angles, using a grid search method over $SO(3)$.

2. *Refinement of secondary structure elements*   If the orientation of the first peptide plane of an m-residue secondary structure element is given, the conformation can be specified uniquely by a sequence of backbone dihedral angles. We refine the fragment by optimizing both the directions of individual NH vectors and also the dihedral angles of the fragment using RDCs alone, while leaving the bond lengths and the six bond angles fixed. Furthermore, the plausible conformations are computed from a DFS-based refinement method, in which every computed (ϕ, ψ) pair is filtered through the favorable Ramachandran regions for the corresponding secondary structure type. Consequently, an optimal conformation is computed from the set of all the plausible conformation vectors using the scoring function defined in Wang and Donald [1].

3. *Backbone structure determination*   We employ a few NOE-derived distances, including the distances between an amide proton and a C^α nucleus or between two C^α nuclei, to compute the relative position of the helix and the single sheet of ubiquitin.

7.1.3   Results

The work [1] has been successfully extended to compute a complete backbone structure [2], including turns and loops (connecting the secondary structure elements) using only NH and CH RDCs in a single medium (i.e., only two RDCs per residue) and two unambiguous NOEs. Here, CH means $C^\alpha H^\alpha$. Figure 7.2a shows the computed structure of ubiquitin backbone without loops, and figure 7.2b compares the computed structure of ubiquitin backbone with loops vs. a reference structure (see plate 2).

A discussion of the one-medium case (exact solutions from NH and CH RDCs in one medium) can be found in [4], where these algorithms were used prospectively to determine the NMR solution structure of the FF domain 2 of human transcription elongation factor CA150 (RNA polymerase II C-terminal domain interacting protein), PDB id: 2kiq [4].

7.2   Determination of Denatured or Disordered Proteins

In this section, we introduce a data-driven algorithm capable of computing a set of structures for denatured or disordered proteins directly from sparse experimental restraints, including the orientational restraints from RDCs, and distance restraints from paramagnetic relaxation enhancement (PRE) experiments.

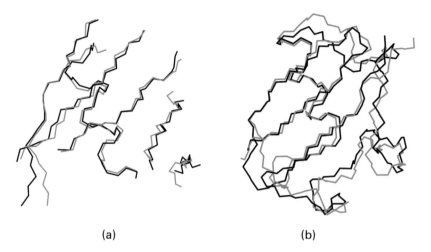

(a)   (b)

Figure 7.2 (plate 2)
(a) Structure of ubiquitin backbone without loops. The ubiquitin backbone structure (blue) was computed using 37 NH and 39 CH RDCs, 12 hydrogen bonds, and four NOEs; (b) structure of ubiquitin backbone with loops. The ubiquitin backbone structure (blue) was computed by extending the algorithm to handle loop regions along the protein backbone. The structure was computed using 59 NH and 58 CH RDCs, 12 H-bonds, and two unambiguous NOEs. A reference structure (from crystallography) is shown (red).

7.2.1   A Probabilistic Interpretation of Restraints in the Denatured State

In contrast to the traditional algorithms for the structure determination of native proteins, Wang et al. [3] employ a *set* of tensors to interpret the RDCs measured in the denatured state. Each tensor in the set represents a cluster of similar denatured structures, and the set of RDCs corresponding to each tensor is sampled from the individual distributions associated with each measured RDC. That is, the experimentally measured RDC value in the denatured state is the expectation, and the different clusters of tensors represent different conformations that are oriented differently in the aligning medium. Similarly, the PRE-derived distance is also considered a random variable, where the measured value is an average over all the possible structures in the denatured state. Hence, the structure determination problem for denatured proteins can be formulated as the computation of a set of conformation vectors, given the distributions of all the RDCs and PREs.

7.2.2   The Algorithm

For a denatured protein, the algorithm in [3] computes a presumably heterogeneous *ensemble* of structures that are consistent with the experimental data within a large range. The input of the algorithm includes the protein sequence, at least two RDCs per residue in a single medium and PREs. For efficiency, a divide-and-conquer strategy is used in the algorithm to compute the ensemble, which consists of six steps (figure 7.3):

1. *Fragment division*   The entire protein sequence is divided into p fragments $F_1, \ldots, F_p$ and $p-1$ linkages $L_1, \ldots, L_{p-1}$, where a linker consists of the residues between two neighboring fragments.

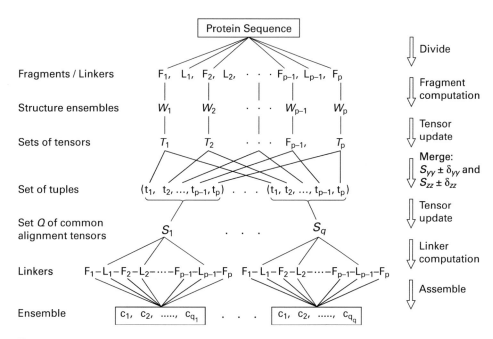

Figure 7.3
Divide-and-conquer strategy. The terms c_i denote conformation vectors for the complete backbone structure.

2. *Fragment computation*   The algorithm computes an ensemble of structures W_i for each fragment F_i, independently.

3. *Tensor computation*   For each structure in ensemble W_i, the corresponding tensor t_i is computed by SVD and saved into a set T_i. The SVD procedure is to minimize the RMSD between the experimental RDC data and the RDC back-computed from the structure using the tensor.

4. *Tensor merge*   All the tensors in the sets T_i are merged into p-tuples, $(t_1, \ldots, t_p)$, where t_i is from the set T_i. All the p tensors in a p-tuple have their S_{yy} and S_{zz} (two of three diagonal elements of the diagonalized Saupe matrix) values in the certain ranges of $[S_{yy} - \delta_{yy}, S_{yy} + \delta_{yy}]$ and $[S_{zz} - \delta_{zz}, S_{zz} + \delta_{zz}]$, where δ_{yy} and δ_{zz} are thresholds.

5. *Tensor update*   For each merged p-tuple, the algorithm computes their common tensor by SVD using the corresponding structures in W_i and all the experimental RDCs for the fragments F_i. Note that the SVD computation can output not only the diagonal elements S_{xx}, S_{yy}, S_{zz}, but also the orientation for each fragment in the common POF as well.

6. *Linker computation and assembly*   The algorithm computes each linker L_i using every common tensor and assembles the corresponding fragments and linkers into complete backbone structures.

7.2.3 Applications to Biological Systems

The algorithm described in Wang and Donald [3] has been applied to compute the structure ensembles of two proteins, an acid-denatured ACBP and a urea-denatured eglin C, from experimental NMR data. For acid-denatured ACBP, an ensemble of 231 structures was computed at pH 2.3, and all the structures have no vdW repulsion larger than 0.1 Å except for a few vdW violations as large as 0.35 Å between the two nearest neighbors of a proline and the proline itself. Further analysis indicates that acid-denatured ACBP is neither random coil nor nativelike. The work [3] also drew a similar conclusion for the urea-denatured eglin C, for which an ensemble of 160 structures was computed in 8 M urea.

References

[1] L. Wang and B. R. Donald. Exact solutions for internuclear vectors and backbone dihedral angles from NH residual dipolar couplings in two media, and their application in a systematic search algorithm for determining protein backbone structure. *Journal of Biomolecular NMR*, 29(3) (2004): 223–242.

[2] L. Wang, R. Mettu, and B. R. Donald. A polynomial-time algorithm for de novo protein backbone structure determination from NMR data. *Journal of Computational Biology*, 13(7) (2006): 1276–1288.

[3] L. Wang and B. R. Donald. A data-driven, systematic search algorithm for structure determination of denatured or disordered proteins. *The Computational Systems Bioinformatics Conference (CSB)*, Stanford, CA (August 2006), pp. 67–78.

[4] J. Zeng, J. Boyles, C. Tripathy, L. Wang, A. Yan, P. Zhou, and B.R. Donald. High-resolution protein structure determination starting with a global fold calculated from exact solutions to the RDC equations. *Journal of Biomolecular NMR*, 45(3) (2009): 265–81.

8 JIGSAW and NMR

In this chapter, we describe the JIGSAW algorithm [1] for automated NMR peak assignment. A main theme is to see how graph theory is used to develop algorithms in structural biology.

8.1 Overview of JIGSAW

JIGSAW is a high-throughput and automated approach for secondary structure determination and resonance assignment from a few, cheap NMR experiments. It applies a graph theoretic framework to represent the atom interactions from the NMR spectra. The contributions of JIGSAW lie in the following parts: First, based on the observation that most protein structures consist of some regular secondary structures, including α-helixes, and β-sheets, and thus result in certain patterns of NMR-measurable distances, JIGSAW shows that the spectral characteristics of secondary structures can be efficiently identified, and useful for the resonance assignment. Second, JIGSAW implies that a good resonance assignment can be obtained by using the fact that the proton chemical shifts of different amino acid types statistically display distinct patterns in the NMR spectra. Third, JIGSAW can identify β-sheets, which are really tertiary structures.

8.2 NMR Spectra Used in JIGSAW

JIGSAW exploits the spectra from the following four NMR experiments: NOESY, HSQC, HNH$^\alpha$, and TOCSY. The NOESY spectra measure the through-space atom interactions, while the other three spectra measure the through-bond atom interactions. We briefly review these four NMR experiments. Figure 8.1 shows the atom interactions that can be captured by these spectra.

1. *HSQC* is used to capture the through-bond interaction of correlated H^N and ^{15}N atoms. Since every residue has a unique H^N- ^{15}N pair on the protein backbone, and thus ideally has distinct frequency signals, the HSQC spectrum serves as the identification of each residue (unassigned), and reference interaction for all other spectra.

2. HNH$^\alpha$ is used to estimate the J coupling constant $^3J_{HNH\alpha}$, which is related to the ϕ torsion angle of the backbone. Thus, we can infer the secondary structure type from an HNH$^\alpha$ spectrum.

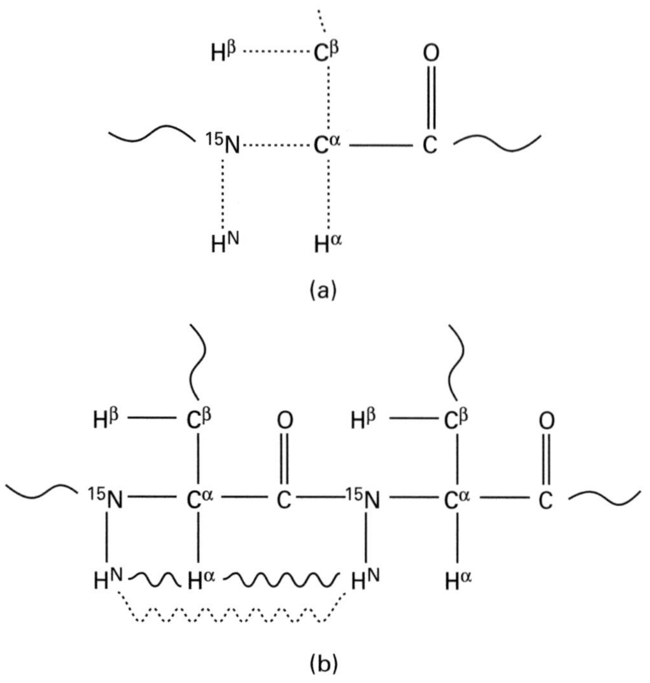

Figure 8.1
NMR interaction graphs showing through-bond and through-space atom interactions in a protein [1]. Atom nomenclature and interactions in a protein. (a) Through-bond interactions shown with dotted lines (HSQC: H^N-^{15}N; HNH$^\alpha$: H^N-^{15}N-H^α; TOCSY: H^N-^{15}N-H^α-H^β-...); (b) through-space interactions in NOESY shown with wavy line ($d_{\alpha N}$ solid and d_{NN} dashed) [1].

3. *TOCSY*   The TOCSY spectrum captures the through-bond interactions of protons on a residue's side-chain. Since the chemical shifts of protons for different amino acids are characteristically different, TOCSY serves as a fingerprint of each amino acid type.

4. *NOESY*   captures the through-space nuclear Overhauser effect (NOE) between an amide proton H^N and a neighboring ^1H in space, which are within a distance less than 6 Å.

8.3   Graph Representation of Atom Interactions in NOESY Spectra

8.3.1   Graph Representation
In JIGSAW, the atom interactions from the NOESY spectra are represented by a directed graph $G = (V, E)$, where V denotes the set of residues (unassigned, but sharing the same atoms H^N and ^{15}N), and E denotes the set of possible NOESY proton interactions between two residues. Because of the noise in the experimental data, an interaction graph has possible erroneous nodes and edges. There are other characteristics (called "labels") associated with the nodes and edges

in the interaction graph, such as secondary structure type label, match score, and atom distance. For more details, please refer to the definition of "NOESY interaction graph" in Bailey-Kellog et al. [1].

Comment Why should we formulate the proton NOE interactions into directed graphs instead of undirected ones? The reason is that, when two protons are connected to different heavy atoms, the NMR pulse sequence will cause the quantum coherence to transfer along different magnetization pathways.

8.3.2 Graph Constraints for Identifying Secondary Structure

Ideally (if there is no noise), the secondary structures, such as α-helix and β-sheet, will display certain graph constraints, as shown in figure 8.2. Such graph constraints should correspond to specific subgraphs in the interaction graph.

8.4 Secondary Structure Pattern Discovery

The first part of JIGSAW is secondary structure pattern discovery. The input is the NOESY interaction graph, and JIGSAW outputs all the subgraphs that possibly form secondary structures, such as α-helix and β-sheet. Notice that if the NOESY is perfect, then we can observe some "local" distance patterns (figure 8.3 a) among vertices. Therefore, these local patterns can help us find a secondary structure quickly. Unfortunately, since there are always some false nodes or edges in the interaction graph due to noise in the experimental data, we need to develop algorithms to identify all the subgraphs that are consistent with the secondary structure graph constraints, taking noise into account. JIGSAW applies the following algorithm (figure 8.3 shows the procedure to find β-sheet secondary structure from the interaction graph.):

Step 1. Identify instances of fragment patterns.

Step 2. Form sequences of consistent fragments.

Step 3. Form sheets of consistent fragments.

Step 4. Identify the best secondary structure graphs from the set of collected possibilities.

Many fragments can be found in step 3, and they can be inconsistent. Therefore, in step 4, we aim at finding a best consistent fragments graph to be the secondary structure. Of course, there can be many consistent fragments graphs. So, to evaluate a fragment graph G^*, a probability is calculated by multiplying the probability of goodness over all the edges. More precisely, for any edge e, let $m(e)$ be the weight of e (which encodes the differences) in the NOESY interaction graph. Let G_σ be the Gaussian prior distribution of width σ. Then, we define probability of the interaction of e to be $P(interaction(e))$ such that

$$P(interaction(e)) = G_\sigma(m(e)).$$

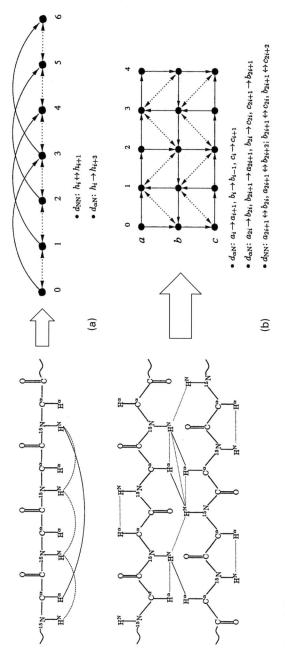

(a)

- d_{NN}: $h_i \leftrightarrow h_{i+1}$
- $d_{\alpha N}$: $h_i \to h_{i+3}$

(b)

- $d_{\alpha N}$: $a_i \to a_{i+1}$, $b_i \to b_{i-1}$, $c_i \to c_{i+1}$
- $d_{\alpha N}$: $a_{2i} \to b_{2i}$, $b_{2i+1} \to a_{2i+1}$, $b_{2i} \to c_{2i}$, $c_{2i+1} \to b_{2i+1}$
- d_{NN}: $a_{2i+1} \leftrightarrow b_{2i}$, $a_{2i+1} \leftrightarrow b_{2i+2}$; $b_{2i+1} \leftrightarrow c_{2i}$, $b_{2i+1} \leftrightarrow c_{2i+2}$

Figure 8.2
Graph constraints [1].

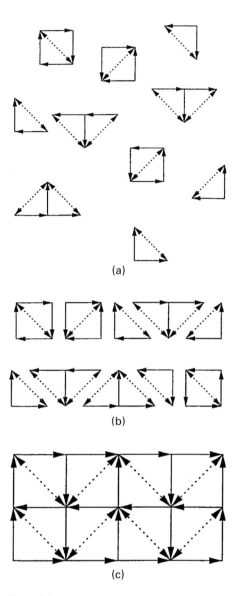

(a)

(b)

(c)

Figure 8.3
Algorithm for identifying β-sheets from the interaction graph [1]. JIGSAW algorithm overview: (a) identify graph fragments; (b) merge them sequentially; and (c) collect them into complete secondary graphs. Only correct fragments are shown here. Graphs from experimental data also generate a large number of incorrect fragments, but mutual inconsistencies prevent them from forming either long sequences or large secondary structure graphs.

Then we normalize all the interaction probabilities of all edges generated for the peak of a given edge e to define the probability for e to be good as $P(good(e))$ such that

$$P(good(e)) = \frac{P(interaction(e))}{\sum_{e' \in C(e)} P(interaction(e'))},$$

where $C(e)$ is the set of all edges generated for the peak of e.

Finally, the *correctness* of a graph G^* is defined to be

$$P(correct(G^*)) = 1 - \prod_{e \in G^*} (1 - P(good(e))).$$

Then we choose the graph G^* with the largest correctness to be the best secondary structure graph.

Although the interaction graph from experimental NOESY spectra unavoidably includes many false edges, it seems that a self-consistent pattern of continuous false edges is unlikely to occur, based on a simple joint probability model. JIGSAW applies the insight that incorrect edges can be mutually inconsistent, while correct edges can consistently reinforce each other.

8.5 Assignment by Alignment of Side-Chain Fingerprints

The BioMagResBank (BMRB) [2] has collected the statistics of observed chemical shifts of the 20 different natural amino acid types, and shown that different amino acid types display distinct fingerprints. The BMRB has been used to predict secondary structure type given chemical shift and amino acid type [2]. The JIGSAW algorithm, on the other hand, predicts amino acid type given chemical shift and secondary structure type. Figure 8.4 shows the mean chemical shifts of the protons for the 20 different amino acid types. JIGSAW applies these distinct "fingerprints" to predict amino acid type, given the chemical shifts from NMR spectra, and secondary structure types discovered. However, since the fingerprint obtained by TOCSY is usually not complete, we need to match the vertex fingerprints (obtained by TOCSY) with the amino acid fingerprints.

The idea of the alignment algorithm is, first, for any fingerprint of residue v, denoted by S_v, and for any BMRB amino acid fingerprint S_a, we define the *match score* by

$$match(S_v, S_a) = \max_{S'_v \subseteq S_v, S'_a \subseteq S_a} partial(S'_v, S'_a),$$

where $partial(S'_v, S'_a)$ is the degree of matching between a subset of S'_v of S_v and a subset S'_a of S_a, defined as follows:

$$partial(S'_v, S'_a) = c_0|S_v - S'_v| + c_1|S_a - S'_a| + c_2 \prod_{p \in S'_v} G_{\sigma_a}(p - d(p)),$$

where d is a partial fingerprint matching between S_v and S_a, that is, a bijection between S'_v and subsets of S'_a.

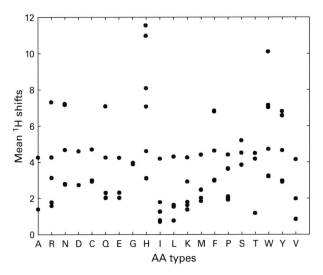

Figure 8.4
The mean proton chemical shifts of the 20 different amino acid types in the BMRB [1]. Cf. figure 3.4.

Then, we define the probability for residue v to be amino acid a by normalizing the match scores between v and all 20 amino acids in A as follows:

$$P(type(v, a)) = \frac{match(S_v, S_a)}{\sum_{b \in A} match(S_v, S_b)}.$$

Now for any sequence of vertices $V = (v_1, \ldots, v_n)$, the beginning position r, and the amino acid sequence $L = (a_1, \ldots, a_{|L|})$, we are ready to define the probability of the alignment of V and L, starting at the rth position of L:

$$P(align(V, L, r)) = \prod_{i=1}^{n} P(type(v_i, a_{r+i-1})).$$

Finally, we define the alignment between V and L to be the position r that maximizes the probability of alignment starting at any position r on L, where $r \leq |L| - |V|$:

$$alignment(V, L) = \arg \max_{r \leq |L| - |V|} P(align(V, L, r)).$$

8.5.1 Experimental Results

The following data were used for testing the performance of JIGSAW: the HSQC, HNH^{α}, 80ms TOCSY, and 3D ^{15}N-edited NOESY experimental spectra of huGrx, CBF-β, and vacGrx proteins. In other tests, synthetic J-coupling constant data were also used in the testing. The results show

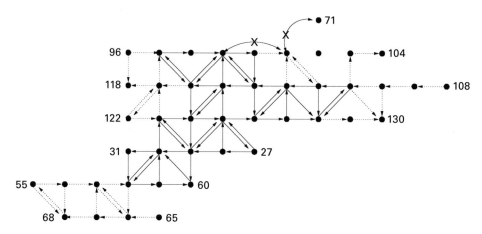

Figure 8.5
β-sheets of CBF-β computed by JIGSAW. Edges: solid = correct; dotted = false negative; X = false positive [1].

that JIGSAW is successful in structure discovery and NMR resonance assignment (see figure 8.5 for an example). For more details on the experimental results, please refer to [1].

References

[1] C. Bailey-Kellogg, A. Widge, J. J. Kelley III, M. J. Berardi, J. H. Bushweller, and B. R. Donald. The NOESY Jigsaw: Automated protein secondary structure and main-chain assignment from sparse, unassigned NMR data. *Journal of Computational Biology* 3–4;7 (2000): 537–558.

[2] Eldon L. Ulrich, Hideo Akutsu, Jurgen F. Doreleijers, Yoko Harano; Yannis E. Ioannidis, Jundong Lin, et al. BioMagResBank. *Nucleic Acids Research* 36 (2007): D402–D408.

9 Peptide Design

This chapter presents methods focusing on peptide design. The three areas of focus are peptide backbone reconstruction from a subset of atom coordinates [1], design of peptides to target transmembrane α-helices [6], and foldamers [7].

9.1 Peptides

Peptides are "short" sequences of amino acids that are linked by peptide bonds. They are important for many biological processes and experiments. One important use of peptides is to create antibodies against full proteins. Peptides can be synthesized and then inserted into an organism to make antibodies against that peptide, resulting in antibodies against proteins containing that peptide sequence. Peptides also act as protein ligands and can be used to probe protein : protein interactions (PPIs). Finally, there are peptides that have antimicrobial properties and have been shown to kill bacteria and other target cells.

It will be useful to clarify the distinction between (a) inhibitors of enzymatic activity versus inhibitors of PPIs, and (b) peptide inhibitors versus small organic molecule inhibitors. While numerous inhibitors of enzymatic activity have been developed, the development of inhibitors of protein:protein interactions has only recently come to the forefront as a viable approach. For example, an approximation to the K^* algorithm (chapter 12) was used by Gorczynski et al. [3] to predict the free energy of protein:ligand interactions to develop novel allosteric small-molecule inhibitors of the binding of two leukemia-associated proteins. Allosteric inhibition of such PPIs presents a number of advantages, including not having to compete for binding with the partner protein. However, there are very few previous examples of such inhibitors. The inhibitors of Gorczynski et al. [3] block the binding of RUNX1 to CBFβ, two proteins whose translocations play a critical role in the development of acute myeloid leukemia and acute lymphocytic leukemia. Treatment of human leukemia cell lines with these inhibitors resulted in changes in morphology indicative of increased differentiation versus untreated cells, providing support for this as a potential therapeutic approach. See chapter 8 for a discussion of CBFβ.

While the study of [3] designed small organic molecules to inhibit PPIs, in this chapter we focus on the design of peptides, which also show considerable promise for inhibition of PPIs.

Furthermore, the design of small peptides builds on the algorithms for the design of proteins, which form a major computational theme in this book (see chapters 10–12).

There is much to gain from understanding peptide interactions and peptide structure. By understanding peptide structure, general rules could be abstracted out that could be applied to protein folding. Pharmaceuticals could be rationally designed not only to take advantage of known peptide properties, but also exhibit new physicochemical characteristics. Also, designed peptides could be used to create novel protein binders and inhibitors.

9.2 Peptide Backbone Reconstruction

9.2.1 Problem Statement

The goal of peptide backbone reconstruction (PBR) is to take as input the coordinates of one type of protein backbone atoms (C^α or C^β) and output the all-atom representation of the protein backbone. Figure 9.1 is a schematic cartoon where the C^α atoms are shown on the left and the all-atom backbone representation of the protein is on the right (see plate 3).

9.2.2 Motivation

There are several current uses for PBR. PBR can be used to enhance low-resolution protein structures. In crystallography, certain atoms scatter better and have a better-known position. By using only those atoms, PBR could potentially be used to enhance the resolution of those structures. PBR could also be used to convert coarse-grained structures into all-atom models. Programs for *ab initio* folding and homology modeling often only output the C^α coordinates for a protein, making PBR applicable. Normal mode analysis (chapters 21, 23, and 48) also generally coarse grains its model to incorporate only C^α atoms, so PBR could be applied when analyzing mode motions.

There are several methods for solving PBR [1]. Some methods use collected fragment libraries and construct the backbone by ranking structures based on energy, homology, or geometry. Other methods perform de novo construction based on statistical position data, molecular dynamics, or

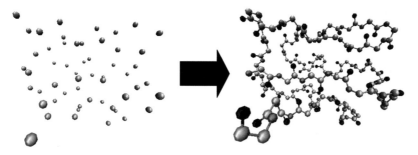

Figure 9.1 (plate 3)
Schematic of going from only C^α atoms to a full backbone model (PDB id: 1DUR). Credit: Kyle Roberts.

Monte Carlo simulations. Unfortunately, none of these current methods can rebuild a structure from the C^β atoms, but only from the C^α atoms.

9.2.3 Algorithm

The algorithm uses dead-end elimination (DEE) and a library of peptide sequences to find the all-atom backbone global energy minimum conformation (GMEC) using a database-derived empirical force field [1]. The algorithm takes as input a library of amino acid peptide sequences and a set of C^α or C^β coordinates. It then overlays all of the peptides in the library onto every input coordinate. Then DEE is used to prune backbone peptides until a GMEC is obtained.

Backbone Fragments   In order to create the peptide backbone fragment library, three-residue backbone fragments were taken from 1,336 random nonredundant PDB structures. All the fragments were clustered by RMSD and duplicates were discarded.

Fragment Overlap   Fragments were overlapped with the input coordinates by minimizing the sum of squared distances. Kearsley's method [4] was adopted, using quaternion algebra to turn this problem into an eigenvalue problem. The advantages to this approach are that the method is not iterative, improper rotations are not produced, and no special cases need to be handled.

Dead-End Elimination   Dead-end elimination [5] was used to prune the backbone conformations until one minimum energy conformation was left. The energy function minimized was

$$E_{total} = \sum_{r=1}^{N} E_{single}(r_i) + \sum_{r=1}^{N} \sum_{s=1}^{N} E_{pair}(r_i, s_j),$$

where r_i is residue r with the backbone conformation i. The original and simple DEE criteria were used in both the singles and pair criteria in order to prune conformations. DEE is discussed further in chapters 11, 12.

Database Force Field   The pairwise energy function used was developed by noting that there is a relationship between the energy between two atom types at a given distance and their probability of occurring at that distance. Thus, there is a relationship between the energy and the radial distribution function. The radial distribution function for all pairwise atom type combinations was approximated by analyzing many structures in the PDB.

Structure Generation   Structures were generated by pruning the aligned peptide backbone fragments using DEE until the GMEC was obtained.

9.2.4 Results

The algorithm was able to obtain structures with RMSDs from the crystal structure ranging from 0.2 to 0.6 Å. The ϕ-ψ angle correlation was on average 0.95 and 0.88, respectively. Also, the computation could be completed in minutes for a protein. When using C^β coordinates as input, the RMSD was always worse than using C^α coordinates. The author attributed this to less constraint at the C^β position. He found that increasing the fragment length could increase the accuracy of the C^β structures.

9.3   Peptides That Target Transmembrane Helices

The idea of Yin et al. [6] was to develop a peptide α-helix that could insert into the membrane and bind to a target membrane α-helix [6]. Some methods exist to design proteins that bind water-soluble proteins, but this problem is much harder since less is known about membrane proteins.

9.3.1   Algorithm

1. *Choose the target alpha helix sequence.* In Yin et al. [6] the integrin αIIb helix was chosen.

2. *Find templates in the PDB that have similar motifs to the target helix.* The αIIb helix was known to have a small-X_3-small motif and a right-handed crossing angle. Helix-helix dimers in the PDB were found with this motif and crossing angle to be used as design templates.

3. *Thread the target sequence onto one of the template helices.* All the amino acid identities of one of the template helices (the"target" helix) were changed to the target sequence (the sequence of αIIb). These amino acids are shown in green in the middle of figure 9.2 (plate 4).

4. *Choose positions that will be mutated on "anti" helix.* Find positions on the other template helix that are close to the helix-helix interface to be considered for redesign. These amino acid positions to be mutated are shown in pink in the middle of figure 9.2 (plate 4).

5. *Mutate those positions to all rotamers and repack.* The "anti" peptide was designed with Monte Carlo simulated annealing. At each step, one residue identity is changed and then the rotamers are optimized with DEE. The energy is computed using a Lennard-Jones potential and membrane depth-dependent knowledge-based potential.

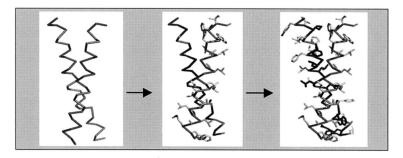

Figure 9.2 (plate 4)
Peptide helix design scheme [6]. Reprinted with permission from AAAS.

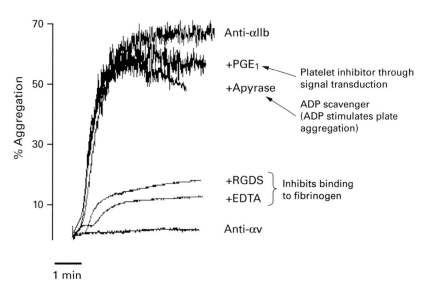

Figure 9.3
Percent platelet aggregation monitored over time with exposure to either anti-αIIb (peptide designed for the target helix) or anti-αv (peptide designed for a homologous target helix) [6].

9.3.2   Results

The authors did many experiments to validate that their designed peptide helix was binding to the target helix. Fluorescence resonance energy transfer was used to confirm that the helices were binding in micelles. TOXCAT was used to analyze the specificity and affinity of the designed peptide for the target peptide. Very good specificity was obtained, and the binding interface seemed to be what they predicted it would be.

One experiment that summarizes these good results is presented in figure 9.3. The role of the target integrin used in this study is to promote platelet aggregation through binding of fibrinogen. This process is naturally induced by ADP. Thus, since the binding of the designed peptide to the target integrin helix would cause the integrin to activate, this binding should promote platelet aggregation as well. As can be seen in figure 9.3, when exposed to anti-αIIb, platelets aggregate confirming that the designed peptide promotes aggregation. The PGE and Apyrase controls show that it is indeed the designed helix causing the aggregation. The RGDS and EDTA controls show that the designed peptide is working as hypothesized. Also, note that the design process is very specific because another designed peptide for a similar target helix (αv) causes no aggregation at all.

9.4   Foldamers

Proteins and RNA are special polymers, in that they adopt a specific compact stable conformation. But biology has been fairly constrained to utilizing only certain monomers to create these structures. In theory, new compounds with stable compact conformations could be created using a variety of monomers. This idea of finding new molecules with compact structure is the study of *foldamers*. A foldamer is "any polymer with a strong tendency to adopt a specific compact conformation" [2].

To create these new polymers with compact conformations, there are three goals that must be achieved. First, new backbone units with suitable folding propensities must be found to be used as monomers. Second, these new backbone units must be given interesting chemical functions. Without interesting function they are of little use biologically. Finally, in order to take full advantage of these foldamers and analyze their properties they need to be produced efficiently.

There are two main uses for foldamers [7]. First, foldamers can be used to test our understanding of protein function. It is possible that our knowledge of proteins has been overfit since we only observe α-amino acids. Creating foldamers makes it possible to test whether or not our knowledge can be generally applied to other systems. Second, foldamers use different building blocks so they could potentially be used for the design of new pharmaceuticals, diagnostic agents, and catalysts.

9.4.1 Types of Monomer Frameworks

At a high level, monomers for the foldamer framework can be classified into two categories: *aromatic* and *aliphatic* (see figure 9.4).

9.4.2 Foldamer Structure

Much analysis has been done on foldamer secondary structure, especially for foldamers incorporating β-amino acids. Foldamers incorporating β-amino acids will form a helix and have hydrogen-bonding patterns similar to α-helices. The composition of the foldamer (incorporation of β or α or both types of amino acids) determines the number of atoms per H-bond ring and the number of residues in the hydrogen-bonding helix spacing as well as the helix polarity [7] (figure 9.5, plate 5).

It turns out that most of our developed intuition for α-amino acids can be transferred to structures with different monomers. For example, adding salt bridges spaced one turn apart will stabilize β-peptides. Also, charged side-chains at the helix ends will stabilize the helix corresponding to the helix's polarity as observed in α-helices.

There has been very limited success in designing foldamer tertiary structure. In one case, a zinc fingerlike motif was built consisting of β-peptides with a β hairpin and 14-helix that could both bind zinc. Also, an octomer was created consisting of β-peptide helices, which maintained the tertiary structure using only noncovalent interactions.

9.4.3 Foldamer Function

There have been many promising designs of foldamers with biological functions. Foldamers were created that could penetrate bacterial cells in a passive process. By varying the length

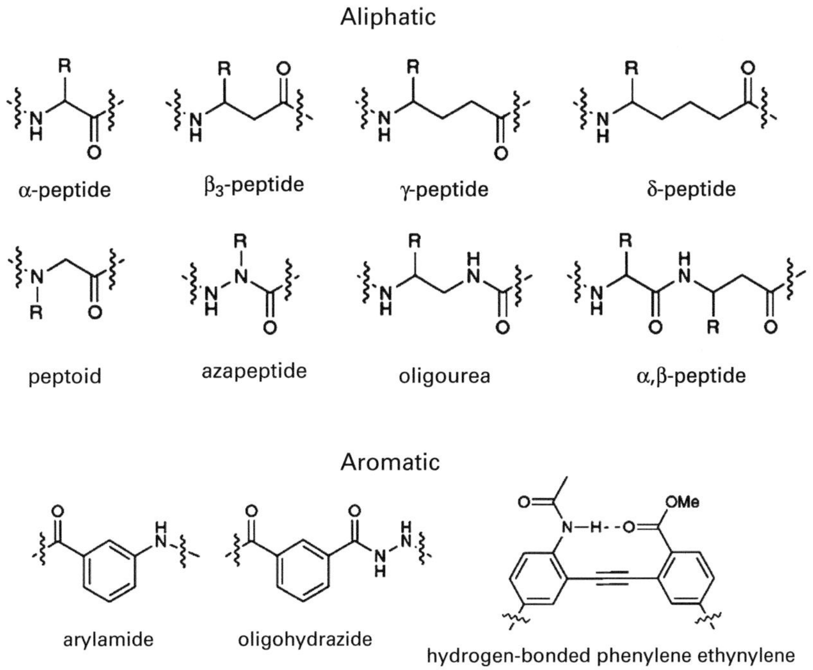

Aliphatic

α-peptide   β₃-peptide   γ-peptide   δ-peptide

peptoid   azapeptide   oligourea   α,β-peptide

Aromatic

arylamide   oligohydrazide

hydrogen-bonded phenylene ethynylene

Figure 9.4
Monomers for foldamer framework are classified into two categories: aliphatic and aromatic [7]. Reprinted by permission from Macmillan publishers Ltd.

and amphilicity of these helices, much insight can be gained about the function of antimicrobial peptides. Many peptides were successfully designed to interrupt protein-protein interactions with very good K_i, even to the nanomolar range. Finally, a foldamer peptide using a combined α/β-amino acid sequence was found to disrupt the Bak-Bcl-x$_L$ interaction with tenfold higher affinity than the native peptide. This is remarkable because previous efforts using only α-amino acid ligands failed.

9.4.4 Foldamer Benefits

Foldamers are more resistant to enzymatic attack than peptides, because the monomers used are non-natural. Hence, enzymes such as proteases are not designed to efficiently break them down. Also, fewer monomeric units for foldamers are needed than for α-amino acid compounds to adopt a well-defined secondary structure. This means that synthetically making these structures could be simpler, but they still should have the needed biological function. Finally, since many backbone monomers exist for foldamers, these backbone monomers can be used as an efficient way to downsize peptides into small molecules, which generally make better drugs.

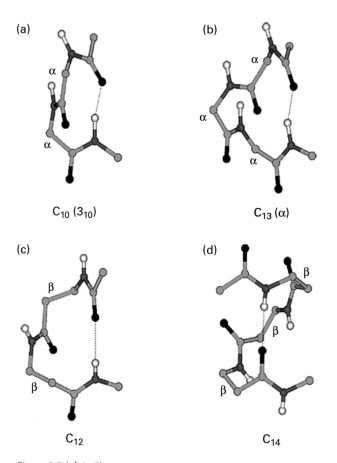

(a)

α

α

α

C_{10} (3_{10})

(b)

α

α

α

C_{13} (α)

(c)

β

β

β

C_{12}

(d)

β

β

β

C_{14}

Figure 9.5 (plate 5)
Secondary structure of α and β amino acid helices [8]. Reprinted with permission. Copyright 2005 American Chemical Society.

References

[1] Stewart A. Adcock. Peptide backbone reconstruction using dead-end elimination and a knowledge-based forcefield. *Journal of Computational Chemistry* 25;1 (2004): 16–27.

[2] S. H. Gellman. Foldamers: A manifesto. *Acc. Chem. Res.* 31(1998): 173–180.

[3] M. J. Gorczynski, J. Grembecka, Y. Zhou, Y. Kong, L. Roudaiya, M. G. Douvas, et al. Allosteric inhibition of the protein-protein interaction between the leukemia-associated proteins RUNX1 and CBFβ. *Chemistry & Biology* 14; 10 (2007).

[4] Simon K. Kearsley On the orthogonal transformation used for structural comparisons. *Acta Crystalographica* A45: (1989) 208–210.

[5] J. Desmet, M. De Maeyer, B. Hazes, and I. Lasters, The Dead End Elimination Theorem and its use in protein side-chain positioning. *Nature* 356(1992): 539–542.

[6] H. Yin, J. S. Slusky, B. W. Berger, R. S. Walters, G. Vilaire, R. I. Litvinov, et al. Computational design of peptides that target transmembrane helices. *Science* 315;5820 (2007 Mar 30): 1817–1822.

[7] C. M. Goodman, S. Choi, S. Shandler, and W. F. DeGrado. Foldamers as versatile frameworks for the design and evolution of function. *Nature Chemical Biology* 3;5 (2007 May): 252–262.

[8] K. Ananda, P. G. Vasudev, A. Sengupta, K. M. Raja, N. Shamala, and P. Balaram. Polypeptide helices in hybrid peptide sequences. *Journal of the American Chemical Society*, 127;47 (2005 Nov 30): 16668–16674.

10 Protein Interface and Active Site Redesign

This chapter describes two quite different but complementary approaches to enzyme redesign: minimal redesign of residues in the active site responsible for catalysis or active site conformation [1, 2], and the use of directed evolution to rescue function in chimeric, domain-swapped enzymes [3].

10.1 Minimalist Active Site Redesign

Enzymes use only a handful of their hundreds of residues to create a local chemical environment customized to lower the activation energy of their reactions. They have traditionally been viewed as fine-tuned to catalyze a single reaction not only with tremendous efficiency, but also with catalytic and substrate specificity. This is true, but in reality many enzymes also display reactive *promiscuity*; they can perform the same reaction on a variety of similar substrates (though still with one overwhelmingly dominating the minor species), or even catalyze a handful of minor reactions in addition to the major one. Redesign by necessity involves undoing some of the specificity evolved for a major reaction and redirecting as much of it as possible in a new direction. Theoretically, this should be easier if the new reaction is already a minor function of the wild-type protein. Promiscuity can be divided into three types: *substrate* (performing the same reaction on different substrates), *product* (yielding different products from the same substrate), and *catalytic* (a combination of both) [1]. In these pioneer days of protein (re)design, substrate and reaction promiscuity offer obvious targets for investigation into the mechanisms by which enzymes can be persuaded to perform new functions.

Often proteins from the same homology family, with remarkably similar folds, use similar catalytic machinery to perform many different tasks—even though their primary sequences may share less than 50% identity [4]. This catalytic variation can be traced to key differences at a handful of residue positions, most often at the active site, that alter either part of the catalytic machinery or the nature (size, local charge, solvent exposure, etc.) of the substrate binding pocket. Enzyme redesign can be seen as an attempt to extend a given homology family by creating new members. Protein engineers have exploited both enzyme homology and promiscuity to devise an array of novel activities from simple, often single-residue active site mutations that leave the

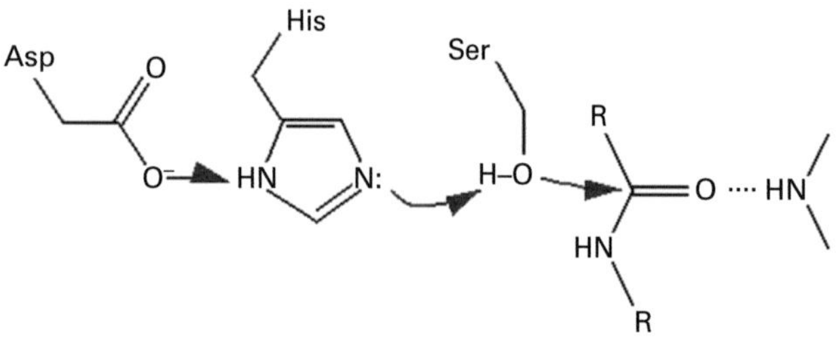

Figure 10.1
The Ser-His-Asp catalytic triad.

wild-type scaffold unaltered. While these novel enzymes rarely approach the efficiency of their evolution-streamlined progenitors, they point the way for future work in design.

10.1.1 Subtilisin

The Ser-His-Asp catalytic triad (figure 10.1) is responsible for the name and activity of serine proteases, which come in two very distinct forms (the result of convergent evolution) that nonetheless share their method of catalysis. The aspartate forms a very strong hydrogen bond to histidine, raising its pK_a, which deprotonates serine, which in turn performs a nucleophilic attack on the carbonyl group of an aromatic residue, converting into a tetrahedral intermediate that is stabilized by the active site residues (the oxyanion hole). This moiety's return to a carbonyl severs the peptide bond.

Subtilisin is a particularly well-studied serine protease. One early experiment used chemical modifications to replace the catalytic serine oxygen of subtilisin with a sulfur, effectively mutating the serine to cysteine [5]. Cysteine proteases are a natural enzyme family possessing an active site geometry similar to that of the serine proteases, so the cysteine variant of subtilisin (dubbed *thiolsubtilisin*) was expected to have an even higher rate of amide bond hydrolysis than the original, given the greater nucleophilicity of sulfur compared to oxygen. In fact, thiolsubtilisin showed a 10^5-fold drop in protease activity, but was able to catalyze the addition of amines to activated esters to form new peptide bonds, a reaction eschewed by subtilisin in favor of hydrolysis. On the premise that the active site of subtilisin was fine-tuned for the alcohol of serine, a proline near the catalytic triad was mutated to alanine to reduce the steric crowding caused by the addition of the bulkier sulfur atom [6]. This double mutant, called subtiligase, lost more protease activity but displayed ten-fold more ligase activity than thiolsubtilisin. Subtiligase operates on the same substrates as subtilisin, showing that the substrate specificity mechanisms of (at least some) enzymes are disjoint from their catalytic machinery. Another subtilisin variant, selenosubtilisin, replaces the sulfur of the cysteine with selenium and acts as both a peroxidase as well as ligase [7]. See figure 10.2.

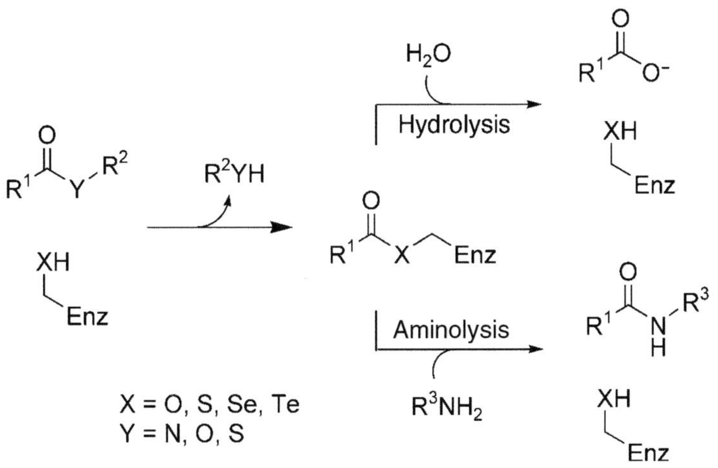

Figure 10.2
Reactions catalyzed by subtilisin and its derivatives. Subtilisin favors hydrolysis, while its synthetic analogs perform aminolysis [1]. Copyright Wiley-VCH Verlag GmbH & Co. KGaA. Reproduced with permission.

10.1.2 Interconverting Homologous Enzymes

Homologous enzymes share the same overall scaffold, and often use the same chemical step to catalyze their various reactions, so comparing their structures can lead to fruitful active-site mutations. L-Ala-D/L-Glu epimerase (AEE), muconate lactonizing enzyme II (MLE II), and ortho-succinylbenzoate synthase (OSBS) are all members of the enolase superfamily that remove a proton from a carboxylate substrate to form an enediolate intermediate stabilized by a Mg^{2+} ligand [8]. AEE and MLE II were mutated to obtain OSBS activity, the former via rational design and the latter using random mutagenesis and screening for the desired activity. In both cases, a single mutation (D297G in AEE and E323G in MLE II) that allowed the bulkier OSBS substrate to fit into the redesigned active sites was enough to confer OSBS activity, albeit at lower rates than OSBS and alongside the original reaction. Directed evolution of the variant enzymes increased their novel catalytic efficiencies by tuning the active site environment toward the new reaction. Still, the novel enzymes did not approach the catalytic efficiency of the original OSBS, a common issue in minimalist active site redesign.

3-alpha-HSD and 5-beta reductase are members of the NADPH-dependent aldo-keto reductase family that use a catalytic tetrad (Tyr-Lys-Asp-Xaa) to catalyze consecutive reactions in steroid metabolism; 3-alpha-HSD's substrate is the product of 5-beta reductase [9]. The identity of the Xaa residue is histidine in 3-alpha-HSD and glutamate in 5-beta reductase. Simply mutating the His of 3-alpha-HSD to Glu is enough to give it 5-beta reductase activity. Similarly, sequence alignment of the monofunctional base-excision repair enzyme MutY with bifunctional repair enzymes suggested that the addition of a single lysine conserved in the bifunctional enzymes could confer strand scission activity on MutY, which it accomplished with only slight effects on the original base excision activity. The addition of a single aspartate was also sufficient to convert the activities of HisA and HisF, involved in histidine biosynthesis, into that of the analogous tryptophan synthesis enzyme TrpF [10]. For HisA, this involved only a change in substrate specificity, since it already catalyzed the analogous reaction in histidine synthesis, but for HisF this single mutation altered both substrate and reaction specificity by enhancing a promiscuous activity that was already present in the wild-type species.

10.1.3 Introduction of Catalytic Machinery

Whereas activity swapping of homologous enzymes uses the catalytic machinery inherent in the original wild types, researchers can also add new activity by replacing the original machinery, while retaining the native substrate specificity. The creation of selenosubtilisin is one such example. This can even work in proteins with no inherent catalytic activity, such as the phospho(serine/threonine/tyrosine)-binding protein STYX. Comparison to a class of phosphatases that binds the same substrates in a similar way suggested that a cysteine necessary for activity in the phosphatases was reduced to a glycine in STYX. Mutation of this gly to cys indeed added phosphatase activity to STYX [11]. In an even more dramatic instance, the Ser-His-Asp catalytic triad was added into the active site of cyclophilin, which normally catalyzes the cis-trans isomerization of proline peptide bonds, to form the novel serine protease cyproase, which maintained

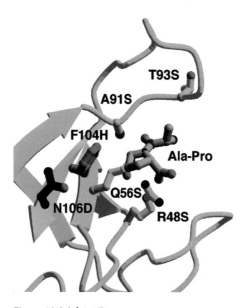

Figure 10.3 (plate 6)
Cyclophilin active site residues mutated in cyproase. Those residues highlighted in yellow were separately mutated to serine in the first round of design. Upon selection of A91S as the best protease, the F104H and N106D mutations were added to reproduce the catalytic triad [2]. Reprinted by permission from Macmillan Publishers Ltd.

the substrate specificity of cyclophilin [2]. Subsequent mutations in the second shell of residues surrounding those of the active site further improved cyproase activity. See figure 10.3 (plate 6).

10.1.4 Removal of Catalytic Nucleophiles

Removal of one reactive group from an active site usually destroys catalytic activity, at least for the main reaction of an enzyme. However, if it does not obliterate substrate binding it can also reveal heretofore blocked secondary reactions, allowing a mechanism previously blocked by nucleophilic attack to proceed. The E54D mutation of glutaconate-CoA-transferase is one such example [12]. The original enzyme transfers CoA from glutaryl-CoA to acetate. The mutant lacks transferase activity because the shorter aspartate carboxyl cannot reach CoA, but does activate a water molecule for hydrolysis. Similarly, removal of the catalytic cysteine in GAPDH by mutation to alanine obliterates phosphorylation, but induces dehydrogenase activity by the addition of water [13]. Finally, removal of the nucleophilic glutamate from the active site of glycosidases prevents hydrolysis but not substrate binding, allowing for the addition of nucleophilic thiosugars (glycoside ligation) [14].

10.1.5 Partitioning of Reaction Intermediates

Enzyme active sites usher reaction intermediates along particular reaction pathways, promoting some and blocking others. Active site redesign can work to alter this partitioning, as in the case of hydrolases and ligases. Both form acyl-enzyme intermediates, with the former transferring the acyl to a water and the latter to a designated acceptor [1]. Mutations favoring one reaction over the other can convert hydrolases to ligases and vice versa. Partitioning redesign can also be as simple as limiting the size of the pocket available to polymerization reactions, such as that catalyzed by farnesyl diphosphate polymerase, in which a single phe to ala mutation alters the length of the major product [15]. See figure 10.4.

Another type of partitioning redesign involves the rescue of enzymes from suicide inhibitors. Organophosphorus compounds such as the nerve agent sarin form tight phosphoester bonds with the catalytic serines of cholinesterases, causing irreversible inhibition. The mutation of an active-site glycine to histidine in human butyrylcholinesterase allows the enzyme to escape this fate by hydrolyzing the bond, either by activating a water molecule for nucleophilic attack on the phosphorus or positioning the inhibitor for such an attack [16].

10.1.6 Controlling Stereo- and Regiochemistry

Rational design can switch the stereo- and regiospecificity of wild-type enzymes by minimal active-site mutations. This can be done by altering the relative orientation of reactive groups and substrates, either by affecting substrate positioning or transferring reactive groups from one side of the active site to another. For example, the Asp170Ser/Thr457Glu double mutant of vanillyl-alcohol oxidase switched its major product from the R to the S form [17], whereas a single mutation in adenylate kinase was sufficient to alter the substrate-binding orientation and subsequently the

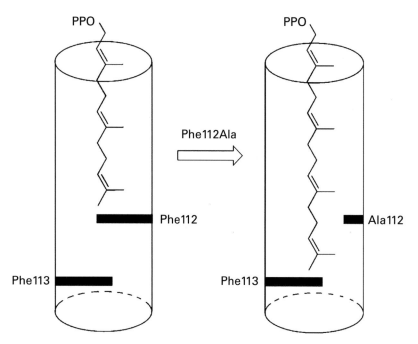

Figure 10.4
Binding pocket size effects the length of polymerization reaction products [1]. Copyright Wiley-VCH Verlag GmbH &
Co. KGaA. Reproduced with permission.

stereospecificity of product production [18]. Similar alterations to active site architecture alter the
regioselectivities of sesquiterpene synthases, in a less well-understood manner, though the effects
of the various mutations on product regioselectivity have been characterized [19]. See figure 10.5
(plate 7).

10.1.7 Improving Promiscuity

Redesign also proceeds by enhancing the rates of promiscuous activities, often furthered by
comparison to homologous enzymes that perform the target's promiscuous reaction as their major
activity (such as in HisF). Original activities in enzymes redesigned to enhance a promiscuous
activity are usually more robust than those in enzymes redesigned to perform a wholly novel
reaction, losing only a few orders of magnitude of catalytic efficiency for the original reaction
while gaining many more for the promiscuous target. Promiscuity enhancement can proceed
by directly altering residues responsible for catalysis or substrate binding (depending on the
type of promiscuity being designed for), or can be more indirect, altering cofactor specificity
or orientation. Altering the bound orientation of flavin in bacterial luciferase changed it from a
monooxygenase to an oxidase [20]; a similar mutation converts l-lactate monooxygenase into

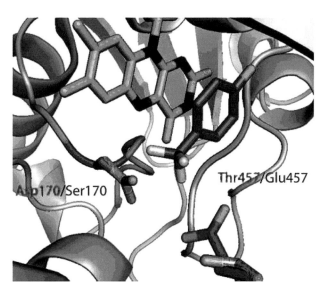

Figure 10.5 (plate 7)
Altering the stereospecificity of vanillyl-alcohol oxidase. Here the flavin cofactor is in orange and a substrate analog in green. Note how the mutations shift the catalytic carboxylate from one side of the substrate to the other [1]. Copyright Wiley-VCH Verlag GmbH & Co. KGaA. Reproduced with permission.

an oxidase [21]. Metals are an important class of cofactor, used in many different enzymes, and altering either the identity of the metal binding site or its orientation to the substrate and/or other catalytic groups can have profound effects on catalysis. For example, altering the primary sequence location of a key histidine in the heme group of myoglobin enhances its promiscuous peroxidase activity 200-fold [22].

Pyridoxal phosphate (PLP) is used as a cofactor in a great variety of reactions, which forge an imine bond between the substrate and PLP that stabilizes carbanion formation on breaking of one of the substrate bonds. *Dunathan's hypothesis* proposes the method that allows this variety: The bond to be broken must be oriented perpendicular to the PLP pyridinium ring. Redesign efforts have exploited this relationship, converting alanine racemase into a D-amino acid aldolase and ornithine decarboxylase into a transaminase, both via single mutations [23, 24]. See figure 10.6.

10.2   Protein Domain Interface Redesign via Directed Evolution

Nonribosomal peptide synthetases (NRPS) are a class of multidomain enzymes that catalyze the construction of small peptides in a manner reminiscent of a factory assemblyline. The adenylation (A) domain of each module activates its substrate amino acid for addition to the growing polypeptide, and is responsible for the majority of specificity in amino acid incorporation. Genetic evidence suggests that A domain rearrangement has occurred in the evolution of several NRPS,

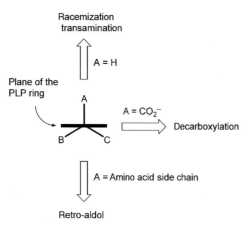

Figure 10.6
Dunathan's hypothesis on PLP-dependent catalysis [1]. Copyright Wiley-VCH Verlag GmbH & Co. KGaA. Reproduced with permission.

and their modular nature makes domain-swapping between homologous NRPS an attractive potential method for the generation of novel antibiotics. Unfortunately, the resultant chimeric proteins display very low or nonfunctional activities, presumably because of the incompatibility of their constituents' domain interfaces. Using directed evolution to rescue protein function in chimeric multidomain proteins offers a solution to this problem.

Walsh and colleagues first attempted domain swapping with the serine-specific A domain of *E. coli* enterobactin, swapping in a serine-specific domain from *P. syringae syringomycin* [3]. The initial chimeric protein showed 30-fold less activity than the original. Following mutagenic polymerase chain reaction (PCR) and growth under screening conditions, the fastest growing clone was selected for another round of mutagenesis and screening. Of the three evolved clones, the best showed 8-fold improvement over the original chimera, within 4-fold of wild-type enterobactin. Later tests focused on the NRPS responsible for the production of the antibiotic andrimid, incorporating an A domain with a new substrate specificity chosen for its promiscuity in the hopes of generating novel antibiotics. Three rounds of mutagenesis and screening later, a mutant was isolated that produced 10.7 more andrimid per unit of culture than the original chimera, only 3-fold less than the wild-type enzyme. These results showed that directed evolution could restore function in chimeric multidomain enzymes. The mutations they incorporated were spread throughout the A domain structure in an unpredictable manner, suggesting the appropriateness of directed evolution for this task as opposed to a more focused rational design approach. It would be interesting to see whether such interfaces could be successfully and reliably redesigned using modern structure-based protein design algorithms, such as those in the dead-end elimination (DEE) family. Such algorithms are introduced in chapter 11. See figure 10.7 (plate 8).

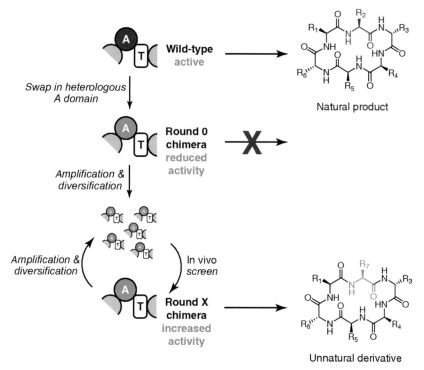

Figure 10.7 (plate 8)
Directed evolution in the rescue of chimeric NRPS activity [3]. Copyright 2007 National Academy of Sciences.

References

[1] M. D. Toscano, K. J. Woycechowsky, and D. Hilvert. Minimalist active-site redesign: Teaching old enzymes new tricks. *Angewandte Chemie International Edition* 46;18 (2007): 3212–3236.

[2] E. Quemeneur, M. Moutiez, J.-B. Charbonnier, and A. Menez. Engineering cyclophilin into a proline-specific endopeptidase. *Nature* 391(1998): 301–304.

[3] M. A. Fischbach, J. R. Lai, E. D. Roche, C. T. Walsh, and D. R. Liu. Directed evolution can rapidly improve the activity of chimeric assembly-line enzymes. *Proceedings of the National Academy of Sciences of the USA* 104;29 (2007): 11951–11956.

[4] M. E. Glasner, J. A. Gerlt, and P. C. Babbitt. Evolution of enzyme superfamilies. *Current Opinion in Chemical Biology* 10(2006): 492–497.

[5] L. Polgar, and M. L. Bender A new enzyme containing a synthetically formed active site: Thiol-subtilisin. *Journal of the American Chemical Society* 1966; 88(13): 3153–3154.

[6] L. Abrahmsen, J. Tom, J. Burnier, K. A. Butcher, A. Kossiakoff, and J. A. Wells. Engineering subtilisin and its substrates for efficient ligation of peptide bonds in aqueous solution. *Biochemistry* 30;17 (1991): 4151–4159.

[7] Z.-P Wu and D. Hilvert. Conversion of a protease into an acyl transferase: Selenolsubtilisin. *Journal of the American Chemical Society* 111;12 (1989): 4513–4514.

[8] D. M. Z. Schmidt, E. C. Mundorff, M. Dojka, E. Bermudez, J. E. Ness, S. Govindarajan, et al. Evolutionary potential of alpha/beta-barrels: Functional promiscuity produced by single substitutions in the enolase. *Biochemistry* 42;28 (2003): 8387–8393.

[9] J. M. Jez, and T. M. Penning. Engineering steroid 5-reductase activity into rat liver 3-hydroxysteroid dehydrogenase. *Biochemistry* 37;27 (1998): 9695–9703.

[10] C. Juergens, A. Strom, D.Wegener, S. Hettwer, M.Wilmanns, and R. Sterner. Directed evolution of a (beta alpha) 8-barrel enzyme to catalyze related reactions in two different metabolic pathways. *Proceedings of the National Academy of Sciences of the USA* 97;18 (2000): 9925.

[11] M. J. Wishart, J. M. Denu, J. A. Williams, and J. E. Dixon. A single mutation converts a novel phosphotyrosine binding, domain into a dual-specificity phosphatase. *Journal of Biological Chemistry* 270;45 (1995): 26782–26785.

[12] M. Mack, and W. Buckel. Conversion of glutaconate CoA-transferase from Acidaminococcus fermentans into an acyl-CoA hydrolase by site-directed mutagenesis. *FEBS Letters* 405;2 (1997): 209–212.

[13] C. Corbier, F. D. Setta, and G. Branlant. A new chemical mechanism catalyzed by a mutated aldehyde dehydrogenase. *Biochemistry* 31;49 (1992): 12532–12535.

[14] M. Jahn, H. Chen, J. Müllegger, J. Marles, R. A. J. Warren, and S. G. Withers. Thioglycosynthases: Double mutant glycosidases that serve as scaffolds for thioglycoside synthesis. *Chemical Communications* (2004): 274–275.

[15] L. C. Tarshis, P. J. Proteau, B. A. Kellogg, J. C. Sacchettini, and C. D. Poulter. Regulation of product chain length by isoprenyl diphosphate synthases. *Proceedings of the National Academy of Sciences of the USA* 93;26 (1996): 15018–15023.

[16] O. Lockridge, R. M. Blong, P. Masson, M.-T. Froment, C. B. Millard, and C. A. Broomfield. A single amino acid substitution, Gly117His, confers phosphotriesterase (organophosphorus acid anhydride hydrolase) activity on human butyrylcholinesterase. *Biochemistry* 36;4 (1997): 786–795.

[17] R. H. H. van den Heuvel, M.W. Fraaije, A. Ferrer, A. Mattevi, and W. J. H. van Berkel. Inversion of stereospecificity of vanillyl-alcohol oxidase. *Proceedings of the National Academy of Sciences of the USA* 97;17 (2000): 9455–9460.

[18] R.-T. Jiang, T. Dahnke, and M.-D. Tsai. Mechanism of adenylate kinase. 10. Reversing phosphorus stereospecificity by site-directed mutagenesis. *Journal of the American Chemical Society* 113;14 (1991): 5485–5486.

[19] D. B. Little, and R. B. Croteau. Alteration of product formation by directed mutagenesis and truncation of the multiple-product sesquiterpene synthases delta-selinene synthase and gamma-humulene synthase. *Archives of Biochemistry and Biophysics* 402;1 (2002): 120–125.

[20] L. Xi, K.-W. Cho, M. E. Herndon, and S.-C. Tu. Elicitation of an oxidase activity in bacterial luciferase by site-directed mutation of a noncatalytic residue. *Journal of Biological Chemistry* 265;8 (1990): 4200–4203.

[21] W. Sun, C. H.Williams, Jr., and V. Massey. Site-directed mutagenesis of glycine 99 to alanine in L-lactate monooxygenase from Mycobacterium smegmatis. *Journal of Biological Chemistry* 271;29 (1996): 17226–17233.

[22] T. Matsui, S.-I. Ozaki, E. Liong, G. N. Phillips, Jr., and Y. Watanabe. Effects of the location of distal histidine in the reaction of myoglobin with hydrogen peroxide. *Journal of Biological Chemistry* 274;5 (1999): 2838–2844.

[23] F. P. Seebeck, and D. Hilvert. Conversion of a PLP-dependent racemase into an aldolase by a single active site mutation. *Journal of the American Chemical Society* 125;34 (2003): 10158–10159.

[24] L. K. Jackson, H. B. Brooks, A. L. Osterman, E. J. Goldsmith, and M.A. Phillips. Altering the Reaction Specificity of Eukaryotic Ornithine Decarboxylase. *Biochemistry* 39;37 (2000): 11247–11257.

11 Computational Protein Design

This chapter introduces the automated protein design and experimental validation of a novel designed sequence, as described in Dahiyat and Mayo [1].

11.1 Introduction

Given a three-dimensional (3D) backbone structure, the *protein design* problem is to find an optimal sequence that satisfies the physical chemical potential functions and stereochemical constraints. Protein design is an "*inverse folding problem*," and fundamental for understanding the protein function.

The term *rotamer* denotes discrete rotational conformations of protein side-chains. Typically these are represented by a finite discretization of the side-chain $\chi_1, \chi_2, \ldots$ dihedral angles. Rotamers are based on observed side-chain conformations from a statistical analysis of high-resolution crystal structures in the PDB. A rotamer can encode a different conformation of the same amino acid side-chain, or a switch in amino acid type. Both are encoded uniformly using a rotamer *library* that contains the low-energy side-chain conformations across different amino acids.

The most basic protein design problem is often viewed as a search for the optimal rotamers to fit on a given protein backbone. Typically, the C^α-C^β bond remains invariant unless the residue is mutated to glycine or proline. The search returning the optimal rotamers yields both side-chain conformations and underlying design sequence. The sequence of the computed rotamers can be obtained by examining the amino acid type of each residue while disregarding its side-chain conformation. However, structural confirmation of a designed structure requires comparing the predicted side-chains (and backbone) versus the experimentally-determined structure by X-ray crystallography or NMR.

11.2 Overview of Methodology

The following is the methodology used in Dahiyat and Mayo [1]:

Given a backbone fold of a target structure, Dahiyat and Mayo [1] first developed an automated side-chain rotamer selection algorithm to (1) screen all possible amino acid sequences, and

(2) find the optimal sequence and side-chain orientations (rotamers). Then experimental validation by using NMR was performed to evaluate the computed optimal sequence/structures.

11.3 Algorithm Design

Input Backbone fold (Zif268), represented by structure coordinates. Here, "Zif" stands for "zinc finger."

Output Optimal sequence (FSD-1). "FSD" stands for "full sequence design"; FSD-1 was the first full-length protein sequence to be designed by computational structure-based algorithms.

Overview

1. The algorithm considers specific interactions between (a) side-chain and backbone and (b) side-chain and side-chain.

2. The algorithm scores a sequence arrangement, based on a van der Waals potential function, solvation, hydrogen bonding, and secondary structure propensity [1].

3. The algorithm considers a discrete set of rotamers, which are all allowed conformers of each side-chain.

4. The algorithm applies a *dead-end elimination* (DEE) algorithm to prune rotamers that are inconsistent with the global minimum energy solution of the system.

Details The inputs of the algorithm are structure coordinates of the target motif's backbone, such as N, C^α, C', and O atoms, and C^α-C^β vectors. The residue positions in the protein structure are partitioned into *core*, *surface*, and *boundary* classes. The set of possible amino acids at the core positions is {Ala, Val, leu, Ile, Phe, Tyr, Trp}. The set of amino acids considered at the surface positions is {Ala, Ser, Thr, His, Asp, Asn, Glu, Gln, Lys, Arg}. The combined set of both core and surface amino acids are considered for the boundary positions.

Note The total number of possible amino acid sequences is equal to the product of possible amino acids at each residue position. For instance, suppose that there are 7 possible amino acids at one core position, and 16 possible amino acids at each of 7 boundary positions, and 10 possible amino acids at each of 18 surface positions. The search space consists of $7 \times 16^7 \times 10^{18} = 1.88 \times 10^{27}$ possible amino acid sequences.

The algorithm is divided into two phases:

Phase 1 (Pruning) The algorithm applies DEE to find and eliminate rotamers that are dead-ending with respect to the global minimum energy conformation (GMEC). A rotamer r at residue position i will be eliminated (i.e., proven to be dead-ending) if there is another rotamer t at the same position such that replacing r by t will always reduce the energy. However, naïvely checking this will still take exponential time. Therefore, the following pruning was applied. Below, i_r denotes rotamer r at sequence position i. Similarly, i_t and j_s denote, respectively, rotamer t at position i, and rotamer s at position j.

DEE Condition   If there exists a rotamer t satisfying

$$E(i_r) - E(i_t) + \sum_j \min_s (E(i_r, j_s) - E(i_t, j_s)) > 0, \tag{11.1}$$

then r will be eliminated, where $E(i_r)$ and $E(i_t)$ represent *self-energies,* that is, energies between the atoms of a single rotamer (e.g., i_r). By convention, and for convenience, we include in the self-energy term the *rotamer-template energies* also. In this context, "template" means the geometric structure of the protein backbone atoms. $E(i_r, j_s)$ and $E(i_t, j_s)$ represent residue *pairwise,* rotamer-rotamer energies for rotamers i_r, i_t, and j_s. The condition in equation (11.1) ensures that replacing r by t will always reduce the energy, regardless of what the rotamers at other residue positions are. The intuition behind equation (11.1) is given in section 11.4.

Note that we have "overloaded" the operator E to represent both self-energies (e.g., $E(i_r)$) and residue-pairwise energies (e.g., $E(i_r, j_s)$). Many protein design algorithms (including most of those in this book) explicitly require that the energy function E be residue-pairwise additive. The DEE algorithms directly exploit this assumption. In general, DEE algorithms could, in principle, be extended to work with residue-k-wise additive energy functions instead, for a small constant $k > 2$. However, parameterizing such energy functions requires care, and can be difficult. In general, "N-body" energy functions (where N is the total number of atoms) such as the Generalized Born/Poisson-Boltzmann solvation models are not amenable to DEE. However, there are approximate pairwise solvation models, and these are discussed in chapter 12.

Different scoring functions E are defined for core, surface, and boundary residues separately. The scoring function for core residues uses "a van der Waals potential to account for steric constraints and an atomic solvation potential favoring the burial and penalizing the exposure of nonpolar surface area" [1]. The surface residues apply a hydrogen-bond potential and secondary structure propensities, and a van der Waals potential. The residues at the boundary positions use a combination of both core and surface scoring functions. The details of these scoring functions, and how they are combined, are sketched in [1], but are discussed at length in a U.S. Patent [7]. Further discussion of empirical molecular mechanics energy and scoring functions is given in chapter 12.

Phase 2 (Enumeration)   For any residue position i, let R_i be the set of remaining rotamers that are not eliminated in Phase 1. The algorithm then enumerates all the combinations of remaining rotamers—that is, $\prod_i R_i$—to find the combination that has the global minimal energy. Enhancements to DEE (e.g., [3, 2]) also prune *pairs* of rotamers that are inconsistent with the GMEC, returning only subsets $R_{ij} \subset R_i \times R_j$ of the pairwise cross products. Rotamer pairs in $R_i \times R_j$ but outside R_{ij} cannot participate in the GMEC.

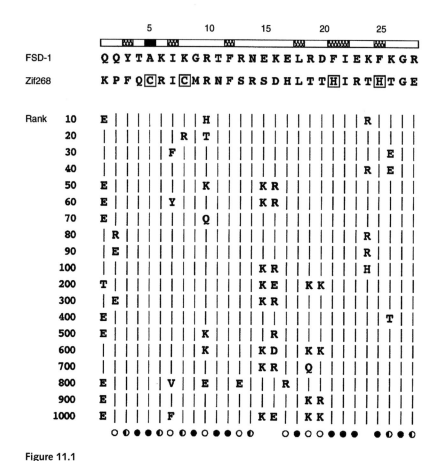

Figure 11.1

Comparison of computed sequence FSD-1 and the target sequence Zif268 [1]. Sequence of FSD-1 aligned with the second zinc finger of Zif268. The bar at the top of the figure shows the residue position and the open bars indicate the 20 surface positions. The alignment matches positions of FSD-1 to the corresponding backbone template positions of Zif268. Of the six identical positions (21 percent) between FSD-1 and Zif268, four are buried (Ile[7], Phe[12], Leu[18], and Ile[22]). The zinc-binding residues of Zif268 are boxed. Representative nonoptimal sequence solutions determined by means of a Monte Carlo simulated annealing protocol are shown with their rank. Vertical lines indicate identity with FSD-1. The symbols at the bottom of the figure show the degree sequence conservation for each residue position computed across the top 1,000 sequences: filled circles indicate more than 99 percent conservation, half-filled circles indicate conservation between 90 and 99 percent, open circles indicate conservation between 50 and 90 percent, and the absence of a symbol indicates highest occurence at each position is identical to the sequence of FSD-1. Single-letter abbreviations for amino acid residues as follows: A, Ala; C, Cys; D, Asp; E, Glu; F, Phe; G, Gly; H, His; I, Ile; K, Lys; L, Leu; M, Met; N, Asn; P, Pro; Q, Gln; R, Arg; S, Ser; T, Thr; V, Val; W, Trp; and Y, Tyr [1]. Reprinted with permission from AAAS.

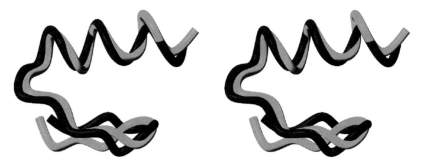

Figure 11.2 (plate 9)
Backbone structure comparison of computed sequence FSD-1 and the target sequence Zif268 [1]. Comparison of the FSD-1 structure (blue) and the design target (red). Stereoview of the best-fit superposition of the restrained energy minimized average NMR structure of FSD-1 and the backbone of Zif268. Residues 3 to 26 are shown. [1]. Reprinted with permission from AAAS.

Phase 1 is provably correct, in that no rotamer will be pruned if it is part of the GMEC. Phase 1 is also polynomial-time. Phase 2 can be made provable using the A^* search algorithm (chapter 12 and [4]). That is, A^* after DEE will guarantee to compute the GMEC. Phase 2 is worst-case exponential-time.

Results   Figure 11.1 shows the comparison of optimal computed sequence FSD-1 and the target sequence Zif268. Figure 11.2 (plate 9) compares the experimentally determined structure of the optimal computed sequence FSD-1 versus the structure of the target backbone sequence Zif268.

11.4   Intuition: Dead-End Elimination

Here is the intuition behind equation (11.1), the Dead-End Elimination (DEE) condition. We repeat it here for clarity:

$$E(i_r) - E(i_t) + \sum_j \min_s (E(i_r, j_s) - E(i_t, j_s)) > 0. \tag{11.1}$$

Recall that lower energy is better; we are searching for the GMEC. The DEE condition (equation 11.1) tells us that we can prune a *candidate* rotamer i_r if certain conditions hold. Those conditions include: the existence of a *competitor* rotamer i_t (i.e., a competitor rotamer t, also at position i) that is better than i_r. But how can we prove that t is better than r? For this calculation, it will be helpful to use the perspective of a *witness* rotamer j_s. In this biophysical modeling problem, the only "perspective" a witness can have on the discrete choice i_r versus i_t is its energetic interaction with the candidate versus the competitor. One of these energies will be more favorable, which implies we may construct a penalty for the choice of rotamer r versus t at position i.

First, the DEE condition contains a local side-chain-backbone penalty encoding the cost of choosing i_r versus i_t. This is $E(i_r) - E(i_t)$. It is independent of j_s.

The DEE condition also includes a pairwise side-chain-side-chain penalty for the cost of choosing i_r versus i_t, from the perspective of j_s. Now, if we *knew* what rotamer s was at position j, then this penalty would simply be $E(i_r, j_s) - E(i_t, j_s)$. Since we don't, all possible rotamers at position j must be considered. The pairwise penalty is built by computing at position j a *lower bound* on the i_r versus i_t penalty, namely, $\min_s(E(i_r, j_s) - E(i_t, j_s))$. The minimization occurs over all *possible* rotamers s at position j. Then a sum is computed of *all* such lower bounds over all residue positions: $\sum_j \min_s(E(i_r, j_s) - E(i_t, j_s))$. If the entire quantity on the left-hand side in equation (11.1) is positive, then rotamer i_r can be pruned, since we have proven it cannot participate in the GMEC.

Finally, the DEE criterion can be efficiently computed, in polynomial time, by enumerating triples of the form (i_r, i_t, j_s). We prove this below.

11.5 Complexity Analysis

Let n denote the number of residues, and r denote the (maximum) number of possible rotamers for each residue.

We first analyze the time complexity of DEE pruning in phase 1. For each rotamer at a specific residue position i, it takes time $O(nr)$ to search all r possible amino acids in all other $n - 1$ positions to find $\sum_j \min_s[E(i_r, j_s) - E(i_t, j_s)]$. Comparisons with other rotamers at the same position i take $r \cdot O(nr) = O(nr^2)$ time. Since we need to consider all possible rotamers at every position i, the total DEE pruning takes $n \cdot r \cdot O(nr^2) = O(n^2 r^3)$ time. So DEE is polynomial time!

Although the pruning step will eliminate many states (that is, many configurations of rotamers) in the search space, it cannot guarantee that the number of the remaining states is small enough for the enumeration to be efficient. Even if there are only two rotamers remaining for each position, the worst-case time to find the state that minimizes the energy is still exponentially large.

Note   In fact, the optimization problem of finding the GMEC in protein design has been proven NP-hard [5], and even NP-hard to approximate [6].

11.6 Experimental Validation: Interplay of Computational Protein Design and NMR

The solution structure for the computed sequence FSD-1 was obtained by using 2D ^{1}H NMR spectroscopy. Sample NMR data, including a NOESY spectrum, are shown in figure 11.3. X-PLOR plus the standard protocols for hybrid distance geometry-simulated annealing were used to calculate the structure. Table 11.1 and Figure 11.4 (plate 10) show an ensemble of 41 structures that are consistent with good geometry and distance constraints within a small tolerance. The structure of FSD-1 was close to the target structure (Zif268), validating the structure-based protein design algorithm using DEE.

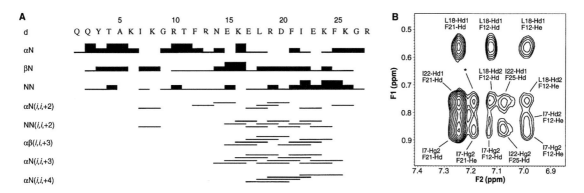

Figure 11.3
NMR data for FSD-1: (a) FSD-1, sequential and short-range NOE connectivities. The "d" denotes a contact between the indicated protons. All adjacent residues are connected by H^α-HN, HN-HN, or H^β-HN NOE crosspeaks, The helix (residues 15 to 26) is well-defined by short-range connections, as is the hairpin turn at residues 7 and 8; (b) 2D ^{1}H NOESY spectrum for the optimal computed sequence FSD-1. Several long-range NOEs from Ile[7] and Phe[12] to the helix help define the fold of the protein. The starred peak has an ambiguous F1 assignment, Ile[22] Hd1 or Leu[18] Hd2 [1]. Reprinted with permission from AAAS.

Table 11.1
NMR structure determination of FSD-1: distance restraints, structural statistics, and atomic root-mean-square (rms) derivations

Distance restraints		
Intraresidue	97	
Sequential	83	
Short range ($|i - j| = 2$ to 5 residues)	59	
Long range ($|i - j| > 5$ residues)	35	
Hydrogen bond	10	
Total	284	
Structural statistics		
rms deviations	$(SA) \pm$ SD	$(SA)_r$
Distance restraints (Å)	0.043 ± 0.003	0.038
Idealized geometry		
Bonds (Å)	0.0041 ± 0.0002	0.0037
Angles (degrees)	0.67 ± 0.02	0.65
Impropers (degrees)	0.53 ± 0.05	0.51
Atomic rms deviations (Å)*		
	$<SA>$ versus $SA \pm$ SD	$<SA>$ versus $(SA)_r \pm$ SD
Backbone	0.54 ± 0.15	0.69 ± 0.16
Backbone + nonpolar side-chains†	0.99 ± 0.17	1.16 ± 0.18
Heavy atoms	1.43 ± 0.20	1.90 ± 0.29

*Atomic rms deviations are for residues 3 to 26, inclusive. Residues 1, 2, 27, and 28 were disordered [ϕ, ψ, angular order parameters (34) < 0.78] and had only sequential $|i - j| = 2$ NOEs. †Nonpolar side-chains are from residues Tyr[3], Ala[5], Ile[7], Phe[12], Leu[18], Phe[21], Ile[22], and Phe[25], which constitute the core of the protein. (SA) are the 41 simulated annealing structures, SA is the average structure before energy minimization, <SA> are the restrained energy minimized average structure, and SD is the standard deviation.
Source: [1]. Reprinted with permission from AAAS.

Figure 11.4 (plate 10)
Empirically determined NMR structure ensemble of FSD-1, including side-chains. Stereoview showing the best-fit superposition of the 41 converged simulated annealing structures from XPLOR. The backbone C^α trace is shown in blue and the side-chain heavy atoms of the hydrophobic residues (Tyr[3], Ala[5], Ile[7], Phe[12], Leu[18], Phe[21], Ile[22], and Phe[25]) are shown in magenta. The amino terminus is at the lower left of the figure and the carboxyl terminus is at the upper right of the figure. The structure consists of two antiparallel strands from positions 3 to 6 (back strand) and 9 to 12 (front strand), with a hairpin turn at residues 7 and 8, followed by a helix from positions 15 to 26. The termini, residues 1, 2, 27, and 28 have very few NOE restraints and are disordered [1]. Reprinted with permission from AAAS.

The structure determination in [1] also represents a simple didactic example of the classic method of NMR protein structure determination in the solution state, based primarily on NOEs. Hence, this experimental study illustrates the biophysical concepts introduced in the preceding chapters on NMR. Although FSD-1 is a small protein, the basic concepts such as sequential and short-range NOEs, NOESY crosspeaks, NOESY assignment, structural ensembles, and the simulated annealing structure determination protocols, are illustrated in this study. For example, note the ambiguous NOESY crosspeak assignment (∗) in figure 11.3b. The NOE patterns exploited by the JIGSAW algorithm (chapter 8) are clearly seen in figure 11.3a.

This example of NMR structure determination in the solution state is a special and restricted case, in which FSD-1 could be synthesized through solid-phase $9H$-fluoren-9-ylmethoxycarbonyl (Fmoc) chemistry. A modern study of a larger protein would typically employ stable isotopic labeling by recombinant protein expression in a bacterial host (such as *E. coli*) followed by protein purification by fast protein liquid chromatography (FPLC), and additional NMR experiments (e.g., triple-resonance, IPAP) for assignments and to measure structural restraints such as RDCs in weakly aligned conditions (chapters 15–18). Nevertheless, these results represent a successful end-to-end study using the techniques we have been discussing, including algorithms for protein design and NMR structural biology.

References

[1] B. I. Dahiyat and S. L. Mayo. De novo protein design: Fully automated sequence selection *Science* 278;5335 (1997 October 3):82.

[2] J. Desmet J, De Maeyer M, and Lasters I. Theoretical and algorithmical optimization of the dead-end elimination theorem. *Pacific Symposium on Biocomputing* (1997):122–133. PubMed PMID: 9390285.

[3] I. Lasters, M. De Maeyer, and J. Desmet. Enhanced dead-end elimination in the search for the global minimum energy conformation of a collection of protein side-chains. *Protein Engineering* (1995) Aug;8;8:815–822.

[4] A. Leach and A. Lemon. Exploring the conformational space of protein side-chains using dead-end elimination and the A^* algorithm. *Proteins*, 33:227–239, 1998.

[5] Niles Pierce and Erik Winfree. Protein design is NP-hard. *Protein Engineering,* 15 (2002):779–782.

[6] Bernard Chazelle, Carl Kingsford, and Mona Singh. A semidefinite programming approach to side-chain positioning with new rounding strategies. *INFORMS Journal on Computing,* 16;4 (2004):380–392.

[7] S.L. Mayo et al. "Apparatus and Method for Automated Protein Design." U.S. Patent 6,269,312 (2001).

12 Nonribosomal Code and K^* Algorithms for Ensemble-Based Protein Design

Protein redesign plays an important role in the realization of novel molecular function and drug design. In this chapter, we introduce an algorithm for the protein redesign problem, which combines a statistical mechanics-derived ensemble-based approach to computing the binding constant with the speed and completeness of a branch-and-bound pruning algorithm [13, 22]. The algorithm has been applied to redesign enzymes [2], and we begin by describing those enzymes' biological function.

12.1 Nonribosomal Peptide Synthetase (NRPS) Enzymes

Nonribosomal peptide synthetase (NRPS) enzymes, usually expressed in microorganisms like bacteria and fungi, complement the traditional ribosomal peptide synthesis pathway. They are also the sources of hundreds of peptidelike products with pharmaceutical properties, including natural antibiotics, antifungals, antivirals, anticancer therapeutics, immunosuppressants, and siderophores. Enzymes of the NRPS pathway have multiple domains with individual functions acting in an assemblyline fashion. The ribosomal machinery is not used. It is commonly believed that the substrate specificity of the NRPS enzymes is dictated primarily by the "gatekeeper" adenylation (A) domain, and recent evidence also indicates that the condensation (C), thiolation (T), and epimerization (E) domains carry some specificity as well.

NRPS enzyme redesign methods can be divided into two main techniques, *domain-swapping* and *active site modification through site-directed mutagenesis*. Domain-swapping techniques modify NRPS enzymes by swapping an adenylation domain of an existing NRPS enzyme for an adenylation domain from a second, different NRPS enzyme. Active site modification through site-directed mutagenesis utilizes structural information of the GrsA-PheA NRPS enzyme. GrsA-PheA is the *phenylalanine (Phe) adenylation domain (PheA) of the of gramicidin synthetase A (GrsA)* enzyme (figure 12.1, plate 11). Gramicidin synthetase has two modules, A and B. Confusingly, the "A" in PheA stands for *adenylation* whereas the "A" in GrsA stands for the A (as opposed to B) *module*.

From sequence alignment of GrsA-PheA with 160 other known adenylation domains, a "signature sequence" can be derived for each adenylation domain by extracting those residues that

Figure 12.1 (plate 11)

(Top) Gramicidin S synthetase is composed of two NRPS proteins, GrsA (3 domains) and GrsB (13 domains). Gramicidin S is produced in an assemblyline manner where two D-Phe-L-Pro-L-Val-L-Orn-L-Leu peptides are joined and cyclized. (A, Adenylation; T, Thiolation (peptidyl carrier protein); E, Epimerization; C, Condensation; TE, Thioesterase). (Bottom) The GrsA-PheA domain controls incorporation of the first amino acid in the synthesis of the antibiotic gramicidin. (Left) The natural gramicidin construct is shown with the incorporated phenylalanine shown in red. By changing the substrate specificity of the GrsA-PheA domain to accept leucine, it may be possible to create a modified gramicidin (right) where the phenylalanines have been replaced by leucine (blue).

align with the structurally determined substrate-binding pocket of the GrsA-PheA crystal structure [27]. The mapping between substrate specificity and the amino acid sequence of NRPS domains is sometimes called the *nonribosomal code*. Structure-based protein design algorithms provide a powerful tool for predicting, deciphering, and extending the nonribosomal code.

12.2 K-star (K^*) Algorithm Basics

Georgiev et al. and Lilien et al. [13, 22] developed an ensemble scoring method K^* (pronounced "K star") to model the protein-ligand binding in active-site mutants. Since it is not currently possible to compute exact partition functions for complex molecular species, K^* approximates these partition functions with the use of rotamerically based conformational ensembles.

Protein-ligand binding for a predicted mutant protein is modeled using the following K^* equation:

$$K^* = \frac{q_{PL}}{q_P q_L}. \tag{12.1}$$

K^* is derived to be an approximation to the true association (binding) constant $K_A = 1/K_D$ by expressing each species' chemical potential as a function of the species partition function q and solving for the equilibrium condition (further details are provided in [22]). To compute exact partition functions for complex molecular species appears to require integrating an exact energy function over a molecule's entire conformational space. We therefore approximate these partition functions with the use of rotamerically based conformational ensembles:

$$q_{PL} = \sum_{b \in B} \exp(-E_b/RT), \quad q_P = \sum_{f \in F} \exp(-E_f/RT), \quad q_L = \sum_{l \in L} \exp(-E_l/RT), \qquad (12.2)$$

where B, F, and L represent rotamer-based ensembles for the bound protein-ligand complex (PL), the free protein (P), and the free ligand (L), respectively, E_s is the energy of conformation s, R is the gas constant, and T is the temperature in Kelvin. The accuracy with which K^* approximates K_A is proportional to the accuracy of the partition function approximation used.

When applying K^* to a protein-ligand system, a number of choices must be made with respect to ensemble generation and single-structure scoring, where *single-structure scoring* is the method by which each individual member of the ensemble is scored. The choices made in ensemble scoring should strike a balance between fidelity to the underlying physical biochemistry and computational feasibility. Lilien et al. [22] first posit a hypothetical brute-force algorithm for mutation search: (1) generate all conformations in each of the molecular ensembles by fixing the protein backbone and using a rotamer library to vary the side-chain conformations; (2) use a numerical minimization procedure with the AMBER energy function [4] (plus the EEF1 [20] solvation model) to energy-minimize each generated conformation [13, 2]; (3) compute each partition function (eq. 12.2) and then combine them to obtain an overall K^* score (eq. 12.1).

However, more efficient algorithms than brute-force are developed in Georgiev et al. and Lilien et al. [22, 13]. To search for the optimal mutation sequence more efficiently, mutation space filters and conformation space pruning are employed. Mutation space filters include two filters: the sequence-space filter, a residue type filter, which restricts the mutation search to include only a subset of amino acids based on compatibility with the target substrate; and the volume filter, which removes mutations that significantly over- or underpack the substrate-bound active site relative to the wildtype. These two filters prune a combinatorial number of conformations from consideration and eliminate the majority of conformations early in the mutation search.

Next, since conformations with large energies are unlikely to be assumed and contribute only a vanishingly small amount to the partition function, it is reasonable to prune high-energy conformations from consideration. Chen et al. and Georgiev et al. [13, 2] employed a *minimized dead-end elimination ("minDEE")* strategy (see chapter 11, and section 12.5) that generates conformations and prunes high-energy conformations and mutations (figure 12.2) to enumerate structures and sequences in gap-free order of energy. By regarding these low-energy conformations as an ensemble of structures, and observing that high-energy structures contribute little to the partition functions, a rigorous threshold for stopping the enumeration can be proven, guaranteeing that a partition function within $(1 - \varepsilon)$ of the correct answer will be computed. This allows

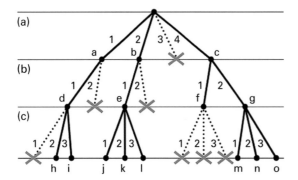

Figure 12.2
An example conformation tree. The rotamers of flexible residue i are represented by the branches at depth i. Internal nodes of a conformation tree represent partially assigned conformations. **x**'s represent nodes of the conformation tree where steric clash has been identified among a partially assigned conformation. All children of **x** nodes are pruned and not considered.

the construction of a provably good ε-approximation algorithm for the binding constant [13]. Figure 12.2 illustrates pruning by a *steric filter*. DEE and minDEE operate by reducing the *branching factor* in the tree, at many nodes at once.

Here is the basic intuition behind the K^* algorithm. When ignoring the pruned conformations, a so-called *intramutation pruning* is then applied during the computation of the partition function for a single mutation. The intramutation pruning algorithm is capable of guaranteeing any desired approximation accuracy to the true partition function, as shown in figure 12.3. That is, a conformation c_{j+1} can be pruned if it satisfies $B(c_{j+1}) \geq -RT \ln(q_j^* \varepsilon - p_j^*)$, and by induction, if all pruned conformations satisfy this condition, then at the end of the computation, q_{j+1}^* will be an ε-approximation to q_n, the true partition function.

The algorithm in figure 12.3 still has to examine every conformation, and therefore provides a merely constant-factor speed-up over a naïve algorithm. The improved algorithm of Georgiev et al. [13] uses the minDEE/A^* algorithm as a preprocessing filter for K^*. A simplified sketch of this algorithm is shown in figure 12.4. MinDEE/A^*/K^* is an efficient algorithm for computing the q_{PL}, q_P, and q_L partition functions, which are used to compute a K^* approximation score for a given mutation sequence. Using the A^* algorithm with minDEE, we can generate the conformations of a rotamerically based ensemble in order of increasing lower bounds on the conformation's minimized energy. We can efficiently compute the lower bound on a conformation's energy as a sum of precomputed pairwise minimum energy terms [13]. As each conformation c is generated from the conformation tree, we compare its lower bound $B(c)$ on the conformational energy to a moving *stop-threshold* and halt the A^* search once $B(c)$ becomes greater than the threshold. The A^* algorithm guarantees that all remaining conformations will have minimized energies above the stop-threshold. Georgiev et al. [13] prove that a partial partition function q^* computed using only those conformations with energies below (i.e., better than) the stop-threshold will lie within a factor of ε of the true partition function q. Note that, by definition, $q \geq q^*$. Thus, q^* is an ε-approximation to q, that is, $q^* \geq (1-\varepsilon)q$.

Let $n \leftarrow$ Number of rotameric conformations
Let $c \leftarrow$ Rotameric conformations
Initialize: $q^* \leftarrow 0, \ p^* \leftarrow 0$
for $j = 1$ to n
    if $\mathrm{B}(c_j) \leq -RT \ln \left(q^* \varepsilon - p^* \right)$
        $q^* \leftarrow q^* + \exp \left(-\texttt{ComputeMinEnergy}(c_j)/RT \right)$
    else
        $p^* \leftarrow p^* + \exp \left(-\mathrm{B}(c_j)/RT \right)$
Return q^*

Figure 12.3
Intramutation pruning, Constant-factor speed-up. Here q^* is the running approximation to the partition function, and p^* is an upper bound on the partition function of the pruned conformations. The function $\mathrm{B}(\cdot)$ computes a lower energy bound for the given conformation. The function ComputeMinEnergy$(\cdot)$ returns the energy of the energy-minimized conformation as computed using steepest-descent minimization and an implementation of the AMBER energy function. Later work used more sophisticated energy functions [13, 2]. At the end, q^* represents an ε-approximation to the true partition function q such that $q \geq q^* \geq (1 - \varepsilon)q$.

Initialize: $n \leftarrow$ Number of rotameric conformations; $q^* \leftarrow 0$
while $(n > 0)$
    $c \leftarrow \texttt{GetNextAStarConf}()$
    if $\mathrm{B}(c) \leq -RT \left(\ln(q^* \rho - k \exp(-E_0/RT)) - \ln n \right)$
        $q^* \leftarrow q^* + \exp \left(-\texttt{ComputeMinEnergy}(c)/RT \right)$
        $n \leftarrow n - 1$
    else Return q^*
if $q^* \rho \geq k \exp(-E_0/RT)$
    Return q^*

Figure 12.4
Efficient partition function computation with energy minimization using the A* search. minDEE/A* provides a combinatorial-factor speed-up. q^* is the running approximation to the partition function. The function $\mathrm{B}(\cdot)$ computes the energy lower bound for the given conformation [13]. The function ComputeMinEnergy$(\cdot)$ returns a conformation's energy after energy minimization. The function GetNextAStarConf() returns the next conformation from the A* search. Define $\rho = \frac{\varepsilon}{1-\varepsilon}$. p^* is the partition function computed over the set P of conformations pruned by minDEE, so that $p^* \leq k \exp(-E_0/RT)$, where $|P| = k$. E_0 is the minimum lower energy bound among all conformations containing at least one pruned rotamer. Upon completion, q^* represents an ε-approximation to the true partition function q, such that $q \geq q^* \geq (1 - \varepsilon)q$. Simplified version of figure 4 from [13].

The application of the minDEE criterion [13] for each rotamer i_r requires that the corresponding minimum energy terms are accessed. Hence, we can piggyback the computation of a lower bound B_{i_r} on the energy of all conformations that contain a pruned rotamer i_r. Let E_0 be the minimum lower energy bound among all conformations containing at least one pruned rotamer,

$$E_0 = \min_{i_r \in S} B_{i_r},$$

where S is the set of *rotamers* pruned by minDEE. E_0 can be precomputed during the minDEE stage and prior to the A^* search. Let p^* be the partition function computed over the set P of *conformations* pruned by minDEE, so that $p^* \leq k \exp(-E_0/RT)$, where $|P| = k$. We can then guarantee an ε-approximation to the full partition function q using the minDEE/A^*/K^* algorithm in figure 12.4. Somewhat surprisingly, the algorithm can ignore a combinatorial number of conformations, and therefore provides a combinatorial-factor speed up, while still guaranteeing to compute an ε-approximation.

We have described *intramutation pruning,* namely, conditions under which a conformation could be pruned when computing a single partition function for a single mutation. What about pruning across mutations?

When using K^* to perform a mutation search, we can bootstrap the pruning condition for improved efficiency (by caching partition functions, we can exploit K^* bounds from other mutations in the same search). Similarly to the development above, pruned conformations are not energy-minimized, thereby saving time in the overall mutation search. Georgiev et al. and Lilien et al. [22, 13] show how, in a mutation search, the partition-function ε-approximation pruning conditions can exploit the partition functions previously computed for other, different mutation sequences evaluated earlier in that search. Therefore, we call this *intermutation pruning:* Georgiev et al. and Lilien et al. [22, 13] derive a lower bound on the partition function that allows us to prune more conformations earlier in the search.

In summary, four levels of approximation are used in computing K^* scores. (1) A rotamer library is used to model side-chain conformations. (2) Sterically disallowed rotamer-based conformations are combinatorially pruned. (3) Intra- and intermutation pruning methods are used to skip evaluation of conformations that are not required to compute ε-approximations to the partition function and K^* values. (4) In computing partition functions and K^* values, the algorithm starts with discrete rotamers and then performs bounded minimization. Finally, in the exposition above, and in many other protein design algorithms, the backbone is fixed. However, this restriction can be lifted, and some backbone flexibility modeled, as we discuss later in this chapter.

12.3　Energy Functions

Any pairwise energy function can be used in K^*. The default K^* energy function includes the Lazaridis-Karplus EEF1 pairwise solvation energy term [20]. A list of some energy functions used in protein design (including K^*) is given in table 12.1. Improvements to protein design

Table 12.1
Some energy functions used in protein design

	FF[a]	vdW[b]	elect[c]	H bond[d]	dihed[e]	bonds[f]	angles[g]	entropy[h]	solv[i]	stats[j]	other[k]
AMBER [4]	A	•	•	implicit[1]	•	•	•				
DREIDING [26]	D	•	•	•	•	•	•				•[2]
CHARMM [25]	C	•	•	implicit[1]	•	•	•				•[3]
Mayo [29]	D	•	•	•	•				•		
Baker [10]	C	•	•	•					•	•[4]	•[5]
Kuhlman [11]	C	•		•				•	•	•[4]	•[5]
Hellinga [19]	C	•	•	•				•	•	•[6]	
Dunbrack [9]	O	•[7]								•[8]	
K*, 2005 [23]	A	•	•	implicit[1]	•			ensembles	•	•[9]	•[5]
K*, 2009 [13]	A	•	•	implicit[1]	•			ensembles	•	•[9]	•[5]
DEE/A* [13]	A	•	•	implicit[1]	•			•[9]	•	•[9]	•[5]
minDEE/A* [13]	A	•	•	implicit[1]	•			•[9]	•	•[9]	•[5]
BD/A* [14]	A	•	•	implicit[1]	•			•[9]	•	•[9]	•[5]
brDEE/A* [15]	A	•	•	implicit[1]	•			•[9]	•	•[9]	•[5]

[a] Force field parameters: A if the corresponding energy function uses the AMBER force field; C—CHARMM; D—DREIDING; and O—other. The "•" in a column for [b] van der Waals, [c] electrostatic, [d] hydrogen bonding, [e] dihedral, [f] bonds, [g] angles, [h] entropy, [i] solvation, and [j] statistical terms, signifies that the corresponding energy function has some representation for the respective energy term. The [k] other column contains terms that cannot be readily classified in the preceding columns. [1] Hydrogen bonding is modeled by the charges and the vdW parameters of the force field; [2] Inversion energies; [3] Urey-Bradley and improper energies; [4] Rotamer free energies modeled by $-\ln \Pr(rot|AA, \phi, \psi)$ and a bonus for close unlike charges; [5] Amino acid reference energies $E_{ref}(AA)$; [6] AA preference for core/surface from PDB; [7] Linear repulsive vdW; [8] Normalized $-\ln \Pr(rot|AA, \phi, \psi)$; [9] Can easily be added; see section 12.3.

energy functions can be implemented, and in some cases they are quite helpful to improve the accuracy of the predictions. Some of these improvements try to incorporate more accurate physics; some introduce statistics from known high-resolution crystal structures; and some use statistical parameter estimation or machine learning techniques to fit the weights and parameters in the energy function terms, so that the results of the design algorithm jibe better with experimental data.

For example, one can use a rotamer probability term such as the (normalized)

$$- \ln \Pr(rot | AA, \phi, \psi) \tag{12.3}$$

score employed by Dunbrack and co-workers [9]; this is the negative natural logarithm of the probability (across a distribution of structures in a database) of a rotamer given the amino acid type and backbone dihedral angles. These probabilities can be obtained from the Richardsons' rotamer library [23] and also from a second library they have provided that has a finer sampling. Other enhancements to energy functions can include: scaling the van der Waals (vdW) radii (which can allow for limited overpacking, and therefore compensate for the lack of backbone flexibility in fixed-backbone redesign [7]), and a distance-dependent dielectric [28].

The original K^* scoring function [22] used the binding energy, without an explicit term to disfavor mutations that destabilize the protein versus the unfolded state. A rigorous solution to this problem requires negative design against the unfolded state [17, 1]. In principle, *Backbone-flexible DEE* (the *BD* algorithm; see section 12.6) can model backbone flexibility for negative design against a range of backbone competitors [14]. While a precise description of the ensemble for the unfolded state can sometimes be constructed from experimental data (see chapter 7), this process is nontrivial. Hence, the selection of charged residues at buried positions is discouraged by excluding polar residues from buried positions and by constraining the amino acid composition (core, surface, and active site compatible with a particular ligand) [22]. Similarly, hydrophobic residues are excluded from surface positions to prevent aggregation [22]. Such sequence constraints are expedient, but a more general approach is desirable [17, 1]. In the newer K^* energy function [13, 2], the solvation term from the EEF1 model [20] implicitly remedies this problem to some extent. One can also introduce an explicit destabilization penalty, which may be necessary when introducing buried polar residues to bind more polar substrates. This may be done by penalizing the burial of polar surface area [6, 5].

Note that minDEE/A* [13] and BD (section 12.6 and [14]) can be used either for GMEC-based design or for ensemble-based design (K^*). An explicit empirical entropy term need not be used for K^*, since the K^* method of ensemble scoring encompasses a measure of conformational entropy through use of the partition function over ligand and side-chain conformations. However, for GMEC-based redesign such as minDEE/A* (without ensembles), the empirical, amino acid–specific entropic penalties suggested by Looger and Hellinga [18] (e.g., for Met residues) can be incorporated. An explicit amino acid reference energy term [10] is not necessary for K^*, since the K^* method of ensemble scoring divides partition functions for the bound and unbound states (which would cause the reference energies to cancel out).

12.4   Redesigning Enzymes with *K**

The work of Chen et al. and Lilien et al. [22, 2] constructed a structural model based on the crystal structure of GrsA-PheA (PDB id: 1AMU), which consists of 9 active site residues (D235, A236, W239, T278, I299, A301, A322, I330, C331), the 30 residues with at least one atom within 8 Å of the active site (termed the *steric shell*), the amino acid substrate, and the AMP cofactor. In structure 1AMU [3], and also in the K^*-predicted structures, residues 235D and 517K make hydrogen bonds to the amino acid backbone of the ligand, thereby stabilizing the substrate in a productive orientation for catalysis.

Flexible residues are represented by rotamers from the Richardsons' rotamer library [23]. Rotameric-based conformations in B, F, and L (eq. 12.2) that survive minDEE are enumerated in order of lower-bound energy by A^*, and are minimized by steepest-descent minimization using the AMBER-EEF1 energy function (section 12.3) and are then combined using equations (12.1) and (12.2).

Several tests were implemented in Lilien et al. [22] to validate the algorithm. One is to find a conformation close to the crystal structure to confirm the rotamer search strategy with minimization and the scoring scheme. Another test is to simulate the biochemical activity assays of L-Phe and L-Leu against wild-type PheA and the T278M/A301G double mutation, and the results qualitatively agreed with the activity assays of Stachelhaus et al. [27]. Later, Chen et al. performed a K^* mutation search to redesign GrsA-PheA to bind and adenylate Leu and several other amino acids instead of Phe. Novel mutants were reported and experimentally validated [2].

In particular, Chen et al. [2] performed a computational structure-based redesign of the GrsA-PheA for a set of noncognate substrates for which the wild-type (WT) enzyme has little or virtually no specificity. Experimental validation of a set of top-ranked computationally predicted enzyme mutants showed significant improvement in the specificity for the target substrates. Chen et al. [2] further developed enhancements to the methodology for computational enzyme redesign, which resulted experimentally in significant additional improvements in the target substrate specificity. The results suggest that structure-based protein design can identify active mutants different from those selected by evolution.

Chen et al. [2] used provable K^* algorithms to redesign GrsA-PheA. They designed mutations to switch the substrate specificity of this enzyme from its cognate substrate phenylalanine, to leucine. The mutant with the highest activity for a noncognate substrate exhibits one-sixth of the WT enzyme:WT substrate activity. That is, the best redesigned enzymes have activity that is up to one-sixth of WT:Phe (k_{cat}/K_M). To do this they employed minDEE and K^* algorithms that not only design mutations to the active site, but also design *distal mutations* that significantly aid in switching the specificity. These algorithms have the possibility to compete *in silico* with directed evolution (an experimental technique that is often used to find distal mutations to restabilize the protein and counterbalance specificity-changing mutations made to the active site). Chen et al. [2] also redesigned the enzyme to adenylate a variety of charged substrates: Arg, Lys, Asp, and

Glu. The latter represent challenges, because the wildtype enzyme has essentially no activity for charged substrates. The designs were experimentally tested by creating the predicted mutant proteins and assaying their activity and binding.

12.5 Minimized Dead-End Elimination (minDEE)

One of the main challenges for protein redesign is the efficient evaluation of a combinatorial number of candidate structures. The modeling of protein flexibility, typically by using a rotamer library of commonly observed low-energy side-chain conformations, further increases the complexity of the redesign problem. A dominant algorithm for protein redesign is Dead-End Elimination (DEE), which prunes the majority of candidate conformations by eliminating rigid rotamers that provably are not part of the global minimum energy conformation (GMEC) (see chapter 11). The identified GMEC consists of rigid rotamers (i.e., rotamers that have not been energy minimized) and is thus referred to as the *rigid-GMEC*. As a postprocessing step, the conformations that survive DEE may be energy minimized. When energy minimization is performed after pruning with DEE, the combined protein design process becomes heuristic, and is no longer provably accurate: A conformation that is pruned using rigid-rotamer energies may subsequently minimize to a lower energy than the rigid-GMEC. That is, the rigid-GMEC and the conformation with the lowest energy among all energy-minimized conformations (the *minimized-GMEC*) are likely to be different. While the traditional ("rigid-rotamer") DEE algorithm succeeds in not pruning rotamers that are part of the rigid-GMEC, it makes no guarantees regarding the identification of the minimized-GMEC. Georgiev et al. [13] derive a provable and efficient DEE-like algorithm called *minimized-DEE (minDEE)* that guarantees that rotamers belonging to the minimized-GMEC will not be pruned, while still pruning a combinatorial number of conformations. MinDEE is sometimes called *minimization-aware dead-end elimination.*

12.5.1 A^* Search and minDEE
An additional combinatorial factor reduction in computational complexity can be achieved through the use of an A^* (pronounced "A star") search algorithm to enumerate conformations in order of increasing energy. In Leach and Lemon [21] an A^* branch and bound algorithm was developed to explore a number of low-energy conformations for a single mutation sequence (i.e., a single protein). In this algorithm, traditional DEE was first used to reduce the number of side-chain conformations, and then surviving conformations were enumerated in order of conformation energy by expanding sorted nodes of a conformation tree. Georgiev et al. [13] modified the A^* algorithm of [21] to compute a partition function using the ideas discussed in section 12.2.

The A^* algorithm deterministically identifies the least-cost path in a tree from the root to a leaf node. When the tree is a *conformation tree* (figure 12.2) and the path costs encode conformation energies, the A^* algorithm can be used to identify minimum-energy conformations [21]. In a conformation tree (figure 12.2), the rotamers of flexible residue i are represented by the branches

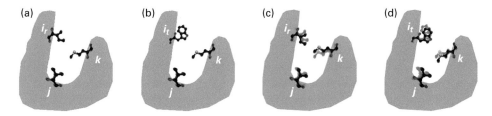

Figure 12.5 (plate 12)
Energy-minimized DEE. Without energy minimization the swapping of rotamer i_r for i_t (panel a to panel b) leaves unchanged the conformations and self and pairwise energies of residues j and k. When energy minimization is allowed, the swapping of rotamer i_r for rotamer i_t (panel c to panel d) may cause the conformations of residues j and k to minimize (i.e., move) to form more energetically favorable interactions (from the faded to the solid conformations in panels c and d).

at depth i. For example, in figure 12.2, residue A has 4 rotamers, residue B has 2 rotamers, and residue C has 3 rotamers. Internal nodes of a conformation tree represent partially-assigned conformations and each leaf node represents a fully-assigned conformation. Let E_w be a scalar energy threshold. The DEE/A^* algorithm of [21] generates all low-energy conformations within a window E_w of the GMEC. Georgiev et al. [13] extended this algorithm (a) for minDEE (i.e., to use minimized rotamers, figure 12.5, plate 12), and (b) to compute provable ε-approximations to the partition functions.

Hence, minDEE is useful not only in identifying the minimized-GMEC, but also as a filter in an ensemble-based scoring and search algorithm, called *minDEE/K**, for protein redesign that exploits energy-minimized conformations. The provable and efficient minimized-DEE algorithm is applicable in protein redesign, protein-ligand binding prediction, and computer-aided drug design.

12.6 Backbone Flexibility in DEE for Protein Design

12.6.1 Continuous Backbone Flexibility DEE

DEE is a powerful algorithm capable of reducing the search space for structure-based protein design by a combinatorial factor. By using a fixed backbone template, a rotamer library, and a potential energy function, traditional DEE identifies and prunes rotamer choices that are provably not part of the GMEC, effectively eliminating the majority of the conformations that must be subsequently enumerated to obtain the GMEC. Since a fixed-backbone model biases the algorithm predictions against protein sequences for which even small backbone movements may result in a significantly enhanced stability, the incorporation of backbone flexibility can improve the accuracy of the design predictions. If explicit backbone flexibility is incorporated into the model, however, the traditional DEE criteria can no longer guarantee that the *flexible-backbone GMEC*, the lowest-energy conformation when the backbone is allowed to flex, will not be pruned. Georgiev and Donald [14] derive a novel DEE pruning criterion, *flexible-backbone DEE (BD)*, that is provably accurate with backbone flexibility, guaranteeing that no rotamers belonging to the flexible-backbone GMEC are pruned (there are also further enhancements to BD for improved pruning efficiency; see [14, 16]). The results from applying BD to redesign the $\beta 1$ domain of protein G and to switch the substrate specificity of the NRPS enzyme GrsA-PheA were compared against the results from previous fixed-backbone DEE algorithms. This showed experimentally that traditional DEE is indeed not provably accurate with backbone flexibility and that BD is capable of generating conformations with significantly lower energies, thus confirming the feasibility of BD.

12.6.2   Backrub DEE

Based on stereochemical intuition, the existence of a subtle backbone motion coupled to rotamer jumps has long been suspected. Such a motion, the "backrub," was confirmed by closely examining the electron density for side-chains modeled as alternates in very high-resolution crystal structures and inferring that the backbone must have shifted between the two conformations to maintain reasonably ideal geometry [8]. It is conservatively estimated that 3% of all residues undergo backrubs, with a large fraction occurring at the protein surface, most likely reacting to bombardment from solvent molecules. In addition to modeling dynamics, [8, 15, 26] show that backrubs can allow rotamer changes. Hence, by deduction, they can accommodate mutations to amino acid types for which no rotamers fit in the original backbone. Therefore, it is reasonable to assume that backrubs may play an evolutionary role. Such an assumption is, of course, impossible to demonstrate from single high-resolution structures and, due to coordinate error on the level of backrub shifts, is also difficult to tease out by comparing otherwise identical-in-sequence point mutant structures. However, one way to address the question is by investigating the effects of *backrubs in protein design [15, 26],* which is essentially a guided form of evolution that contributes to our knowledge of the determinants of protein packing and folding. Backrubs enable a provable algorithm to design proteins with low energies [15]; therefore, we can be confident that they may also contribute on an evolutionary time scale [8].

The *Backrub* [8] is a small but kinematically efficient side-chain–coupled local backbone motion frequently observed in atomic-resolution crystal structures of proteins. A backrub shifts the C^α-C^β orientation of a given side-chain by rigid-body dipeptide rotation plus smaller individual rotations of the two peptides, with virtually no change in the rest of the protein. Backrubs can therefore provide a biophysically realistic model of local backbone flexibility for structure-based protein design. Georgiev et al. [15] present a combinatorial search algorithm for protein design that incorporates an automated procedure for local backbone flexibility via backrub motions. This derives a DEE-based criterion for pruning candidate rotamers that, in contrast to previous DEE algorithms, is provably accurate with backrub motions. The backrub-based DEE algorithm ("brDEE") successfully predicts alternate side-chain conformations from ≤ 0.9-Å resolution structures, confirming the suitability of the automated backrub procedure. Finally, the application of the algorithm to redesign two different proteins was shown to identify a large number of lower-energy conformations and mutation sequences that would have been ignored by a rigid-backbone model.

12.7 Application to Negative Design

It is intriguing to speculate how protein redesign algorithms based on K^* can be used for negative design. An illuminating example focuses on prospective prediction of resistance mutations. These mutations arise in protein targets to resist antimicrobial and antibiotic drugs, antivirals, and even in cancer cells to evade antineoplastic and chemotherapy agents.

Drug resistance resulting from mutations to the target is an unfortunately common phenomenon that limits the lifetime of many of the most successful drugs. In contrast to the investigation of mutations *after* clinical exposure, it would be powerful to be able to incorporate strategies early in the development process to predict and overcome the effects of possible resistance mutations. Frey et al. [29] describe the successful *in silico* prediction of resistance mutations to a novel antifolate inhibitor of *Staphylococcus aureus* (MRSA) dihydrofolate reductase (DHFR). This was the first prospective computational prediction of resistance mutations in a drug target, thus opening the possibility of overcoming drug resistance early in the iterative drug discovery process.

The K^* algorithm predicted potential resistance mutations in DHFR using *positive design* to maintain catalytic function and *negative design* to interfere with binding of a lead inhibitor [29]. Enzyme inhibition assays showed that three of the four highly ranked predicted mutants are active yet display lower affinity (18-, 9-, and 13-fold) for the inhibitor. A crystal structure of the top-ranked mutant enzyme validated the predicted conformations of the mutated residues and the structural basis of the loss of potency.

• This study demonstrates K^* can do negative design. Negative design is difficult because much greater conformational flexibility must be modeled.

• Deterministic algorithms were used to enumerate top mutants in gap-free order. All tested predictions (#1, #3, #7, and #9) were in the top 10 predictions (out of 1173 mutants and 10^{12} conformations), and the top-ranked mutant (#1) performed best (followed by #3 and #7).

• Except for the selection of four of the top ten mutants, there was no human intervention or manual selection; the entire design was automated.

• Experimentally measured enzyme activities had precisely the same rankings predicted by the algorithm.

• While the negative designs were based on a crystal structure of the DHFR:inhibitor complex, the positive designs were based on a *model* (structure prediction) of MRSA DHFR bound to NADPH and dihydrofolate. The success of the algorithm in designing to a model (as opposed to a crystal structure) is notable.

• A crystal structure of the top-ranked mutant (#1) proved the algorithm's structural predictions were highly accurate, explaining the affinity loss as predicted.

This study used the protein design algorithms discussed in this chapter (minDEE/K^*), and the predictions were then tested prospectively. The ability of these provable, ensemble-based protein design algorithms to predict resistance mutations could be incorporated in a lead design strategy against any target that is susceptible to mutational resistance. The accuracy of the algorithms suggests the possibility to analogously predict future mutational resistance to other targets.

There were several extensions made to K^* to handle the negative design. K^* uses both intra- and intermutation pruning, as described earlier in this chapter [13]. The intra- and intermutation pruning algorithms ensure provably accurate K^* scores for the best-scoring mutants and structures. This suffices for positive design, in which good-scoring structures and mutants are of interest. However for negative design, we are interested in poorly scoring structures and mutants as well. For poorly scoring structures and mutants, the intermutation pruning algorithm described earlier does not guarantee accurate scores [13]. For this reason, some modifications were undertaken to extend the algorithm for provably accurate negative design. Specifically, either changing γ (which controls the intermutation pruning fraction) or disabling the intermutation pruning may be necessary for provably accurate negative design. Frey et al. [29] tried both ways. In practice they have found that the negative design predictions are often still adequate even with the intermutation pruning enabled. These changes to the K^* parameters are described in the open-source code provided with [29].

12.8   Discussion

Algorithms for determining protein structures using NMR or X-ray crystallography comprise a family of techniques for *analyzing* biological data and biological problems. However, this chapter described another family of algorithms and their application to *synthetic biology* problems such as a protein engineering. These ensemble-based scoring and search algorithms for protein design have been applied to modify the substrate specificity of an antibiotic-producing enzyme in the non-ribosomal peptide synthetase (NRPS) pathway [2], and to predict resistance mutations in MRSA DHFR [29]; The latter were confirmed with crystal structures.

Realization of novel molecular function requires the ability to alter molecular complex formation. Enzymatic function can be altered by changing enzyme-substrate interactions via modification of an enzyme's active site. A redesigned enzyme may either perform a novel reaction on its native substrates or its native reaction on novel substrates. A number of computational approaches have been developed to address the combinatorial nature of the protein redesign problem. These approaches typically search for the GMEC among an exponential number of protein conformations. In contrast, new algorithms for protein redesign combine a statistical mechanics-derived ensemble-based approach to computing the binding constant with the speed and completeness of a branch-and-bound pruning algorithm [22, 2, 13]. In addition, efficient deterministic approximation algorithms have been obtained, capable of approximating biophysical scoring functions to arbitrary precision [2, 13]. These techniques include provable ε-approximation algorithms for estimating partition functions to model binding and protein flexibility.

This lecture only covers the basic ideas behind the K^* algorithm. Since 2005, there has been rapid progress on improved versions of the K^* and DEE algorithms, and the designs have been experimentally tested. For a general introduction to this line of research, see Chen et al. [2]. Recent provable algorithms include minDEE [13]; continuous backbone flexibility in DEE (BD) [14]; backrub motions in DEE (brDEE) [15]; and divide-and-conquer splitting (DACS) [16]. For K^* implementation, see Chen et al. and Georgiev et al. [13, 2]. For the general idea of K^*, also see Lilien et al. [22]. Three of these extensions and improvements—minDEE, BD, and brDEE—were discussed earlier. MinDEE/K^* and BD were used to obtain novel enzyme and DHFR resistance designs that were experimentally verified [2].

The algorithms in chapters 11 and 12 are provable and complete, which allows an engineering segregation of the search procedure from the model. The *input model* typically includes initial protein backbone structures, a library of rigid rotamers, an energy function, and models of continuous or discrete flexibility for the backbones and rotamers. The search procedure rank-orders, gap-free, the best solutions in sequence and conformational space. We expect some of these solutions to be improved designs, but the unsuccessful designs will be useful to improve the model or the algorithm. In contrast, if a heuristic (as opposed to provable) design algorithm is used, we have no idea whether the failures arose from inadequate optimization or from flaws in the model (energy function, structure, flexibility, rotamer library, etc.). Thus, there is a danger that the failures will be used to overfit the energy function. In particular, without a provable algorithm, we do not know whether the original model can successfully discriminate the decoys, since the heuristic technique is not guaranteed to return the optimal answers for the model nor to compute a list in the order ranked by the model. If a perfectly good energy function is overfit to handle spurious predictions (artifacts of the algorithm, not of the model) then the process of improving the model has failed, and the additional experimental data will be misused to effectively "break" a good model to overfit the putative outliers. Hence, with heuristic design algorithms the operation of tuning the model to match the data can be fundamentally unsound. On the other hand, with a provable algorithm, we remove one "variable" from the experiment (the algorithm), and we know that any discrepancies between the experimental results and the algorithm's predictions must be blamed solely on inadequacies of the model (and not the algorithm). In this manner, by using provable algorithms it is substantially more straightforward to improve the model given new experimental data, and these improvements or "tuning" are sound.

Recent work (e.g., chapters 11, 12) has extended the boundaries of what is possible with provable algorithms, in a series of end-to-end studies going from mathematics to novel algorithms to software to prospective experimental validation including binding, kinetics, stability, in vivo assays, and NMR and crystal structures. The algorithms were experimentally demonstrated to have remarkable accuracy and predictive power. Computational protein designs can only be as good as the (necessarily approximate) input model. But provable algorithms can avoid the undersampling, overfitting, and local minima that may cause heuristic algorithms to predict solutions far worse than the model. Provable algorithms guarantee that the predictions are exactly as good as the model, and as described above, this enables a powerful and principled path to improving the model based on experimental feedback.

Exercise Explain why more conformational flexibility must be modeled for negative design (as opposed to positive design). How much more conformational flexibility was modeled for negative design (versus positive design) in the study of Frey et al. [29]? Will that about be sufficient, in general? How could additional conformational flexibility be modeled?

References

[1] Daniel N. Bolon, Robert A. Grant, Tania A. Baker, and Robert T. Sauer. Specificity versus stability in computational protein design. *Proceedings of the National Academy Sciences (U S A)*, 102;36(Sep 2005):12724–12729.

[2] C. Chen, I. Georgiev, A.C. Anderson, and B.R. Donald. Computational structure-based redesign of enzyme activity. *Proceedings of the National Academy of Sciences (USA)*, 106;10(2009): 3764–3769.

[3] E. Conti, T. Stachelhaus, M. Marahiel, and P. Brick. Structural basis for the activation of phenylalanine in the non-ribosomal biosynthesis of Gramicidin S. *The EMBO Journal* 16 (1997):4174–4183.

[4] W. Cornell, P. Cieplak, C. Bayly, I. Gould, K. Merz, D. Ferguson, et al. A second generation force field for the simulation of proteins, nucleic acids and organic molecules. *Journal of the American Chemical Society* 117(1995):5179–5197.

[5] B. I. Dahiyat, D. B. Gordon, and S. L. Mayo. Automated design of the surface positions of protein helices. *Protein Science* 6;6(Jun 1997):1333–1337.

[6] B. I. Dahiyat and S. L. Mayo. Protein design automation. *Protein Science* 5;5(May 1996):895–903.

[7] B. I. Dahiyat and S. L. Mayo. Probing the role of packing specificity in protein design. *Proceedings of the National Academy of Sciences (USA)* 94(1997):10172–10177.

[8] I. W. Davis, W. B. Arendall 3rd, D. C. Richardson and P. S. Richardson. The backrub motion: how protein backbone shrugs when a side-chain dances. Structure. 2006 Feb;14(2):265–274.

[9] A. A. Canutescu, A. A. Shelenkov, and R. L. Dunbrack Jr. A graph-theory algorithm for rapid protein side-chain prediction. *Protein Science* 12(2003):2001–2014.

[10] B. Kuhlman and D. Baker. Native protein sequences are close to optimal for their structures. *Proceedings of the National Academy of Sciences (USA)* 97(2000):10383–10388.

[11] X. Hu and B. Kuhlman. Protein design simulations suggest that side-chain conformational entropy is not a strong determinant of amino acid environmental preferences. *PROTEINS: Structure, Function, and Bioinformatics*, 62(2006):739–748.

[12] G. D. Friedland, A. J. Linares, C. A. Smith, and T. Kortemme. A simple model of backbone flexibility improves modeling of side-chain conformational variability. *Journal of Molecular Biology* 380;4(2008 Jul):757–774.

[13] I. Georgiev, R. Lilien, and B. R. Donald. The minimized dead-end elimination criterion and its application to protein redesign in a hybrid scoring and search algorithm for computing partition functions over molecular ensembles. *Journal of Computational Chemistry* 29;10(2008):1527–1542.

[14] I. Georgiev and B. R. Donald. Dead-end elimination with backbone flexibility. *Bioinformatics* 23;13(2007): i85–94. Proceedings of the International Conference on Intelligent Systems for Molecular Biology (ISMB), Vienna, Austria (2007).

[15] I. Georgiev, D. Keedy, J. S. Richardson, D. C. Richardson, and B. R. Donald. Algorithm for backrub motions in protein design. *Bioinformatics* 24;13(2008):i196–204. Proceedings of the International Conference on Intelligent Systems for Molecular Biology (ISMB), Toronto, Canada (2008).

[16] I. Georgiev, R. Lilien, and B. R. Donald. Improved pruning algorithms and divide-and-conquer strategies for dead-end elimination, with application to protein design. *Bioinformatics*, 22;14(2006):e174–183, Proceedings of the International Conference on Intelligent Systems for Molecular Biology (ISMB), Fortaleza, Brazil (2006).

[17] James J. Havranek and Pehr B. Harbury. Automated design of specificity in molecular recognition. *Nature Structural Biology*, 10;1(2003):45–52.

[18] L. Looger and H. Hellinga. Generalized dead-end elimination algorithms make large-scale protein side-chain structure prediction tractable: Implications for protein design and structural genomics. *Journal Molecular Biology* 307(2001):429–445.

[19] H. Hellinga and F. Richards. Construction of new ligand binding sites in proteins of known structure: I. Computer-aided modeling of sites with pre-defined geometry. *Journal Molecular Biology* 222(1991):763–785.

[20] T. Lazaridis and M. Karplus. Effective energy function for proteins in solution. *PROTEINS: Structure, Function, and Genetics* 35(1999):133–152.

[21] A. Leach and A. Lemon. Exploring the conformational space of protein side-chains using dead-end elimination and the A^* algorithm. *Proteins: Structure, Function, and Bioinformatics* 33(1998):227–239.

[22] R. H. Lilien, B. W. Stevens, A. C. Anderson, and B. R. Donald. A novel ensemble-based scoring and search algorithm for protein redesign and its application to modify the substrate specificity of the gramicidin synthetase a phenylalanine adenylation enzyme. *Journal of Computational Biology* 12;6(2005 Jul–Aug):740–761.

[23] S. Lovell, J. Word, J. Richardson, and D. Richardson. The penultimate rotamer library. *Proteins: Structure, Function, and Bioinformatics* 40(2000):389–408.

[24] A. D. MacKerell Jr., D. Bashford, M. Bellott, R. L. Dunbrack Jr., J. D. Evanseck, M. J. Field, et al. All-atom empirical potential for molecular modeling and dynamics studies of proteins. *The Journal of Physical Chemistry B*, 102(1998):3586–3616.

[25] S. L. Mayo, B. D. Olafson, and W. A. Goddard III. DREIDING: A generic force field for molecular simulations. *The Journal of Physical Chemistry* 94(1990):8897–8909.

[26] C. A. Smith, and T. Kortemme. Backrub-like backbone simulation recapitulates natural protein conformational variability and improves mutant side-chain prediction. *Journal of Molecular Biology* 380;4(2008 Jul 18):742–756.

[27] T. Stachelhaus, H. Mootz, and M. Marahiel. The specificity-conferring code of adenylation domains in nonribosomal peptide synthetases. *Chemical Biology* 6(1999):493–505.

[28] Eric S. Zollars, Shannon A. Marshall, and Stephen L. Mayo. Simple electrostatic model improves designed protein sequences. *Protein Science* 15(2006):2014–2018.

[29] Frey K, Georgiev I, Donald BR, Anderson A. Predicting resistance mutations using protein design algorithms. *Proceedings of the National Academy of Sciences (PNAS)* USA. 2010;107(31):13707–12.

13 RDCs in NMR Structural Biology

This chapter continues our discussion of the residual dipolar coupling method in solution NMR spectroscopy, and then presents some related computational topics.

13.1 Residual Dipolar Couplings

The *nuclear Overhauser effect* (NOE) in NMR spectroscopy provides a local distance constraint, and the *scalar coupling* provides a local dihedral angle measurement. Both NOEs and scalar couplings have limitations for measuring structural information between atoms far apart. On the other hand, *residual dipolar couplings* (RDCs) provide global orientational restraints between remote internuclear vectors, and thus give a potential solution to these limitations.

Residual dipolar coupling, measuring anisotropic spin interaction, between spins i and j is described as follows:

$$D_{ij} = \frac{\mu_0 \gamma_i \gamma_j \hbar}{4\pi^2 \langle r_{ij}^3 \rangle} \left\langle \frac{3\cos^2 \theta - 1}{2} \right\rangle, \tag{13.1}$$

where γ_i and γ_j denote the gyromagnetic ratios of spins i and j, and r_{ij} denotes the internuclear distance, and θ denotes the angle between the internuclear vector and the axis of the external magnetic field.

When we only consider the coupling between covalently bound spins, the internuclear distance is essentially a constant. Thus, the coupling D_{ij} between spin i and j depends only on the orientation of the corresponding internuclear vector, that is, the orientational information of the internuclear vector related to the molecular coordinate frame can be obtained from the residual dipolar couplings.

Each experimentally measured RDC can be formulated by the following equation [1]:

$$d = D_{\max} v^T S v, \tag{13.2}$$

where $D_{\max}$ is a constant, and v is the internuclear vector orientation relative to an arbitrary coordinate frame, and S is the 3×3 *Saupe order matrix* [1]. Moreover, S is a real-valued, symmetric, traceless, rank 2 tensor with 5 degrees of freedom.

The tensor S can be diagonalized according to the following equation:

$$S = U \Delta U^T, \tag{13.3}$$

where $U \in SO(3)$ is a 3×3 rotation matrix, representing the *principal order frame* (POF), and Δ is a 3×3 diagonal and traceless matrix consisting of the eigenvalues of S. Note that $U^T = U^{-1}$.

13.2 Computational Topics Related to RDCs

13.2.1 Assignment Problem

Problem Statement   *Given the experimental RDCs d_i and the internuclear bond vectors v_j of a known protein structure, how can we assign each RDC d_i to its corresponding internuclear vector v_j?*

We could try to assign each pair of d_i and v_j with weight $w_{ij} = |d_i - D_{max} v_j^T S v_j|$. Then the assignment problem can be formulated into a *maximum bipartite matching*, which can be solved in $O(n^3)$ time by using the *Kuhn-Munkres matching* algorithm. Here, n denotes the number of residues.

Notes
Suppose that we know $k \geq 5$ pairs of unambiguous assignments between d_i and v_j, that is, we are given the following algebraic equation:

$$\begin{pmatrix} d_1 \\ \cdot \\ \cdot \\ \cdot \\ d_k \end{pmatrix} = D_{max} V^T S V, \tag{13.4}$$

where V is derived from $(v_1, \ldots, v_k)$. Given $(d_1, \ldots, d_k)^T$ and V, we can find the tensor S in equation (13.4) using *singular value decomposition* (SVD).

13.2.2 Structure Determination Problem

Problem Statement   *Given an alignment tensor S and experimental RDC data, how can we find conformation vectors such that the backbone (ϕ, φ) angles best fit the experimental RDCs [7]?*

Suppose that we are given the experimental RDC data in two alignment media, that is, we have two alignment tensors, which give two orientational constraints on each internuclear vector of a certain type, such as NH. Since each RDC in one medium yields a quartic constraint, the corresponding bond vector must lie on the intersection of two conic curves [7].

Let (x, y, z) denote the coordinates of the unit bond vector in a principal order frame of one medium. Based on two RDC constraints and the rotation transformation between two principal order frames of two media, we can obtain a polynomial in x in which only the degrees of 0, 2, 4, 8 appear. Thus, we obtain a quartic in x^2, and there are only a finite number of exact solutions for each internuclear bond vector, which generates a finite and discrete set for the final conformation space. This is because univariate quartic polynomials can be solved exactly, in closed form. Based on the computed NH (or other) bond vectors, we can calculate the (ϕ, ψ) torsion angles by using inverse kinematics [7].

13.2.3   Estimation of Alignment Tensor without Assignments

Problem Statement   *Suppose we know the eigenvalues of the alignment tensor S. Then given unassigned RDC data and a known protein structure, how can we estimate the alignment tensor without knowing assignments?*

When the internuclear vectors are distributed reasonably uniformly in orientation (chapter 6), the magnitude and rhombicity of a molecular alignment tensor can be obtained without assignment from a histogram of experimental RDCs [8]. This immediately yields an estimate of the eigenvalues. This technique is sometimes called the "*powder pattern.*"

Suppose that the eigenvalues Δ can be estimated using the powder pattern [1, 3]. Let U be the rotation matrix of the principal order frame, and D be the set of unassigned experimental RDC data. Let $B(U)$ denote the set of back-computed RDCs, that is, $B(U) = E^T SE = (E^T (U^T \Delta U)E)$, where E is the set of internuclear bond vectors. Now we are going to find

$$\underset{U \in SO(3)}{\text{argmin}} \; KL(D, B(U)), \tag{13.5}$$

where $KL(\cdot)$ stands for the Kullback-Leibler distance. This approach finds the rotation that brings the two distributions (experimental and predicted) into best agreement.

When U is obtained, the alignment tensor can be computed by equation (13.3).

Alternative Approach   We can also estimate the alignment tensor in an alternative way, that is, we can find the optimal tensor that makes the unassigned bond vectors, computed from the experimental RDCs, best match the geometric pattern of known structure bond vectors. The geometric matching distance can be measured by Hausdorff distance. This approach appears to require more RDCs per residue, or more aligning media.

13.2.4   Structural Homology Detection

Problem Statement   *Given the unassigned experimental RDC data from a protein of unknown structure, can we apply the algorithm in section 13.2.3 and rank the known structures in a database based on their value of KL in equation (13.5)?*

Given a known model (structure in PDB format), we can back-compute the RDCs based on the estimated result of the alignment tensor in section 13.2.3. By measuring the Kullback-Leibler distance between simulated RDCs and experimental RDC data, we can compare the distributions of RDCs and see how close the unknown structure is to any known structure.

Alternative Approach   We can also measure the structural homology in another way. We first compute the unassigned bond vectors based on the alignment tensor obtained in section 13.2.3. Then we can rank the known structures in the database by measuring the geometric matching distance, such as Hausdorff distance, between the orientational distributions of unassigned bond vectors and known structures. This approach appears to require more RDCs per residue, or more aligning media.

In the next chapters, we will discuss some solutions to these algorithmic problems in more detail.

References

[1] J. A. Losonczi, M. Andrec, M. W. F. Fischer, and J. H. Prestegard. Order matrix analysis of residual dipolar couplings using singular value decomposition. *Journal of Magnetic Resonance* 138 (1999): 334–342.

[2] C. J. Langmead and B. R. Donald. An expectation/maximization nuclear vector replacement algorithm for automated NMR resonance assignments. *Journal of Biomolecular NMR,* 29;2(2004):111–138.

[3] C. J. Langmead, A. K. Yan, R. H. Lilien, L. Wang, and B. R. Donald. A polynomial-time nuclear vector replacement algorithm for automated NMR resonance assignments. *RECOMB* (2003): 176–187; *Journal Computational Biology* 11;2–3(2004):277–298.

[4] C. J. Langmead and B. R. Donald. 3D structural homology detection via unassigned residual dipolar couplings. *Proceedings of IEEE Computational Systems Bioinformatics Conference (CSB).* Stanford University, Palo Alto (August 10, 2003) pp. 209–217.

[5] C. J. Langmead and B. R. Donald. High-throughput 3D structural homology detection via NMR resonance assignment. *Proceedings of IEEE Computational Systems Bioinformatics Conference (CSB).* Stanford University Palo Alta, (August, 2004) pp. 278–289.

[6] C. J. Langmead and B. R. Donald. An improved nuclear vector replacement algorithm for nuclear magnetic resonance assignment. *Tech. Rep. TR2004-494,* Dartmouth Department of Computer Science, 2004.

[7] L. Wang and B. R. Donald. Exact solutions for internuclear vectors and backbone dihedral angles from NH residual dipolar couplings in two media, and their application in a systematic search algorithm for determining protein backbone structure. *Journal of Biomolecular NMR,* 29;3(2004):223–242.

[8] G. M. Clore, A. M. Gronenborn, and A. Bax. A robust method for determining the magnitude of the fully asymmetric alignment tensor of oriented macromolecules in the absence of structural information. *Journal of Magnetic Resonance* 133(1998):216–221.

14 Nuclear Vector Replacement

This lecture presents the nuclear vector replacement (NVR) algorithm, and its applications in automated NMR resonance assignments [1, 2] and three-dimensional (3D) structural homology detection [3, 4]. The expectation-maximization approach is introduced, which complements bipartite matching as a useful framework for many structural problems.

NVR stands for "Nuclear Vector Replacement." It is named by analogy to "molecular replacement" in X-ray crystallography (chapters 47 and 48).

14.1 Experimental Input

NVR is an algorithm for computing resonance assignments given a putative 3D structural model of the target protein. This model is typically a PDB format structure. The following data are processed in the NVR algorithm: unassigned chemical shifts in H^N-^{15}N HSQC spectra, H^N-^{15}N RDCs in two media, amide exchange data, and unassigned peaks in 3D ^{15}N-NOESY spectra. The amide exchange information provides probabilistic classifications for the peaks in the HSQC corresponding to non-hydrogen-bonded, solvent-accessible backbone amide protons. All the H^N-^{15}N RDC, H-D exchange and ^{15}N-NOESY data serve as the geometric constraints on assignment. Table 14.1 shows the information content and roles of the input data.

14.2 Nuclear Vector Replacement

The NVR algorithm consists of three phases: *tensor estimation*, *resonance assignment*, and *structure refinement*, as shown in figure 14.1. In the first phase, NVR estimates the alignment tensors in both media. In the second phase, NVR applies the estimated tensors, and uses an iterative computational process, such as a Bayesian or expectation/maximization framework, to search the optimal resonance assignments. After resonance assignments, the assigned RDCs are used to refine the structure of the model by using the algorithm in Wang and Donald [6]. In the following subsections, we will present more details on the tensor estimation and resonance assignment phases.

Table 14.1
Information contents and roles of the input data [1]

Experiment/Data	Information Content	Role
H^N-^{15}N HSQC	H^N, ^{15}N chemical shifts	Backbone resonances, cross-referencing NOESY
H^N-^{15}N RDC (in 2 media)	Restraints on amide bond vector orientation	Tensor determination, resonance assignment,
H-D exchange HSQC	Identifies solvent exposed amide protons	Tensor determination
H^N-^{15}N HSQC-NOESY	Distance restraints between spin systems	Tensor determination, resonance assignment
^{15}N TOCSY	Side-chain chemical shifts	Tensor determination, resonance assignment
Backbone structure	Tertiary structure	Tensor determination, resonance assignment
Chemical shift predictions	Restraints on assignment	Tensor determination, resonance assignment

Reprinted from [1]. Copyright 2004 IEEE.

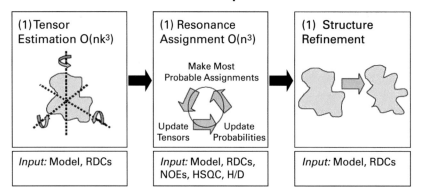

Nuclear Vector Replacement

Figure 14.1
Nuclear vector replacement [2].

14.2.1   Tensor Estimation

An alignment tensor S is a symmetric and traceless 3×3 matrix with five degrees of freedom, for example, three Euler angles α, β and γ, and the axial D_a and rhombic D_r components of an ellipsoid scaling the dipolar couplings. The alignment tensor can be diagonalized as follows:

$$S = V \Sigma V^T$$

where $V \in SO(3)$ is a 3×3 rotation matrix that defines the *principle order frame*, and Σ is a 3×3 diagonal and traceless matrix containing the eigenvalues of S and encoding D_a and D_r.

$SO(3)$ denotes the *special orthogonal group* of real 3×3 invertible matrices. It is the intersection of the *orthogonal group* $O(3)$ of orthogonal matrices, and the *special linear group* $SL(3)$ of matrices with determinant $+1$. $SO(3)$ is isomorphic to the group of three-dimensional rotations.

The diagonal elements of Σ can be obtained based on the powder pattern method. Given Σ, a rotation matrix $V(\alpha, \beta, \gamma)$, corresponding to an alignment tensor S, is found such that the Kullback-Leibler distance is minimized between the distribution of unassigned experimental RDCs and the distribution of the back-computed RDCs.

Time Complexity   The eigenvalues of S are computed in $O(nk^3)$ time, where n is the number of experimental RDCs and k is the resolution of the search-grid over $SO(3)$. The rotation matrix is calculated in $O(nk^3)$ time. Given Σ and V, constructing S takes only $O(1)$ time. Thus, the total running time for tensor estimation is $O(nk^3)$. The Euler angle representation of three-dimensional rotations $V \in SO(3)$ is familiar to many students; however, it suffers from singularities. A preferred representation of rotations is unit *quaternions*. Geometrically, quaternions may be represented by the three-dimensional sphere S^3, which is a double cover of $SO(3)$. That is, if $q \in S^3$ is a quaternion, then we have a smooth map $\pi : S^3 \to SO(3)$ such that $\pi(q)$ is the rotation by q and $\pi(q) = \pi(-q)$. The quaternions, like the Euler angles, form a three-dimensional space (S^3) and therefore can be searched using three parameters (similar to α, β, γ) to construct a grid over S^3. Only half of S^3 must be searched, since $\pi(q) = \pi(-q)$.

14.2.2   Resonance Assignment

Based on the estimated tensors S_1 and S_2 in two alignment media, the back-computed RDCs are obtained according to the following equation:

$$d = D_{\max} v^T S_i \, v, \quad (i = 1, 2), \tag{14.1}$$

where $D_{\max}$ is the dipolar interaction constant, and v is the internuclear vector orientation.

Let Q be the set of HSQC peaks, R be the set of residues, D_m be the set of experimental RDCs in the medium m, and B_m be the back-computed RDCs in the medium m, based on S_m. Let M_m be the assignment probability matrix for the medium m such that

$$M_m(q, r) = \Pr(q \mapsto r | S_m) = N(d_m(q) - b_m(r, S_m), \mu_m, \sigma_m), \tag{14.2}$$

where $q \in Q, r \in R, d_m(q) \in D_m, b_m(r, S_m) \in B_m$, and $N(d_m(q) - b_m(r, S_m), \mu_m, \sigma_m)$ denotes the normal distribution function with the observed variable $d_m(q) - b_m(r, S_m)$, the mean μ_m, and the standard deviation σ_m. Each entry in M_m will be set to zero if the assignment $q \mapsto r$ violates some geometric constraint, such as the d_{NN}s from ^{15}N-NOESY spectra, or amide exchange. The d_{NN}s provide distance constraints between peaks that can be correlated to the residues in the model, while the H-D exchange information identifies fast-exchanging amide protons that are likely to be solvent-exposed and non-hydrogen-bonded, and can be correlated to the structure model.

Equation (14.2) is used to compute the assignment probability at each iteration. The most likely assignment in each medium is considered. Let $r_1(q), r_2(q) \in R$ be the most likely assignments of peak q in media 1 and 2. The assignment $q \mapsto r$ is added into the master list of assignments if both following conditions are satisfied:

$$r_1(q) = r_2(q)$$

and

$$r_m(q) \neq r_m(k), \quad \forall k \in Q, \ k \neq q.$$

It is possible that the second condition is not satisfied. In this case, a maximum bipartite matching is chosen.

Suppose that peak q has been assigned to r, then q and r are removed from the consideration of next iterations. At the end of each iteration, the alignment tensors S_1 and S_2 are refined based on the master list of assignments and model. At the same time, the assignment probabilities are renormalized and updated.

Time Complexity   It takes $O(n^2)$ time to construct the probability matrices. Updating the alignment tensors S_1 and S_2 takes $O(n^2)$ time. Since at least one residue is assigned at each iteration, there are at most $O(n)$ iterations. Thus, all assignments are guaranteed to be finished in $O(n^3)$ time. The time complexity has been improved to be $O(n^{5/2} \log(cn))$ in Langmead and Donald [5], where c is the maximum edge-weight in an integer-weighted bipartite graph.

14.3 An Expectation/Maximization NVR Algorithm

Langmead and Donald [1] applied an expectation/maximization NVR algorithm for automated NMR resonance assignments. In the expectation/maximization framework, the observed variables include chemical shifts, d_{NN}s, amide exchange rates. The hidden variables include the correct resonance assignments and the correct alignment tensors. Let X be the set of observed variables, and Y be the set of hidden variables. Let Θ denote the set of all assignments obtained so far. Initially, Θ is set to be empty, and then grows as the algorithm finds more assignments. The Expectation (E) step of the EM algorithm computes the following expectation:

$$E(\Theta \cup \Theta' | \Theta) = E(\log \Pr(X, Y | \Theta \cup \Theta')),$$

where Θ' denotes the set of new assignments that is disjoint with the previous set Θ. The Maximization (M) step computes the new assignment Θ^* that has the maximum likelihood:

$$\Theta^* = \underset{\Theta'}{\operatorname{argmax}} \, E(\Theta \cup \Theta' | \Theta).$$

At the same time, the algorithm updates the master list of assignment, that is, $\Theta \leftarrow \Theta \cup \Theta^*$. Also, the alignment tensors are recomputed at the end of each iteration, based on updated assignments in Θ. The algorithm terminates when all peaks have been assigned.

14.4   3D Structural Homology Detection via NVR

Given unassigned RDCs of a target protein and a set of known models (protein structures in PDB format) in a database, we measure the 3D structural homology between the target protein and each known model by computing the likelihood of the assignment from the experimental RDCs to the model. We run the NVR algorithm for each model in the database, and then rank them by the likelihoods of the assignments.

Time Complexity   Suppose that there are in total k models in the database. Since the NVR algorithm runs in $O(n^{5/2} \log(cn))$ time, processing all k proteins takes $O(kn^{5/2} \log(cn))$. Finally, sorting all k models according to their likelihoods of assignments takes $O(k \log k)$ time. Thus, the total required time is $O(kn^{5/2} \log(cn) + k \log k)$.

14.5   Matching Modulo a Group, and Clustering Modulo a Group

NVR for automated NMR resonance assignments can be viewed as a *matching problem* on a quotient space of (bond-vector) orientations, induced by a quadratic form ξ on the 2-sphere S^2. ξ is given by the familiar RDC tensor equation (eq. 14.3):

$$d = \xi(v, v) = D_{\max} v^T S v, \tag{14.3}$$

from above. Using the alignment tensor S, ξ is parameterized by the Lie group $SO(3) \times \mathbb{R}^2$. S is traceless and symmetric. When the eigenvalues of S can be experimentally estimated (as is often the case), the degrees of freedom reduce to $SO(3)$. Hence, 3D structural homology detection from *unassigned* NMR data (enabling rapid fold determination) can be performed by combinatorial optimization, searching over $SO(3)$ to minimize a functional that compares distributions generated by ξ's image of the bond vectors from putative database protein models (chapter 13). These algorithms are developed further in chapters 15–18.

Using the techniques in chapters 4, 13, and 15–18, the RDC equations can be combined with protein kinematics to reduce the problem of protein structure determination from RDCs, to a search over the (discrete and finite number of) roots of low-degree univariate polynomial equations. These techniques have been developed for NH RDCs in two media, and for NH and CH (i.e., C^α-H^α) RDCs in one medium. They can be extended to use alternative RDCs (on different bond vectors), additional RDCs, or additional aligning media.

In some cases, such as NH RDCs in two media, or NH and CH RDCs in one medium, solving the resulting polynomial equations (generically) results in multiple (albeit finite and discrete) roots. Hence, the structure determination algorithm must typically perform a discrete and combinatorial search for the correct root, i.e., the true solution.

In 3D structural homology detection, there are two ambiguities in matching: which bond vector, and which root. The following toy example illustrates the concept. We consider the case of NH

RDCs in two media. Suppose we have NH bond vectors in a putative structural model from only two residues, Arg25 and Phe53. Call these v_1 and v_2, and without loss of generality consider them to be unit vectors lying on S^2. Now, suppose from the NMR data there are two unassigned residues A and B, from which we compute roots $w_1, \ldots, w_4$ and $u_1, \ldots, u_4$ to the RDC equations. These roots define possible NH bond vectors for the two unassigned residues A and B. Thus, we will think of the NH bond vector of residue A as lying in the set $W = \{ w_1, \ldots, w_4 \}$. The NH bond vector of residue B lies in the set $U = \{ u_1, \ldots, u_4 \}$. Note that U and W each have four points (in this example) and therefore are discrete and finite subsets of S^2. A *legal matching* has the form $((v_1, w_i), (v_2, u_j))$ or $((v_1, u_i), (v_2, w_j))$, for $i, j \in \{ 1, \ldots, 4 \}$. But all matches of the form $((v_1, w_i), (v_2, w_j))$ and $((v_1, u_i), (v_2, u_j))$ are *illegal* since in the first case w_i and w_j are roots of the same polynomial, and hence correspond to two possible but mutually exclusive choices for *same* bond vector based on the data. Similarly, u_i and u_j are both roots of a (different) polynomial, and correspond to mutually exclusive choices for a single bond vector in the data. If U and W are viewed as sets, then the legal matchings have a simple "XOR" (exclusive OR) structure. However, U and W have a group structure, since each is constructed from the roots of a polynomial. The group-theoretic viewpoint gives the problem a fascinating mathematical structure, that can likely be exploited by advanced algorithms. We call this problem *matching modulo a group*. It is worth explaining the terminology. If G is the group of U, then the intuition is, to say "u_i is the same as u_j modulo G" signalizes that "u_i and u_j are the same except for differences accounted for or explained by G." The word *modulo* is mathematical shorthand to denote equivalence in the orbit space S^2/G. S^2/G is the quotient of the group action G on the manifold S^2.

To summarize, in NVR, bond vector orientations from a putative structural model can be *matched* against the multiple roots of the RDC equations. Only one root-match per model bond vector can be correct. This problem may be viewed as *matching modulo a group*, where the group encodes the multiplicity of the roots, and is related to the Galois group of the RDC-kinematic polynomials derived by Wang and Donald (chapters 16, 17). While the problems of *matching* and *clustering* have received a great deal of attention in computer science, machine vision, and computational biology, the problems of *matching modulo a group* and *clustering modulo a group* are relatively new, and are of considerable biological and computational importance. *Clustering modulo a group* will be discussed later, in chapter 47.

References

[1] C. J. Langmead and B. R. Donald. An expectation/maximization nuclear vector replacement algorithm for automated NMR resonance assignments. *Journal of Biomolecular NMR* 29;2(2004):111–138.

[2] C. J. Langmead, A. K. Yan, R. H. Lilien, L. Wang, and B. R. Donald. A polynomial-time nuclear vector replacement algorithm for automated NMR resonance assignments. *RECOMB* (2003): 176–187. *Journal Computational Biology* 11;2–3(2004):277–298.

[3] C. J. Langmead and B. R. Donald. 3D structural homology detection via unassigned residual dipolar couplings. *Proceedings of IEEE Computational Systems Bioinformatics Conference (CSB),* Stanford University, Palo Alto (August 10, 2003), pp. 209–217.

[4] C. J. Langmead and B. R. Donald. High-throughput 3D structural homology detection via NMR resonance assignment. *Proceedings of IEEE Computational Systems Bioinformatics Conference (CSB),* Stanford, (August, 2004), pp. 278–289.

[5] C. J. Langmead and B. R. Donald. An improved nuclear vector replacement algorithm for nuclear magnetic resonance assignment. *Tech. Rep. TR2004-494,* Dartmouth Department of Computer Science, 2004.

[6] L. Wang and B. R. Donald. Exact solutions for internuclear vectors and backbone dihedral angles from NH residual dipolar couplings in two media, and their application in a systematic search algorithm for determining protein backbone structure. *Journal of Biomolecular NMR,* 29;3(2004):223–242.

15–18 Short Course: Automated NMR Assignment and Protein Structure Determination Using Sparse Residual Dipolar Couplings

Bruce R. Donald and Jeffrey Martin

15.1 Introduction

By now, the reader will have learned some of the basics, both of NMR physics and the computational algorithms required to interpret the data. Hence, it is possible to pull together some of these concepts in order to build a comprehensive picture of the algorithms and mathematics that can be employed to assign spectra and determine solution-state NMR structures. This chapter addresses a special case, but an important one, namely, that of using sparse data obtained mostly from residual dipolar couplings (RDCs). The subject is mathematically rich, and fertile ground for students and researchers who wish to understand and develop new methodology.

We have organized this material into a longer chapter, which is presented as a "short course"— a discrete module within a semester-long class, or even the bulk of a shorter class in a typical medical school teaching schedule. The exposition is at approximately the same level as the other chapters in this book, but a longer narrative allows us to get a running start, and persistently attack the key computational biophysical problems with a systematic perspective.

15.1.1 Motivation

The introduction of RDCs for protein structure determination in the late 1990s energized development of NMR methods. Robust automation of the complete NMR structure determination procedure has been a longstanding goal, and RDC-based algorithms may increase the consistency and reliability of NMR structural studies. It has also been recognized that structure determination based primarily on orientational restraints could be quicker and more accurate than traditional distance-restraint-based methods. Furthermore, NMR is increasingly important in applications where structural information is already available, so that methods which effectively automate NMR assignment of known structures would also be a substantial contribution.

Since RDCs are measured in a global coordinate frame, they enable molecular replacement–like methods that perform assignments using structural priors. Furthermore, recent methods for structure determination have exploited novel RDC equations, which combine RDC data and protein kinematics. Under fairly mild assumptions, the dihedral torsional angles of a protein can be analytically expressed as roots of these low-degree polynomials. Solving these equations exactly has enabled a departure from earlier stochastic methods, and led to linear-time, combinatorially precise algorithms for NMR structure determination. These algorithms are optimal in terms of combinatorial (but not algebraic) complexity, and show how structural data can be used to produce a deterministic, optimal solution for the protein structure in polynomial time.

The coefficients of the RDC equations are determined by the data. An RDC error bound therefore defines a range of coefficients, which, in turn, yield a range of roots representing the structural dihedral angles. Hence, the RDC equations define an analytical relationship between the RDC error distribution, and the coordinate error of the ensemble of structures that satisfy the experimental restraints. Precise methods that relate the experimental error to the coordinate error of the computed structures therefore appear within reach. In this chapter, we describe these and other recent advances in NMR assignment and structure determination based on sparse dipolar couplings.

15.1.2   Glossary of Abbreviations

We use the following abbreviations in this chapter:

C' = carbonyl carbon

$CH = C^\alpha\text{-}H^\alpha$

DOF = degrees of freedom

FF2 = FF domain 2 of human transcription elongation factor CA150 (RNA polymerase II C-terminal domain interacting protein)

hSRI = human Set2-Rpb1 interacting domain

HSQC = heteronuclear single quantum coherence spectroscopy

MD = molecular dynamics

NMR = nuclear magnetic resonance

NOE = nuclear Overhauser effect

PDB = protein Data Bank

POF = principal order frame

pol η = zinc finger domain of the human DNA Y-polymerase Eta

ppm = parts per million

RDC = residual dipolar coupling

RMSD = root mean square deviation

SA = simulated annealing

SSE = secondary structure element

vdW = van der Waals

WPS = well-packed satisfying

15.1.3 Background

While automation is revolutionizing many aspects of biology, the determination of three-dimensional (3D) protein structure remains a harder, more expensive task. Novel algorithms and computational methods in biomolecular NMR are necessary to apply modern techniques such as structure-based drug design and structural proteomics on a much larger scale. Traditional (semi-) automated approaches to protein structure determination through NMR spectroscopy require a large number of experiments and substantial spectrometer time, making them difficult to fully automate. A chief bottleneck in determinating 3D protein structures by NMR is the assignment of chemical shifts and NOE restraints in a biopolymer.

The introduction of RDCs for protein structure determination enabled novel attacks on the assignment problem, to expedite high-throughput NMR structure determination. Similarly, it is difficult to determine protein structures accurately using only *sparse data*. New algorithms have been developed to handle the increased spectral complexity encountered for larger proteins, and sparser information content obtained either in a high-throughput setting, or for larger or difficult proteins. The overall goal is to minimize the number and types of NMR experiments that must be performed and the amount of human effort required to interpret the experimental results, while still producing an accurate analysis of the protein structure.

This short course is tempered by our recent experiences in automated assignments [79, 82, 83, 118, 153, 174], novel algorithms for protein structure determination [152, 156, 117, 89, 110, 151, 155, 154], characterization of protein complexes [118, 99] and membrane proteins [117], and fold recognition using only unassigned NMR data [82, 83, 78, 80]. Recent algorithms for automated assignment and structure determination based on sparse dipolar couplings represent a departure from the stochastic methods frequently employed by the NMR community (e.g., simulated annealing/molecular dynamics (SA/MD), Monte Carlo (MC), etc.) A corollary is that such stochastic methods, now routinely employed in NMR structure determination pipelines [60, 53, 91, 64], should be reconsidered in light of their inability to ensure identification of the unique or globally optimal structural models consistent with a set of NMR observations. In this vein, this short course focuses on *sparse data*. While SA/MD may perform adequately in a data-rich, highly constrained setting, it is difficult to determine protein structures accurately using only sparse data. Sparse data arises not only in high-throughput settings, but also for larger proteins, membrane proteins [117], symmetric protein complexes [118], and difficult systems including denatured or disordered proteins [154]. Sparse-data algorithms require *guarantees of completeness* to ensure that solutions are not missed and local minima are evaded.

We caution that in the context of NMR, "high-throughput" is relative, and currently not as rapid as, for example, gene sequencing or even crystallography. Hence the term "batch mode" may be more appropriate. The challenge is to develop new algorithms and computer systems to exploit sparse NMR data, demonstrating the large amount of information available in a few key spectra, and how it can be extracted using a blend of combinatorial and geometric algorithms. More-over, because of their (relative) experimental simplicity, we hypothesize that the computational advantages offered by such approaches should ultimately obtain an integrated system in which automated assignment and calculation of the global fold could be performed at rates comparable to current-day protein screening for structural genomics using ^{15}N-edited heteronuclear single quantum coherence spectroscopy (^{15}N-HSQC).

This chapter describes how sparse dipolar couplings can be exploited to address key computational bottlenecks in NMR structural biology. The past few years have yielded rapid progress in automated assignments, novel algorithms for protein structure determination, characterization of protein complexes and membrane proteins, and fold recognition using only unassigned NMR data. We discuss recent algorithms that assist these advances, including (1) *sparse-data algorithms for protein structure determination from RDCs* using exact solutions and systematic search; (2) RDC-based molecular replacement–like techniques for structure-based assignment; (3) *structure determination of membrane proteins and complexes,* especially symmetric oligomers, enabled by RDCs; and (4) *automated assignment of NOE restraints* in both monomers and complexes, based on backbones computed primarily using sparse RDC restraints.

These define the four main themes in this chapter.

1. It is difficult to determine protein structures accurately using only sparse data. Sparse data arises not only in high-throughput settings, but also for larger proteins, membrane proteins, and symmetric protein complexes. For de novo structure determination, there are now roots-of-polynomials approaches to compute exact solutions, by systematic search, for internuclear bond vectors and backbone dihedral angles using as few as 2 recorded RDCs per residue (for example, NH in two media, or NH and H^{α}-C^{α} in one medium [177]). By combining systematic search with exact solutions, it is possible to efficiently compute accurate backbone structures using less NMR data than in traditional approaches.

De novo structure determination from sparse dipolar couplings can exploit structure equations derived by Wang and Donald [152, 151]. These include a quartic equation to compute the internuclear (e.g., bond) vectors from as few as 2 recorded RDCs per residue, and quadratic equations to subsequently compute protein backbone (ϕ, ψ) angles *exactly* [152, 151]. The structure equations make it possible to compute, exactly and in constant time, the backbone (ϕ, ψ) angles for a residue from very sparse RDCs. Simulated annealing, molecular dynamics, energy minimization, and distance geometry are not required, since the structure is computed exactly from the data. Novel algorithms build on these exact solutions, to perform protein structure determination, using mostly RDCs but also sparse NOEs. For example, the RDC-EXACT algorithm employs a systematic search with provable pruning, to determine the conformation of helices, strands, and loops and to compute their orientations using exclusively the angular restraints from RDCs [152, 156, 177]. Then, the algorithm uses very sparse distance restraints between these computed segments of structure, to determine the global fold.

2. Algorithms using sparse dipolar couplings can accelerate protein NMR assignment and structure determination by exploiting a priori structural information. By analogy, in X-ray crystallography, the molecular replacement (MR) technique allows solution of the crystallographic phase problem when a "close" or homologous structural model is known, thereby facilitating rapid

structure determination. In contrast, a key bottleneck in NMR structural biology is the *assignment problem* — the mapping of spectral peaks to tuples of interacting atoms in a protein. For example, peaks in a 3D nuclear Overhauser enhancement spectroscopy (NOESY) experiment establish distance restraints on a protein's structure by identifying pairs of protons interacting through space. An automated procedure for rapidly determining NMR assignments given a homologous structure, can similarly accelerate structure determination. Moreover, even when the structure has already been determined by crystallography or computational homology modeling, NMR assignments are valuable because NMR can be used to probe protein-protein interactions and protein-ligand binding (via chemical shift mapping or line-broadening), and dynamics (via, e.g., nuclear spin relaxation). Molecular replacement-like approaches for structure-based assignment of resonances and NOEs, including structure-based assignment (SBA) algorithms, can be applied when a homologous protein is known. Moreover, to find structural homologs, it is possible to apply (filter) modules of SBA to a structure database (as opposed to a single structure). This technique performs rapid fold recognition by correlating structural geometry versus distributions of *unassigned* NMR data, enabling detection of homologous structures *before* assignments [82, 83, 78, 145, 97, 80]. Hence, the algorithm finds candidate homologs using only unassigned spectra; then SBA algorithms perform assignments given the structural homolog.

Several algorithms have been proposed for structure-based assignment using RDCs [4, 3, 63, 82, 83, 79, 67, 64]. For example, NVR [79] exploits RDCs to perform structure-based assignment (backbone and NOEs) of proteins when a homologous structure is known, and requires only ^{15}N-labeling. NVR was a step in developing a MR–like method for NMR (useful because many NMR studies, especially in drug design and pharmacology, are of homologous proteins). NVR exploits a priori structural information. Automated procedures for rapidly determining NMR assignments given a homologous structure, can accelerate structure determination, since assignments must generally be obtained before NOEs, chemical shifts, RDCs, and scalar couplings can be employed for structure determination/refinement. NVR offers a high-throughput mechanism for the required assignment process. However, the spectral assignment produced by NVR is itself an important product: even when the structure has already been determined by X-ray crystallography or computational homology modeling, NMR assignments are valuable for structure-activity relation (SAR) by NMR [133, 54] and chemical shift mapping [18], which compare assigned NMR spectra for an isolated protein and a protein:ligand or protein:protein complex. Both are used in high-throughput drug activity screening to determine binding modes. Assignments are also necessary to determine the residues implicated in the dynamics data from nuclear spin relaxation (e.g. [113, 112, 71]). Building on NVR [79], the algorithm GD [78] performs rapid fold recognition (via geometric hashing against a protein structural database) to correlate distributions of *unassigned* NMR data. GD exploits novel approaches for alignment tensor estimation from *unassigned* RDCs [78, 82, 83], to perform maximum-likelihood resonance and NOE assignment [80] (in the NVR framework) against the PDB, to detect the fold even *before* assignments.

In contrast to traditional methods, the set of NMR experiments required by NVR and RDC-EXACT is smaller, and requires less spectrometer time. While these algorithms have exploited uniform ^{13}C-/^{15}N-labeling [151, 156], NVR and RDC-EXACT have been successfully applied to experimental spectra from different proteins using only ^{15}N-labeling, a cheaper process than ^{13}C labeling (cf. Wüthrich [166]: "A big asset with regard to future practical applications ... [is] ... straightforward, inexpensive experimentation. This applies to the isotope labeling scheme as well as to the NMR spectroscopy ...").

3. We will describe recent algorithms that assist in determining membrane protein structures. In such systems, RDCs serve several functions. First, RDCs enable accurate subunit backbone structure calculation in complexes. Second, in symmetric homo-oligomers, the RDCs aid in determining the symmetry axis [2, 168]. These two advantages enable complete algorithms for NOE assignment and structure determination that overcome limitations of the simulated annealing/molecular dynamics (SA/MD) methodology when the data are sparse. Several methods, including SYMBRANE and AMBIPACK, cast the problem of structure determination for symmetric homo-oligomers (such as many membrane proteins) into a systematic search of symmetry configuration space, automatically assigning NOEs and handling NOE ambiguity while provably characterizing the uncertainty in the structural ensemble [117, 118].

Membrane proteins present experimental and computational challenges. Structural studies can be difficult if a protein is hard to crystallize (for X-ray) or is not well-behaved in an artificial membrane (for NMR). Many membrane proteins are symmetric oligomers. In an n-mer, identical electronic environments obtain identical chemical shifts, thereby boosting each signal approximately n-fold. However, it is not possible to distinguish signals from symmetric atoms in different subunits. The ambiguity in intersubunit NOEs (with *identical* chemical shifts) adds to the usual chemical shift ambiguity in assigning NOEs in monomers (with 'merely' similar shifts). While the latter could, in principle, be resolved experimentally (for example, using 4D NOESY [21]), the former is *inherent* in the symmetry and cannot currently be resolved by experimental methods: computational solutions are required. On the other hand, the symmetry (which is known to exist from the signal overlap) can be used as an *explicit* kinematic constraint during structure determination. SYMBRANE is both *complete,* in that it evaluates all possible conformations, and *data-driven,* in that it evaluates conformations separately for consistency with experimental data and for quality of packing. Completeness ensures that the algorithm does not miss the native conformation, and being data-driven enables it to assess the structural precision possible from data alone. SYMBRANE performs a branch-and-bound search in the symmetry configuration space. It eliminates structures inconsistent with intersubunit NOEs, and then identifies conformations representing every consistent, well-packed structure. SYMBRANE has been used to determine the complete ensemble of NMR structures of unphosphorylated human cardiac phospholamban [117], a pentameric membrane protein. SYMBRANE addresses some of the challenges of protein complex determination, larger proteins, and the difficulties arising from symmetry and NOE ambiguity. By running SYMBRANE using different priors (starting structures) encoding the putative oligomeric number, one can determine, solely from NMR data, the maximum-likelihood oligomeric state.

4. Since accurate protein backbone structures can be computed from RDCs, these backbones can then be used to bootstrap NOE assignment. Novel techniques, including the algorithms TRIANGLE [153] and HANA [174], exploit the accurate, high-throughput backbone structures obtained exactly using sparse RDCs. NOE assignment can be difficult to fully automate, and structure determination of symmetric membrane proteins by NMR can be challenging. We discuss how, by *combining* these two difficult problems, recent results indicate an algorithm that solves both simultaneously, and enjoys guarantees on its completeness and complexity [117, 118].

An overview of the major steps to automated assignment and structure determination using sparse RDCs is given in figure 15.1. The figure suggests how these algorithms and software tools could be developed into a set of integrated programs for automated fold recognition, assignment, monomeric and oligomeric structure determination. For each of the modules in the figure, there are algorithms and implementations reported by several groups working on NMR methodology. One example for each module is shown in the figure, and should be interpreted as a representative for a class of algorithms (described later) with similar function. *Fold recognition* [82, 83, 78, 145, 97, 80, 128] denotes correlation of *unassigned* NMR distributions (e.g., RDCs) against a database of known folds. *Structure-based assignment (SBA)* [4, 3, 63, 82, 83, 79, 67, 64] denotes automated assignment given *priors* on the putative structure(s) of the protein. Note that, like sparse data and completeness, *SBA* is a crosscutting theme: NVR uses priors on putative homologs (detected by GD) to assign resonances (and unambiguous H^N-H^N NOEs). A number of algorithms for protein structure determination are based on exact solutions to the RDC equations [122, 159, 152, 151, 156, 154, 155, 174]. RDC-EXACT is one such algorithm [151, 152, 156]. While the algorithms of [38, 69, 167] are not exact, it is likely that a roots-of-polynomials exact solutions version of these algorithms could be derived, although possibly not in closed-form. TRIANGLE uses backbone structure (computed by RDC-EXACT) to assign ambiguous backbone and side-chain NOEs. Several algorithms exist to determine the structure of symmetric homooligomers using a combination of RDCs, NOEs, and other NMR data [150, 117, 118, 129, 168]. SYMBRANE [117, 118] and AMBIPACK [150] exploit the subunit (monomer) structure to assign intermolecular NOEs and determine the complex structure. Finally, note that assignment (NVR) and Fold Recognition (GD) operate entirely on unassigned data. Structure Determination by RDC-EXACT operates on assigned data.

This chapter concentrates on the information content of the NMR experiments, and the methods for assignment and structure determination, with an emphasis, where possible, on provable algorithms with guarantees of soundness, completeness, and complexity bounds. A number of excellent articles have appeared on the experimental aspects of RDCs; we recommend Weidong and Wang [160] for a good introduction to RDCs and the interplay between experimental and computational challenges.

Rather than describing a competition between computer programs, this chapter tries to evaluate the strengths and weaknesses of the underlying ideas (algorithms). There are several reasons. First, we believe no one will be using the same programs in ten years (and if we are, that would reflect poorly on the field). However, the underlying mathematical relationships between the data and the

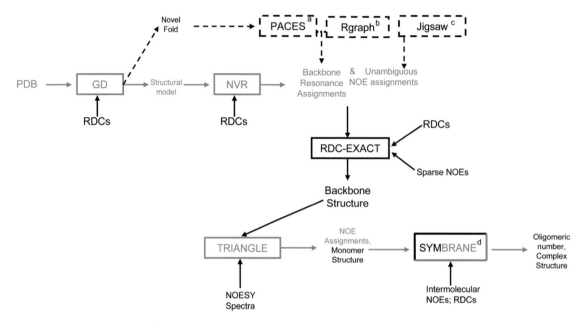

Figure 15.1

Overview of the major modules in automated NMR assignment and protein structure determination using sparse dipolar coupling constraints. One example for each module is shown in the figure, and should be interpreted as a representative for a class of algorithms (described in the text) with similar function. GD performs rapid fold determination from unassigned NMR data. NVR performs structure-based assignment. RDC-EXACT determines backbone structure de novo, from 2 RDCs per residue plus sparse NOEs. TRIANGLE assigns the NOESY spectra, allowing determination of a high-resolution monomer structure. SYMBRANE assigns intermolecular NOEs and determines the oligomeric number and complex structure. Each of these modules takes as input NMR data that can be collected in a high-throughput fashion. The major data sources are shown; complete descriptions of the data requirements are in the text. The solid arrows show the data flow for the molecular replacement-like method for NMR. Dashes show an alternative pathway for de novo assignment and structure determination in the case where a completely novel fold is detected (by GD) from unassigned NMR data. [a]PACES, [b]RGRAPH, and [c]JIGSAW are *ab initio* assignment algorithms [22, 105, 149, 70, 148, 12]. Right solid arrows show the data flow from *structure to assignments*. Downward arrows show the data flow from *assignments to structure*. [d]SYMBRANE simultaneously performs assignment and structure determination.

structures should prove enduring, warranting a characterization of the completeness, soundness, and complexity of structure determination algorithms exploiting sparse dipolar couplings.[1]

16.1   The Power of Exact Solutions

Let us consider an analogy. A point-mass p is fired from a cannon with velocity $\mathbf{v}$, where $\mathbf{v}$ is a tangent vector to Euclidean three-dimensional space $\mathbb{R}^3$. Assuming Newtonian dynamics, when, and where, will it hit the ground?

---

[1] This chapter is based on our review article in *Progress in Nuclear Magnetic Resonance Spectroscopy* (2009).

This problem can be solved by numerical forward-integration of the dynamical equations of motion, or by random guessing (also known as Monte Carlo sampling), simulated annealing, neural networks, genetic algorithms, systematic grid search, or a host of other techniques. However, the following simple technique, from middle-school physics, suffices. The trajectory of the mass p is given by a quadratic equation in one scalar variable (time). By solving this equation simultaneously with the plane of the ground, $z = 0$, the solution to our problem may be calculated exactly, in closed form, and in constant-time (using only a constant number of computer operations). In this case, "closed-form" means using only the field operations (addition, subtraction, multiplication, and division) plus calculating roots $\sqrt[j]{\cdot}$. In this case, $j \leq 2$.

A similar trick is available to assist in protein structure determination, (figures 16.1–16.6, plates 1, 13–15) when we have measured dipolar couplings (figure 16.5). A simplified example will be helpful to understand the idea. The example arises in protein structure determination from RDCs measured in one medium, using exact solutions.

Suppose we have recorded RDCs (figures 16.1–16.3) in a single alignment medium for NH and C^{α}-H^{α} bond vectors (figure 16.6, plate 1), and that secondary structure regions have been identified using either chemical shifts, short-range NOEs, or scalar coupling experiments such as HNH^{α} or J-doubling to measure the ϕ bond angles. Consider the simplified problem of computing the orientation and conformation of a secondary structure element (helix or strand) h containing

Residual Dipolar Couplings

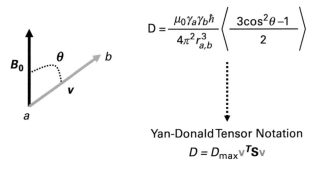

$$D = \frac{\mu_0 \gamma_a \gamma_b \hbar}{4\pi^2 r_{a,b}^3} \left\langle \frac{3\cos^2\theta - 1}{2} \right\rangle$$

Yan-Donald Tensor Notation

$$D = D_{max} \mathbf{v}^T \mathbf{S} \mathbf{v}$$

Figure 16.1
The scalar component of the residual dipolar coupling; $\hbar$ is Planck's constant, μ_0 is the magnetic permeability of vacuum, γ_a and γ_b are the gyromagnetic ratios of two nuclei a and b, and θ is the angle between the external magnetic field $\mathbf{B}_0$ and the internuclear vector $\mathbf{v}$ (from a to b) in the weakly aligned anisotropic phase; $\left\langle \frac{3\cos^2\theta - 1}{2} \right\rangle$ is the ensemble average of the second Legendre polynomial of $\cos\theta$; $r_{a,b}$ is the distance between nuclei a and b. Here, a and b are assumed to be covalently bonded, and therefore the ensemble average $\langle r_{a,b}^3 \rangle$ in the denominator is replaced with the single scalar $r_{a,b}^3$. In classical solution-state NMR, proteins tumble rapidly and isotropically and therefore the dipolar couplings average to zero. RDCs are measured by introducing a dilute alignment medium, which biases the orientational distribution of the protein so that dipolar couplings can be measured. In contrast to NOEs, whose magnitude is proportional to the interatomic distance to the inverse sixth power, RDCs are proportional to $1/r_{a,b}^3$. The alignment tensor $\mathbf{S}$ represents the molecular alignment in the anisotropic phase. It is convenient to express the residual dipolar coupling in Yan-Donald tensor notation [82, 83], as $D_{max} \mathbf{v}^T \mathbf{S} \mathbf{v}$.

$$D = D_{max}\mathbf{v}^T\mathbf{S}\mathbf{v}$$

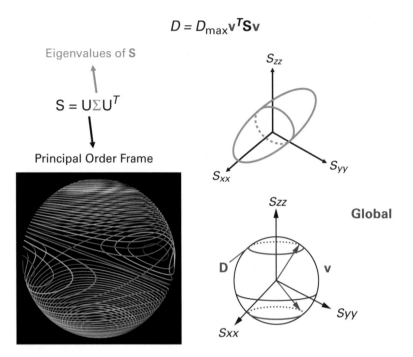

Figure 16.2 (plate 13)
The alignment tensor **S** is a symmetric second-rank tensor that may be represented by a real-valued 3×3 matrix that is symmetric and traceless. Hence **S** has 5 degrees of freedom and may be decomposed using singular value decomposition (SVD) into a rotation matrix U, called the *principal order frame,* and a diagonal matrix Σ encoding its eigenvalues. The principal axes of U encode the eigenvectors. For a fixed experimental RDC D, the possible orientations of the corresponding internuclear vector **v** must lie on one of two *RDC curves* on the two-dimensional sphere S^2. Each curve is the intersection of an ellipsoidal cone with S^2. (Lower left) RDCs curves are shown spaced at 1 Hz intervals. Credit: Vincent Chen and Tony Yan.

k residues, and that a good estimate of the alignment tensor is available. As described in several studies [152, 151, 156, 177], an initial estimate of the alignment tensor may be obtained by fitting parametric ideal helical geometry to RDCs from a secondary structure element such as a helix. The alignment tensor can be subsequently refined in an iterative fashion [152, 177].

Henceforth we will simply refer to the C^α-H^α bond as a CH bond, and to a C^α-H^α RDC as a CH RDC. Let us assume standard protein backbone geometry (figure 16.6); our example proceeds by analogy with the mathematical concept of strong induction. Assume we have already computed the structure of the first $i - 1 < k$ residues of h starting at the N-terminus. In this case, the $(i - 1)^{st}$ peptide plane (between residues $i - 1$ and i) is known (figure 16.6). As the ith ϕ dihedral angle, ϕ_i rotates, the orientation of the ith C^α-H^α bond vector will move in a circle (figures 16.6 and 16.7). Under any change of coordinate system, this circle will transform to an ellipse, E, on the two-dimensional sphere S^2. Such an ellipse is shown in green in figure 16.7 (plate 16).

Residual Dipolar Coupling

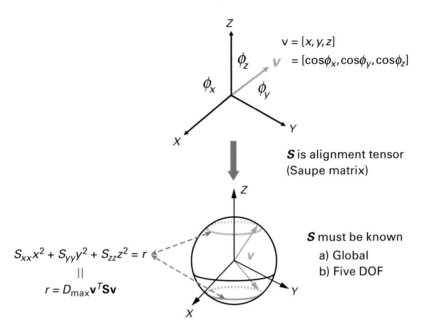

$$v = [x, y, z]$$
$$v = [\cos\phi_x, \cos\phi_y, \cos\phi_z]$$

S is alignment tensor
(Saupe matrix)

$$S_{xx}x^2 + S_{yy}y^2 + S_{zz}z^2 = r$$
$$\|$$
$$r = D_{max}\mathbf{v}^T\mathbf{S}\mathbf{v}$$

S must be known
a) Global
b) Five DOF

Figure 16.3 (plate 14)
It is convenient to express the internuclear vector as a unit vector of the form **v**, corresponding to its direction cosines. The RDC r can be expressed in Yan-Donald tensor notation [82, 83], as $r = D_{max}\mathbf{v}^T\mathbf{S}\mathbf{v}$, or in a principal order frame that diagonalizes the alignment tensor, namely $r = S_{xx}x^2 + S_{yy}y^2 + S_{zz}z^2$. Here, S_{xx}, S_{yy} and S_{zz} are the three diagonal elements of a diagonalized Saupe matrix **S** (the alignment tensor), and x, y and z are, respectively, the x, y, z-components of the unit vector **v** in a principal order frame (POF) which diagonalizes **S**.

For each RDC r, the dipolar coupling is given by the top equation in figure 16.1 (sec [140, 142]). It is convenient to express the residual dipolar coupling in Yan-Donald tensor notation [82, 83], as

$$r = D_{max}\mathbf{v}^T\mathbf{S}\mathbf{v}, \tag{16.1}$$

where D_{max} is the dipolar interaction constant, **v** is the internuclear vector orientation relative to an arbitrary substructure frame, and **S** is the 3×3 *Saupe order matrix* [130] (figure 16.1). **S** is a symmetric, traceless, rank 2 tensor with 5 degrees of freedom, which describes the average substructure alignment in the weakly aligned anisotropic phase (figures 16.1 and 16.2). The measurement of five or more independent RDCs in substructures of known geometry allows determination of **S** [92].

Now, if the RDC has been measured for the CH bond vector in residue i, then the RDC equation (16.1) constrains the bond vector orientation to lie on one of two curves $R = R_1 \cup R_2$.

RDCs in two different media

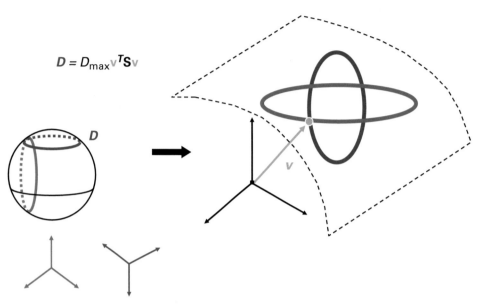

$$D = D_{max} \mathbf{v}^T \mathbf{S} \mathbf{v}$$

2 Principal Order Frames

Figure 16.4 (plate 15)
A cartoon of the geometry and algebra of RDCs measured in two independent aligning media. If the two principal order frames (POFs) are independent, then the internuclear unit vector $\mathbf{v}$ is constrained to simultaneously lie on the blue RDC curve (from the blue POF) and the red RDC curve (from the red POF). Generically, the blue and red RDC curves will intersect at 0, 2, 4, or 8 points. Here, only one of the two RDC curves is shown for each POF. Suppose that r is the red RDC and that the diagonalized red POF can be represented as (S_{xx}, S_{yy}, S_{zz}). Let $u = 1 - 2\left(\frac{x}{a}\right)^2$, where x is the x-component of $\mathbf{v}$ and $a^2 = (r - S_{zz})/(S_{xx} - S_{zz})$; see equation (18.5) below and [152, p. 238]. The discrete points corresponding to the RDC curve intersections are calculated exactly by solving a quartic polynomial equation in u, of the form $f_4 u^4 + f_3 u^3 + f_2 u^2 + f_1 u + f_0 = 0$ [152], which is also a quartic polynomial equation in x^2.

Each of these curves is the intersection of S^2 with an ellipsoidal cone. One such curve is shown in orange in figure 16.7.

Therefore, the ϕ_i angles that simultaneously satisfy protein backbone kinematics and the CH RDC data are given by the intersection of curves E and R, shown as the green and orange ellipses, respectively, in figure 16.7. Generically, this intersection will be a set of 0, 2, or 4 points (the 1-point solution is nongeneric), as shown in figure 16.7. Wang and Donald showed that these points are the roots of a quartic univariate polynomial equation [151, 152, 177] and hence may be computed exactly and in closed form. (Technically, we compute exact solutions for the sine and cosine of this angle ϕ_i, which completely determine the angle ϕ_i, which, if desired, can then be computed numerically using the 2-argument arctangent function ATAN2).

Key Ingredients

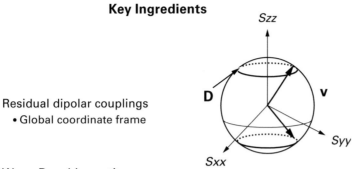

Residual dipolar couplings
 • Global coordinate frame

Wang-Donald equations
(exact solutions to NH, C^α-H^α orientations and (ψ, ϕ) angles):

$$f_4 u^4 + f_3 u^3 + f_2 u^2 + f_1 u + f_0 = 0$$

Optimization via algebraic geometry
 • Conformations of secondary structure elements
 • Global fold

$$\exists x^* : \forall x : f(x^*) \leq f(x)$$

Figure 16.5
Key ingredients to a structure determination algorithm exploiting exact solutions and systematic search.

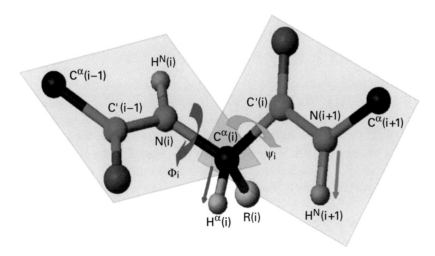

Figure 16.6
Protein backbone kinematics and RDC restraints. See plate 1.

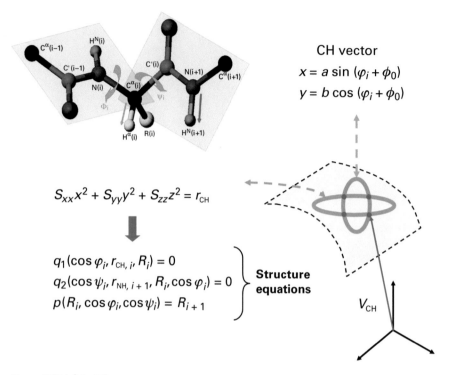

CH vector

$$x = a \sin (\varphi_i + \phi_0)$$
$$y = b \cos (\varphi_i + \phi_0)$$

$$S_{xx}x^2 + S_{yy}y^2 + S_{zz}z^2 = r_{CH}$$

$$q_1(\cos \varphi_i, r_{CH,\,i}, R_i) = 0$$
$$q_2(\cos \psi_i, r_{NH,\,i+1}, R_i, \cos \varphi_i) = 0 \left.\vphantom{\begin{matrix}a\\b\\c\end{matrix}}\right\} \text{Structure equations}$$
$$p(R_i, \cos \varphi_i, \cos \psi_i) = R_{i+1}$$

V_{CH}

Figure 16.7 (plate 16)
Given NH and C^α-H^α RDCs measured in one medium, the Wang-Donald Structure Equations yield exact solutions for the (ϕ, ψ) backbone dihedral angles.

Now, a solution is chosen for ϕ_i from among these multiple exact solutions, and the procedure continues along the polypeptide chain. With ϕ_i fixed, we find ourselves in a symmetrical situation. As the ψ_i bond angle rotates, the orientation of the NH vector of residue $i + 1$ will move in an ellipse E' (figure 16.7). Similarly, if the RDC has been measured for this NH bond, then its orientation will similarly be constrained to lie on curves R' on S^2. Therefore, the orientation of the $(i + 1)^{st}$ NH bond vector must lie on the intersection of the curves E' and R', and the ψ_i angles satisfying this constraint can similarly be solved for exactly and in closed form, as done previously for ϕ_i.

Again, a solution is chosen for ψ_i from among these multiple exact solutions, which defines the ith peptide plane (between residues i and $i + 1$), and the procedure continues along the polypeptide chain. Equations describing the geometric procedure above were introduced in [151]; see [177] (Supplementary Information) for an intuitive derivation.

Every exact solution precisely satisfies the data. Since there are multiple solutions for each backbone dihedral angle a choice must be made, and this defines a discrete combinatorial search

(φ, ψ) Angles in Conformation Tree

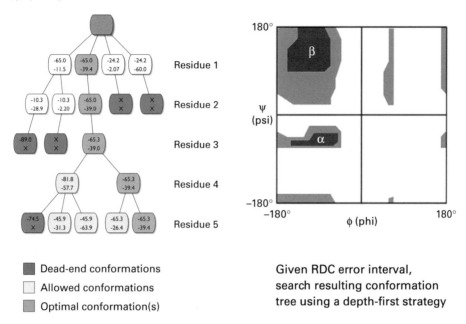

Figure 16.8 (plate 17)
A *conformation tree* is a data structure used in depth-first search over the exact solutions (roots of polynomials) with backtracking or A* search, to optimally compute the backbone dihedral angles that *globally* best fit the RDC data and an empirical scoring function.

for the structure of the secondary structure element h. A scoring function is used to choose the correct root, and hence the correct backbone dihedral angle. The scoring function can use the Ramachandran diagram, molecular mechanics energies, and any of the usual components of an empirical scoring function [152, 151, 156, 154, 177]. By structuring the search into a conformation tree [152, 88, 156, 45] and using a depth-first search with backtracking [152], or A* search [45], the optimal solution over the entire secondary structure element h can be found (figure 16.8, plate 17). Henceforth, we will call such a secondary structure element h a *fragment*.

The RDC RMSD term in the scoring function calculates the sum of the squared differences between the experimental RDCs and the back-computed RDCs over all k residues of the secondary structure element h [152]. Minimizing the scoring function over the combinatorial number of choices of the polynomial roots representing the backbone dihedral angles, will yield in the structure that optimally fits the data [152, 156, 177]. Unlike some traditional methods (SA/MD, MC, etc.) that can only compute local minima, this technique is guaranteed to compute the globally optimal solution for h. We discuss this point next.

16.1.1   Computing the Globally Optimal Solution

It is important to note that the choice of backbone dihedral angle is not made locally solely using the RDC information for that residue. Rather, the scoring function includes an RDC RMSD term, *so that the global optimum is computed over the entire fragment* [152, 156, 177]. By *global optimal* we mean the minimum of the scoring function, where the minimum is taken over all (ϕ, ψ) angles in the fragment that are zeroes of the structure equations. While a grid search over all (discretized) (ϕ, ψ) angles is not computationally feasible, a complete search with full backtracking that considers all possible exact solutions for all possible dihedral angles is possible over secondary structure elements of up to about 20 residues [152, 156, 174, 177]. Typical scoring functions over this tree search have included terms for RDC RMSD, Ramachandran suitability, and hydrogen bonds [152, 156, 174, 177] or van der Waals packing [154], but any empirical molecular mechanics energy function would be feasible. Note that although an exhaustive search over the entire tree is theoretically necessary in the worst case, in practice, combinatorial speed-up can be obtained since, when a node is pruned, the entire subtree below it is eliminated [152, 156, 177]; (see figure 16.8).

This gives us a procedure to compute the structure of h that optimally fits the data under the scoring function. Now, the procedure is exponential in k, the length of h. This exponential dependence provides a combinatorial obstruction to simply proceeding along the polypeptide chain for the entire protein. To overcome this problem, the following algorithm is used. The protein is partitioned into secondary structure regions. The orientation and conformation of each secondary structure element is solved using the techniques described earlier. Each may be solved independently and in parallel since, under suitable assumptions about the dynamics, they all share the same alignment tensor. This allows the algorithm to divide and conquer: for a protein with n residues, there could, in principle, be at most n secondary structure elements. However, each will have only constant length ($k = O(1)$). Therefore the problem is divided into a series of $\Theta(n)$ subproblems, each of constant size. Each of these can be solved in constant time since the exponential of a constant is also a constant.

When RDCs are recorded in a single medium, there is a fourfold orientational ambiguity between a pair of secondary structure elements. This cannot be disambiguated solely using the RDCs. However, not all combinations need to be tried. The secondary structure elements can be assembled sequentially using sparse NOEs to pack them together (figure 16.9, plate 18). For example, if the secondary structure elements, whose orientations (up to the symmetry of the dipolar operator) and conformations have been optimally determined are $(h_1, h_2, \ldots, h_m)$, then the algorithm would first pack h_2 to dock with h_1, and then pack h_3 to dock with the packed substructure (h_1, h_2), and then pack h_4 to dock with the packed substructure (h_1, h_2, h_3), and so forth. Each of these packing operations can determine the optimal packing including the orientational ambiguity. This may be done using a complete algorithm as described by several researchers [117, 152, 156, 174, 177]. Note that, although there could be $\Theta(n)$ secondary structure elements, the packing and assembly problem is not exponential since it is transformed into a linear-sized sequence of constant-sized packing problems. The requirements on the NOEs

Fold via sparse NOEs

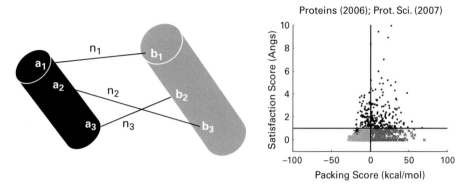

Using NOEs, find a translation* x minimizing:

$$\sum_{i=1}^{\ell} \left(\|a_i - b_i + x\| - n_i \right)^2 \qquad \begin{array}{l} \text{*4-fold discrete} \\ \text{orientational} \\ \text{symmetry (1 medium)} \end{array}$$

Figure 16.9 (plate 18)
The orientations and conformations of secondary structure elements (SSEs) can be calculated using sparse RDCs. Then the SSEs are packed using sparse NOEs. Packings are scored separately for data fit and molecular mechanics energies [117] to avoid bias. The packing by NOEs also disambiguates the discrete 4-fold orientational degeneracy due to the symmetry of the dipolar operator.

are fairly mild since the conformation of the secondary structure elements is determined up to translation (and the fourfold discrete orientational degeneracy). This means that the translation between the oriented secondary structure elements is not mathematically determined using RDCs alone. Therefore, a small number of sparse NOEs will suffice to pack the secondary structure elements [152, 177].

Once the global fold has been determined by packing together the secondary structure elements based on the RDCs and sparse NOEs, loops must be determined to connect them. This problem is similar to the kinematic loop closure problem in X-ray crystallography. The similarity arises because once the core structure of the secondary structure elements has been computed, it defines orientations and positions for the helices and strands. Models of loops must be built that close these kinematic gaps, to connect the secondary structure elements. The kinematic loop closure problem is combined with the RDC restraints that are measured for the loop residues, to compute an ensemble of loops that simultaneously satisfies the closed-chain kinematics and the polynomial equations arising from the RDCs [152, 174].

Finally, it is possible to model error in the input RDCs. In the simplest method, a distribution is placed over the input data, and that distribution can be sampled [152, 151]. This sampling results in a set of perturbed RDCs. The preceding combinatorially precise, exact algorithms are run for different sets of the sampled RDCs, resulting in different solutions to the structure. Out of this ensemble of solutions, the maximum-likelihood solution can be computed [152, 151, 156, 177]. Alternatively, an ensemble of structures that fit the data can be returned [154]. In the case that sampling is undesirable, it is possible, in principle, to use algebraic algorithms (polynomial arithmetic) to push the RDC error intervals through the RDC equations, and obtain a representation of the probability density function over backbone dihedral angles [156].

In general, when different RDCs (at least two) have been measured per residue, a similar algebraic and kinematic derivation holds to obtain exact solutions. The case for NH RDCs recorded in two media is shown in figure 16.4. In all cases, the coefficients of the RDC equations are determined by the data [151]. An RDC error bound therefore defines a range of coefficients, which, in turn, yield a range of roots representing the structural dihedral angles. Hence, the RDC equations define an analytical relationship between the RDC error distribution and the coordinate error of the ensemble of structures that satisfy the experimental restraints [156]. Precise methods that relate the experimental error to the coordinate error of the computed structures therefore appear within reach.

Of course, whenever exact solutions exist, there usually are also excellent numerical algorithms [37] (as opposed to exact algorithms) that stably solve the same analytical equations (18.6 and 18.9 below) not exactly, but up to the accuracy of the floating-point representation. In our motivating example of a point-mass fired from a cannon, these numerical algorithms include (for example), the eponymous Newton's method. Such techniques, born in the field of numerical analysis and scientific computation [37], enable provably good approximation algorithms for our structure determination problem.

16.1.2 Limitations and Extensions

The approach described above assumes that dynamics can be neglected, although recent studies indicate that modest dynamic averaging can be tolerated, albeit with reduced accuracy in the determined orientations of internuclear bond vectors [69]. In addition, it is assumed that the alignment tensor can be estimated initially by fitting parametric secondary structure geometry (helices and β-strands) to the RDCs to obtain the alignment [152], and that the alignment parameters can be optimized in an iterative fashion by alternating the roots-of-polynomials exact solutions approach to structure determination (given an alignment tensor) with fitting the alignment tensor to RDCs and the just-determined nascent partial structure (using SVD) [152, 151, 156]. While good results have generally been obtained from this methodology [152, 151, 156, 174, 177], if inaccurate tensors are used, the resulting structures may have inaccuracies [69]. We observe that the RDCs are scaled by the order parameter S. Suppose order parameters S^2 are measured for the same bond vectors as the RDCs, using, for example, relaxation experiments. In this case, neglecting dynamics outside the timescale of the dynamics measurements, one may heuristically assume that when S^2 is sufficiently uniform (i.e., the core of the protein is largely rigid), then the dynamic averaging due to S in the RDC measurement is safe to use for structure determination.

17.1 NMR Structure Determination Algorithms Using Sparse RDCs

Several papers [152, 151, 156, 154, 155, 174, 177] make contributions (see list on p. 146) to the method of determining protein structures by solution NMR spectroscopy using RDCs as the main restraints. These contributions may be valuable not only to the NMR community in particular and structural genomics in general, but also to structural biologists more broadly. This is because in both experimental and computational structural biology, exact computational methods have been, for the most part, elusive to date. Second, rigorous comparisons of structures derived from NMR versus X-ray crystallography are made possible by these techniques, and these comparisons should be of general interest.

One algorithm, RDC-EXACT (figure 16.5), requires the following experimental NMR data: (a) two RDCs of backbone internuclear vectors per residue (e.g., assigned NH RDCs in two media or NH and CH RDCs in a single medium); (b) identified α-helices and β-sheets with known hydrogen bonds (H-bonds) between paired strands; and (c) a few NOE distance restraints. The implementation [152, 151, 156, 174, 177] uses this experimental data, and allows for missing data as well. In contrast to NOE assignment, RDCs can be recorded and assigned on the order of hours. Additionally, it is relatively straightforward to rapidly obtain the few (three or four), unambiguous NOEs required for the packing algorithm (see section 16.1) from a standard NOESY spectrum, or by using, for example, the labeling strategy of Kay and coworkers [40]. The secondary structure types of residues along the backbone can be determined by NMR from experimentally recorded scalar coupling HNH$^\alpha$ [16, p. 524–528] data, or J-doubling [31] data for larger proteins (these experiments report on the ϕ backbone angles). NMR chemical shifts [163, 165, 164, 94, 25] or automated assignment [12] can also be used. Hydrogen bonds can be determined by NMR from experimentally recorded data [24, 157], or, for example, by using backbone resonance assignment programs such as JIGSAW [12]. In the remainder of this chapter, we discuss the algorithm RDC-EXACT assuming that we are given assigned NH RDCs in two media (figure 16.4). However, the results also hold for the case of NH and CH RDCs in one medium with slight modifications to the equations in section 18.1.1 as shown in Wang and Donald [151, 177], and as illustrated in section 16.1.

Most traditional algorithms focus on using NOE restraints to determine protein structure. This approach has been shown to be NP-hard [131], essentially due to the local nature of the constraints. The most notable characteristic of NP-hard problems is that no fast solution to them is known [162]; that is, the time required to solve the problem using any currently known algorithm increases very quickly as the size of the problem grows. As a result, the time required to *provably* solve even moderately large versions of many of these problems becomes prohibitive using any currently available amount of computing power. Here, this implies that no algorithm for computing a structure that globally satisfies a dense network of NOE constraints can be mathematically *proven* to produce a satisfactory solution in a reasonable time. This is an undesirable property for structure determination software. In particular, in the case of algorithms such as SA/MD and Monte Carlo, no guarantees of soundness, efficiency, or completeness can be made. In contrast, it is remarkable that by primarily using RDCs instead of NOEs, provably polynomial-time algorithms can be obtained that have guarantees of soundness, efficiency, and completeness.

In practice, approaches such as molecular dynamics and simulated annealing [15, 52], which lack both combinatorial precision and guarantees on running time and solution quality, are used routinely for structure determination. Several structure determination approaches do use RDCs, along with other experimental restraints such as chemical shifts or sparse NOEs [6, 32, 46, 62, 125, 139], yet remain heuristic in nature, without guarantees on solution quality or running time. Unlike previous approaches ([14] is a notable exception), which have either no theoretical guarantees [15, 52, 6, 32, 46, 62, 125, 139], or run in worst-case exponential time [131, 29, 28, 55, 56], recent methodology has shown that it is possible to exploit RDC data, which gives global restraints on the orientation of internuclear bond vectors, in conjunction with very sparse NOE data, to obtain an algorithm that runs in polynomial time and provably computes the structure that agrees best with the experimental data.

These results are consistent with earlier observations [140, 1, 139, 38, 120, 6, 32, 46, 62, 125, 159] that, empirically, RDCs increase the speed and accuracy of biomacromolecular structure determination: RDC-EXACT formally quantifies the complexity-theoretic benefits of employing globally referenced angular data on internuclear bond vectors. The main contributions of this work were as follows:

1. To derive low-degree polynomial equations that can be solved *exactly* and in *constant time*, to determine backbone (ϕ, ψ) angles from experimentally recorded RDCs. Only two RDCs per residue are required. For example, after measuring RDCs corresponding to a single internuclear vector **v** in two different aligning media, the easily-computable exact solutions eliminate the need for one-dimensional grid search previously employed [159] to compute the direction of **v** or two-dimensional grid search [46, 139, 158, 93] to compute (ϕ, ψ) angles. The main results also hold for the case of NH and CH RDCs in one medium with slight modifications to the Wang-Donald equations in section 18.1.1, as shown in several studies [151, 156, 174, 177] (see section 16.1). Furthermore, these equations are very general and can be extended to compute other backbone and side-chain dihedral angles. The method can be applied *mutatis mutandis* to derive similar equations for computing dihedral angles from RDCs in nucleic acids.

2. The first NMR structure determination algorithm that simultaneously uses exact solutions, systematic search, and only two RDCs per residue. (A systematic search is a search over all possible conformations (solutions) that employs a provable pruning strategy that guarantees pruned conformations need not be considered further).

3. The first *combinatorially precise, polynomial-time* algorithm for structure determination using RDCs, secondary structure type, and very sparse NOEs.

4. The first provably polynomial-time algorithm for de novo backbone protein structure determination solely from experimental data (of any kind).

Table 17.1
Results of RDC-EXACT

Protein[a]	α/β residues[b]	RDCs[c]	Type of RDCs[d]	Hydrogen bonds[e]	NOEs[f]	RMSD[g]
Ubiquitin	39/75	78	NH in two media	12	4	1.23 Å
Ubiquitin	41/75	76	NH, CH in one medium	12	4	0.97 Å
Dini	41/81	75	NH in two media	6	9	1.55 Å
Dini	41/81	80	NH, $C^{\alpha}C'$ in one medium	6	9	1.35 Å
Protein G	29/56	53	NH in two media	9	4	0.98 Å
Protein G	33/56	61	NH, $C^{\alpha}C'$ in one medium	9	4	1.30 Å

(a) experimental RDC data for ubiquitin (PDB ID: 1D3Z), Dini (PDB ID: 1GHH), and Protein G (PDB ID: 3GB1) were taken from the Protein Data Bank (PDB). (b) Number of residues in α-helices or β-sheets, versus the total number of residues. (c) The total number of experimental RDCs (note that RDCs are missing for some residues). (d) RDCs from different experimental datasets (for different bond vectors) were used. (e) Number of hydrogen bonds used. (f) Number of NOEs used. (g) RMSD (for C^{α}, N, and C' backbone atoms) between the oriented and translated secondary structure elements (excluding loop regions) computed by RDC-EXACT to reference structures: ubiquitin to a high-resolution X-ray structure (PDB ID:1UBQ); Dini to an NMR structure (PDB ID: 1GHH); and Protein G to an NMR structure (PDB ID: 3GB1).

5. An implementation of the algorithm that is competitive in terms of empirical accuracy and speed, but requires much less data than, previous NMR structure determination techniques.

6. Testing and results of the algorithm on protein NMR data.

Representative results from RDC-EXACT, including RMSD to high-resolution crystal structures and NMR structures, are shown in table 17.1 and figure 17.1 (plate 19). In addition to these studies several blind tests of RDC-EXACT were performed, in which the structure was not known ahead of time [174]. NMR data were recorded for the ubiquitin-binding zinc finger domain of the human DNA Y-polymerase eta (polη). Zeng et al. [174] used NH and H^{α}-C^{α} RDCs recorded in one medium, with typically 10–15% missing data (but up to 32%) in one secondary structure region, plus 9 NOEs between the helix and β-strands. The structure of polη was then computed by RDC-EXACT, and is shown in figure 17.2L (plate 20). The RDC-EXACT structure [174, 177] was compared with the structure being determined (by conventional techniques) in Pei Zhou's laboratory (figure 17.2C). The core structure (helix and sheet) computed by RDC-EXACT was 1.28 Å RMSD from the Zhou lab structure (figure 17.2C). In a second test, the same suite of NMR data [174, 177] was obtained for a second protein, the human Set2-Rpb1 interacting (SRI) domain. The global fold of human SRI was determined by RDC-EXACT, similarly using NH and H^{α}-C^{α} RDCs recorded in one medium, plus sparse NOEs. The resulting core structure (a 3-helix bundle) had an RMSD of 1.61 Å to the reference structure (see figure 17.2R). Both reference structures were determined using traditional methods (XPLOR [15]) requiring a much larger set of experimental spectra. The accuracy of RDC-EXACT on these blind tests is comparable to the accuracy achieved in the retrospective studies (figure 17.1 and table 17.1). The ability of RDC-EXACT to determine the global fold of polη and human SRI with reasonable accuracy,

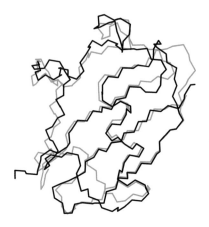

Reference[a]	Program	Technique[b]	Restraints per Residue[c]	Accuracy[d]
Brown Lab [46]	X-plor	MD/SA	6 RDCs	1.45 Å
Blackledge Lab [62]	SCULPTOR	MD/SA	11 RDCs,	1.00 Å
Bax Lab [32]	MFR	Database	10 RDCs, 5 Chemical shifts	1.21 Å
Baker Lab [125]	RosettaNMR	DataBase/MC	3 RDCs, 5 Chemical shifts	1.65 Å
Baker Lab [125]	RosettaNMR	DataBase/MC	1 RDC	2.75 Å
Donald Lab[e]	RDC-EXACT	Exact equations	2 RDCs	1.45 Å

Figure 17.1 (plate 19)

Top: Structure of ubiquitin backbone with loops. The ubiquitin backbone structure (blue) was computed by extending RDC-EXACT to handle loop regions along the protein backbone [156]. The structure was computed using 59 NH and 58 CH RDCs (117 out of 137 possible RDCs; 20 are missing), 12 H-bonds, and 2 unambiguous NOEs. The structure has a backbone RMSD of 1.45 Å with the high-resolution X-ray structure (PDB ID: 1UBQ, in magenta) [147]. Bottom: Comparison of RDC-EXACT with previous approaches. (a) References to previously computed ubiquitin backbone structures (including loop regions), (b) algorithmic technique; (c) data requirements; (d) backbone RMSD of structure (for C^α, N, and C' backbone atoms) compared to the X-ray structure (1UBQ). The structure computed by RDC-EXACT includes loops and turns, as shown at top. [e]References [152, 151, 156, 155].

using only a minimal suite of experiments, that could be collected in a high-throughput fashion, supports the feasibility of the exact solutions approach for structure determination.

7. RDC-EXACT can compute β-sheets from RDC data alone, which fundamentally extends previous methods [38] targeting only entirely helical proteins. Unlike α-helices, β-strands are often twisted in globular proteins so it is important to refine them accurately from RDC data. RDC-EXACT can determine the backbone structures of proteins consisting of either α-helices, β-sheets, or both, and thus has wider application since most proteins have both α-helices and β-sheets.

8. RDC-EXACT was the first demonstration that the conformations and orientations of both α-helices and β-strands can be computed accurately and efficiently using exclusively RDCs measured on a single bond vector type (NH) in only two aligning media [152]. Similar accuracies and efficiency were obtained using only NH and CH RDCs in one medium [151, 156, 174, 177].

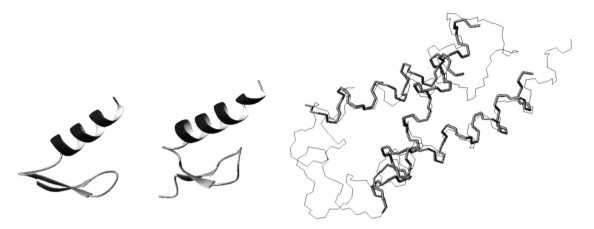

Figure 17.2 (plate 20)
Left: The global fold of polη, computed by RDC-EXACT using 2 RDCs per residue measured in one medium plus 9 NOEs between the helix and β-strands. Center: Comparison of polη to the reference structure, PDB id: 2I5O. The RMSD of the secondary structure elements is 1.28 Å. Right: Global fold of human SRI, computed by RDC-EXACT (thick lines) using two RDCs per residue measured in one medium plus sparse NOEs. The reference structure is shown in thin lines (PDB id: 2A7O [86]).

With a minimum number of additional distance restraints a three-dimensional structure could be computed consequently [152, 156, 174, 177].

Structure determination using sparse data is an underconstrained problem. Additional constraint may be obtained using structure prediction [73, 126] or homology modeling [121]. The former reduces to decoy detection, pruned by NMR data. The second reduces to biasing the structure determination using the PDB. In both approaches, sparse RDCs can be employed, but, compared with conventional protocols, the resulting structures obtain their authority less from the data and more from modeling or homology. In contrast, the exact solutions technique admits algorithms that can extract more structural information from less NMR data, than had been previously exploited. This can be done using a combination of computer algebra, computational geometry, and statistical methods. Compared with previous algorithms for computing backbone structures using RDCs, RDC-EXACT achieves similar accuracies but requires less data, relies less on statistics from the PDB and does not depend on molecular dynamics or simulated annealing (figures 17.1 and 17.2; table 17.1). Since RDCs can be acquired and assigned much more quickly than NOEs in general, the results show it is possible to compute structures rapidly and inexpensively using mainly RDC restraints.

17.2 Nuclear Vector Replacement for Automated NMR Assignment and Structure Determination

High-throughput NMR structural biology can play an important role in structural genomics. Recent results have generalized and extended structure-based assignment algorithms such as

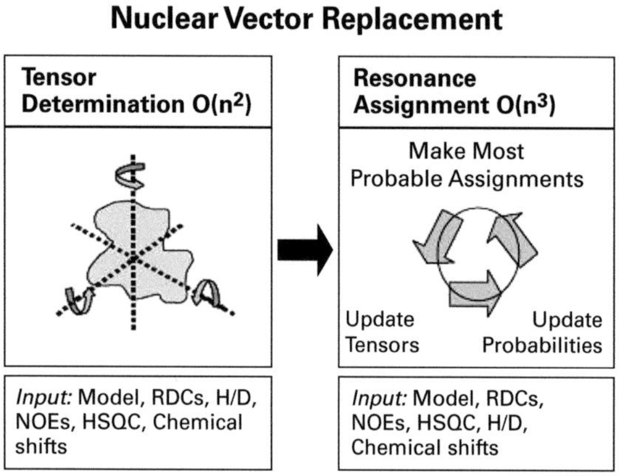

Nuclear Vector Replacement

Tensor Determination $O(n^2)$	Resonance Assignment $O(n^3)$
	Make Most Probable Assignments
	Update Tensors Update Probabilities
Input: Model, RDCs, H/D, NOEs, HSQC, Chemical shifts	*Input:* Model, RDCs, NOEs, HSQC, H/D, Chemical shifts

Experiment/ Data	Information Content	Role in NVR
$H^N\text{-}^{15}N$ HSQC	$H^N, ^{15}N$ Chemical shifts	Backbone resonances, cross-referencing NOESY
$H^N\text{-}^{15}N$ RDC (in 2 media)	Restraints on amide bond vector orientation	Tensor determination, resonance assignment
H-D exchange HSQC	Identifies solvent exposed amide protons	Tensor determination
$H^N\text{-}^{15}N$ HSQC-NOESY	Distance restraints between spin systems	Tensor determination, resonance assignment
Backbone structure	Tertiary structure	Tensor determination, resonance assignment, chemical shift prediction
Chemical shift predictions	Restraints on assignment	Tensor determination, resonance assignment

Figure 17.3
Nuclear vector replacement. Left: Schematic of the NVR algorithm for resonance assignment using EM. The NVR algorithm takes as input a model of the target protein and several unassigned spectra, including the ^{15}N-HSQC, $H^N\text{-}^{15}N$ RDC, ^{15}N-HSQC-NOESY, and an H-D exchange-HSQC to measure amide exchange rates. In the first phase, NVR computes the alignment tensors for both media using chemical shift prediction, d_{NN}s, H-D exchange-exchange rates and the EM algorithm. This step takes time $O(n^2)$, where n is the number of residues. In the second phase, chemical shift predictions, d_{NN}s, RDCs in two media and the EM algorithm are used to assign all remaining peaks. This entire process runs in minutes, and is guaranteed to converge in time $O(n^3)$. NVR Experiment Suite: Right: The 5 *unassigned* NMR spectra used by NVR to perform resonance assignment. The HSQC provides the backbone resonances to be assigned. $H^N\text{-}$^{15}N RDC data in two media provide independent, global restraints on the orientation of each backbone amide bond vector. The H-D exchange HSQC identifies fast-exchanging amide protons. These amide protons are likely to be solvent-exposed and non-hydrogen-bonded and can be correlated to the structural model. A sparse number of unambiguous, unassigned d_{NN}s can be obtained from the NOESY. These d_{NN}s provide distance constraints between spin systems which can be correlated to the structural model. Chemical shift predictions are used as a probabilistic constraint on assignment.

JIGSAW [12], to obtain an automated procedure for high-throughput NMR resonance assignment for a protein of known structure, or of an homologous structure [79, 82, 83, 78, 8]. *Nuclear vector replacement* (NVR) uses *expectation/maximization* (EM) to compute assignments (figure 17.3L). NVR is an RDC-based algorithm, which computes assignments that correlate experimentally measured RDCs, chemical shifts, $H^N\text{-}H^N$ NOEs (which are called d_{NN}s) and amide exchange rates to a given a priori 3D backbone structural model. The algorithm requires only uniform ^{15}N-labeling of the protein, and processes unassigned $H^N\text{-}^{15}N$ HSQC spectra, $H^N\text{-}^{15}N$ RDCs, and sparse d_{NN}s, all of which can be acquired in a fraction of the time needed to record the traditional suite of experiments used to perform resonance assignments (figure 17.3R). NVR could form the basis for "Molecular Replacement (MR) by NMR." RDCs provide *global* orientational restraints on internuclear bond vectors (see equation16.1 and figures 16.1 and 16.2). Once the alignment tensor **S** has been determined, RDCs may be simulated (back-calculated) given any other internuclear vector $\mathbf{v}_i$. In particular, suppose an $(H^N, {}^{15}N)$ peak i in an H^N-^{15}N HSQC (subsequently termed simply "HSQC") spectrum is assigned to residue j of a protein, whose crystal structure is known. Let r_i be the measured RDC value corresponding to this peak. Then the RDC r_i is assigned to amide bond vector $\mathbf{v}_j$ of a known structure, and we should expect that $r_i \approx D_{\max} \mathbf{v}_j^T \mathbf{S} \mathbf{v}_j$ (although noise, dynamics, crystal contacts in the structural model, and other experimental factors will cause deviations from this ideal).

SBA approaches use unassigned NMR data, such as RDCs. Note that, in contrast, *assigned* RDCs have also been employed by a variety of structure refinement [19] and structure determination methods [62, 6, 159], including orientation and placement of secondary structure to determine protein folds [38], pruning an homologous structural database [7, 96], de novo structure determination [125], in combination with a sparse set of assigned NOEs to determine the global fold [100], and a method for fold determination that selects heptapeptide database fragments best fitting the assigned RDC data [32]. Bax and coworkers termed their technique "molecular fragment replacement" [32], by analogy with X-ray crystallography MR techniques. *Unassigned* RDCs have been successfully used to expedite resonance assignments [176, 32, 139].

The idea of correlating unassigned experimentally measured RDCs with bond vector orientations from a known structure was first proposed by Al-Hashimi and Patel [4] and subsequently demonstrated by Al-Hashimi et al. [3] who considered permutations of assignments for RNA, and also in reference [63]. Brüschweiler and coworkers [63] successfully applied RDC-based maximum bipartite matching to structure-based resonance assignment. Their technique requires RDCs from several different bond types which, in turn, requires ^{13}C-labeling of the protein and triple resonance experiments. NVR builds on these works and offers some improvements in terms of isotopic labeling, spectrometer time, accuracy, and computational complexity. NVR algorithms have addressed the following hypothesis: *Are backbone amide RDCs and $d_{NN}s$ sufficient for performing resonance assignment?* Like the techniques of Hus et al. [63], NVR calls optimal bipartite matching as a subroutine, but within an Expectation/Maximization (EM) framework that offers some benefits, described later. Previous methods (and later algorithms [67, 64]) required ^{13}C-labeling and RDCs from many different internuclear vectors (for example, C'-^{15}N, C'-H^N, C^α-H^α, etc.). NVR uses a different algorithm and requires only amide bond vector RDCs, no triple-resonance experiments, and no ^{13}C-labeling. Moreover, NVR is more efficient. The combinatorial complexity of the assignment problem is a function of the number n of residues (or bases in a nucleic acid) to be assigned, and, if a rotation search is required, the resolution k^3 of a rotation-space grid over the Lie group $SO(3)$ of 3D rotations. The time-complexity of the RNA-assignment method, named CAP [3], grows exponentially with n. In particular, CAP performs an exhaustive search over all permutations, making it difficult to scale up to larger RNAs. The method presented in Hus et al. [63] runs in time $O(In^3)$, where $O(n^3)$ is the complexity of bipartite matching [76] and I is the number of times that the bipartite matching algorithm is called. I may be bounded by $O(k^3)$, the time to search for the principal order frame (POF) over $SO(3)$. Thus, the full time-complexity of the algorithm presented in Hus et al. [63] is $O(k^3n^3)$. Version 1.0 of NVR [82, 83, 78] also performed a discrete grid search for the POF over $SO(3)$, but used a more efficient algorithm with time-complexity $O(nk^3)$. Once the POF has been computed, resonance assignments are made in time $O(n^3)$. Thus, the total running time of NVR Version 1.0 [82, 83] is less: $O(nk^3 + n^3)$. Zweckstetter and Bax [175] estimated alignment tensors (but not assignments) using permutations of assignments on a subset of the residues identified using either selective labeling or $^{13}C^{\alpha/\beta}$ chemical shifts. If m residues can be thus identified a priori, then this method provides an $O(nm^6)$ tensor estimation algorithm that searches over all possible assignment permutations.

NVR Version 2.0 [79] requires neither a search over assignment permutations, nor an explicit rotation search over $SO(3)$. Rather, EM [33] is used to correlate the chemical shifts of the H^N-^{15}N HSQC resonance peaks with the structural model. In practice, the application of EM on the chemical shift data is sufficient to uniquely assign a small number of resonance peaks, and directly determine the alignment tensor by singular value decomposition (SVD) (see figure 16.2). NVR 2.0 eliminates the rotation grid-search over $SO(3)$, and hence any complexity dependence on a grid or its resolution k, running in $O(n^3)$ time, scaling easily to proteins in the middle NMR size range ($n = 56$ to 129 residues) [79]. Moreover, NVR elegantly handles missing data (both resonances and RDCs).

NVR adopts a sparse-data, or minimalist approach [12], demonstrating the large amount of information available in a few key spectra. By eliminating the need for triple resonance

Table 17.2

Backbone Amide Resonance Assignment Accuracy using NVR

PDB ID[a]	Accuracy[b]	PDB ID[a]	Accuracy[b]
1G6J [10]	100%	1GB1 [50]	100%
1UBI [123]	100%	2GB1 [50]	96%
1UBQ [147]	100%	1PGB [39]	100%
1UD7 [66]	100%		
(i) Ubiquitin		(ii) SPG	

PDB ID[a]	Accuracy[b]	PDB ID[a]	Accuracy[b]
193L [146]	100%	1LYZ [35]	100%
1AKI [9]	100%	2LYZ [35]	100%
1AZF [90]	100%	3LYZ [35]	100%
1BGI [106]	100%	4LYZ [35]	100%
1H87 [47]	100%	5LYZ [35]	100%
1LSC [77]	100%	6LYZ [35]	100%
1LSE [77]	100%		
(iii) Lysozyme		(iv) Lysozyme	

Accuracies report the percentage of correctly assigned backbone HSQC peaks. [a]Structural model used. [b]Accuracy of NVR on the NMR data shown in figure 17.3 R for ubiquitin (i), SPG (ii), and lysozyme (iii-iv). The 96% accuracy for 2GB1 reflects a single incorrect assignment.

experiments, NVR saves spectrometer time. The required data (figure 17.3R) can be acquired in about one day of spectrometer time using a cryoprobe. NVR runs in minutes and efficiently assigns the (H^N,^{15}N) backbone resonances as well as the sparse d_{NN}s from the 3D ^{15}N-NOESY spectrum. NVR was tested on NMR data from 3 proteins using 20 different alternative structures, all determined either by X-ray crystallography or by *different* NMR experiments (without RDCs) (table 17.2). When NVR was run on NMR data from the 76-residue protein, human ubiquitin (matched to four structures, including one mutant/homolog), it achieved 100% assignment accuracy. Similarly good results were obtained in experiments with the 56-residue streptococcal protein G (SPG) (99%) and the 129-residue hen lysozyme (100%) when they were matched by NVR to 16 3D structural models. Table 17.1 summarizes the performance of NVR using alternative structures of ubiquitin, SPG, and lysozyme, none of which were refined using RDCs. 1UD7 is a mutant form of ubiquitin where 7 hydrophobic core residues have been altered (I3V, V5L, I13V, L15V, I23F, V26F, L67I). It was chosen to test the effectiveness of NVR when the model is a (close) homolog of the target protein. This success in assigning the mutant 1UD7, suggests that NVR could be applied more broadly to assign spectra based on homologous structures [79, 8]. Thus, NVR could play a role in structural genomics.

17.3 Protein Fold Determination via *Unassigned* Residual Dipolar Couplings

Sequence homology can be used to predict the fold of a protein, yielding important clues as to its function. However, it is possible for two dissimilar amino acid sequences to fold to the "same" tertiary structure. For example, the RMSD between the human ubiquitin structure (PDB Id: 1D3Z) and the structure of the Ubx domain from human Fas-associated factor 1 (Faf1; PDB Id: 1H8C) is quite small (1.9 Å), yet they have only 16% sequence identity. Detecting structural homology given low sequence identity poses a difficult challenge for sequence-based homology predictors. Is there a set of fast, cheap experiments that can be analyzed to rapidly compute 3D structural homology? A new method for homology detection has been developed, called GD, that exploits high-throughput solution-state NMR. This algorithm extends the NVR technique to perform protein 3D structural homology detection, demonstrating that NVR and its generalization GD, are able to identify structural homologies between remote amino acid sequences from a database of structural models. The first paper on protein fold determination using *unassigned* RDCs was published in *RECOMB* in April 2003 [82]. Other papers on this topic include [78] and [83, 80]. One goal of structural genomics is the identification of new protein folds. GD is an automated procedure for detecting 3D structural homologies from sparse, *unassigned* protein NMR data, and could aid in prioritizing unknown proteins for structure determination. GD identifies the 3D structural models in a protein structural database whose "*unassigned* geometries" best fit the unassigned experimental NMR data. It does not use sequence information and is thus not limited by sequence homology. GD can also be used to confirm or refute structural predictions made by other techniques such as protein threading or sequence homology.

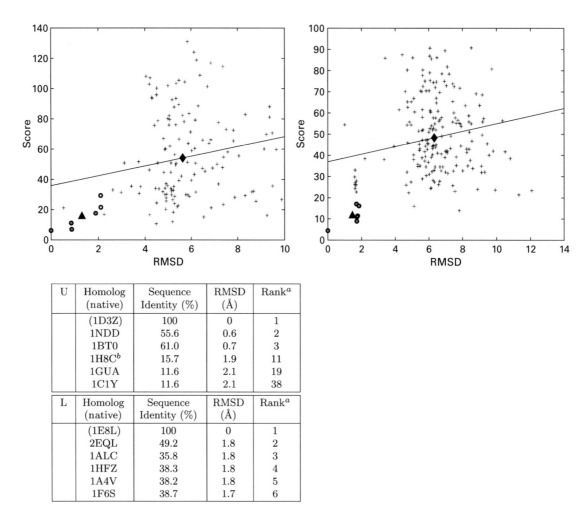

U	Homolog (native)	Sequence Identity (%)	RMSD (Å)	Rank[a]
	(1D3Z)	100	0	1
	1NDD	55.6	0.6	2
	1BT0	61.0	0.7	3
	1H8C[b]	15.7	1.9	11
	1GUA	11.6	2.1	19
	1C1Y	11.6	2.1	38

L	Homolog (native)	Sequence Identity (%)	RMSD (Å)	Rank[a]
	(1E8L)	100	0	1
	2EQL	49.2	1.8	2
	1ALC	35.8	1.8	3
	1HFZ	38.3	1.8	4
	1A4V	38.2	1.8	5
	1F6S	38.7	1.7	6

Figure 17.4

GD, Representative results. GD was tested on unassigned H^N-^{15}N RDCs for 5 proteins [78]; representative scatter plots of RMSD vs. the score computed by GD are shown for human ubiquitin (top left) and hen lysozyme (top right). Only proteins within 10% of the target protein's length are plotted. Circles are data points for the native structure (1D3Z for ubiquitin; 1E8L for lysozyme) and five homologous structures (tables U and L). The + signs are the data points associated with nonhomologous proteins. The diamond is the 2D mean of the +'s while the triangle is the 2D mean of the circles. The trend line shows the correlation between the score computed by GD and RMSD for all the data points. The scores associated with the native fold and the 5 homologs are statistically significantly lower than the scores of unrelated proteins (p-values $< 2.7 \times 10^{-5}$). Tables: GD results for (U) ubiquitin and (L) lysozyme. The sequence identity and RMSD of these 2 test proteins with their respective top 5 homologs are shown. [a]The rank of each model, out of 2,456 proteins in the database, using the score computed by GD. [b]The Ubx domain from human Faf1 (see section 17.3) was ranked 11th.

GD runs in $O(pnk^3)$ time, where p is the number of proteins in the database (2,456 in [82, 78, 83]), n is the number of residues in the target protein, and k is the resolution of a rotation search. GD requires only uniform ^{15}N-labeling of the protein and processes unassigned H^N-^{15}N RDCs, which can be acquired rapidly. Experiments on NMR data from 5 different proteins demonstrated that GD identifies closely related protein folds, despite low-sequence homology between the target protein and the computed model [82, 78, 83, 80]. The overall rankings of the top predicted homologs are good (figure 17.4). Human Faf1 (1H8C) (discussed earlier) was identified as a structural homolog of ubiquitin (figure 17.4U). GD does best on lysozyme, where the native structure and 5 homologous structures occupied the top 6 places (figure 17.4L).

GD [82] represented the first systematic algorithm to exploit an unassigned "fingerprint" of RDCs for rapid protein fold determination. After Langmead et al. [82], a similar idea was independently proposed and extended by number of researchers including Prestegard [145, 128], Baker [97], [98], and coworkers. GD was later extended to demonstrate high-throughput inference of protein-protein interfaces using only unassigned NMR data [99], which should be valuable in structural proteomics.

17.4 Automated NOE Assignment Using a Rotamer Library Ensemble and RDCs

Despite recent advances in RDC-based structure determination (see sections 16.1–17.1), NOE distance restraints are still important for computing a complete three-dimensional solution structure including side-chain conformations. In general, NOE restraints must be assigned before they can be used in a structure determination program. NOE assignment is very time-consuming to do manually, challenging to fully automate, and has become a key bottleneck for high-throughput NMR structure determination. The difficulty in automated NOE assignment is *ambiguity*: there can be tens of possible different assignments for an NOE peak based solely on its chemical shifts. Most automated NOE assignment approaches [103, 57, 59, 68, 60, 53, 91] rely on an ensemble of structures, computed from a subset of all the NOEs, to iteratively filter ambiguous assignments. Despite this progress, there is room for improvement since previous methods require quite high-quality input data. For example, they typically require initializing the assignment/structure determination/assignment cycle with a large number of unambiguous or manually assigned NOE peaks (e.g., >5 NOEs per residue in [103]), > 85% complete resonance assignments (including side-chains), almost complete aromatic side-chain assignments, a low percentage of noise peaks, and only small errors in chemical shifts (≤ 0.03 ppm for 3D NOESY spectra). Moreover, previous algorithms are heuristic in nature, provide no guarantees on solution quality or running time, and consume many hours to weeks of computation time.

Sparse dipolar couplings have enabled the development of new NOE assignment algorithms, including TRIANGLE [153] and HANA [174, 177]. An in-depth understanding of HANA [177] would require a review of the minimum Hausdorff distance [37] and Chernoff tail bounds [174], which are beyond the scope of this book. Therefore, we describe the conceptually simpler TRIANGLE algorithm; since the information content of the input data is identical, TRIANGLE will illustrate the basic paradigm. The interested reader can consult Zeng et al. [177] for a description of HANA, which extends TRIANGLE and is more general and robust. TRIANGLE begins with known resonance assignments and an accurate backbone structure computed using RDC-EXACT (sections 16.1– 17.1; [152, 151, 156]). In principle, another structure determination algorithm could be used instead. However, RDC-EXACT is the only algorithm that can compute a complete protein backbone structure de novo using only two RDCs per residue. TRIANGLE uses the backbone structure determined by RDC-EXACT to bootstrap the automated assignment of NOEs: the assignment proceeds by *filtering* the experimentally measured NOEs based on consistency with the backbone structure.

One novel feature of TRIANGLE is the use of a rotamer-library–derived ensemble of intraresidue vectors between the backbone atoms and side-chain protons to reduce the ambiguity in the assignment of NOE restraints involving side-chain protons, especially aliphatic protons. The rotamer database was built from ultra-high- resolution structures (<1.0 Å) in the PDB [153]. TRIANGLE merges this ensemble of intraresidue vectors, together with internuclear vectors from the computed backbone structure. For example, consider the putative assignment of an NOE to an (H^N, H^δ) pair of protons. The *triangle relationship* (figure 17.5L, plate 21) defines a decision procedure to filter NOE assignments by *fusing* information from RDCs ($\mathbf{v}_{N\beta}$), rotamer modeling ($\mathbf{v}_{\beta\delta}$), and NOEs ($|\mathbf{v}_{N\delta}|$). More details of the TRIANGLE algorithm are provided in other studies [153, 174] and section 18.1.3.

NMR Structures   NMR vs. X-ray Structure

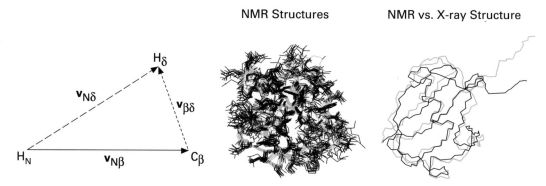

Figure 17.5 (plate 21)

Left: The triangle relationship, defines a decision procedure to filter NOE assignments by fusing information from structure ($\mathbf{v}_{N\beta}$), modeling ($\mathbf{v}_{\beta\delta}$), and experiment ($|\mathbf{v}_{N\delta}|$). An accurate backbone structure is first computed using only 2 RDCs per residue (section 17.1). The vector $\mathbf{v}_{N\beta}$ is computed from this backbone structure. $\mathbf{v}_{\beta\delta}$ is a rotamer ensemble–based intraresidue vector mined from the PDB. The computed length of $\mathbf{v}_{N\delta}$ is compared with the *experimental* NOE distance d_N (measured using NOE cross-peak intensities) to filter ambiguous NOE assignments. The complexity of NOE assignment is $O(n^2 \log n)$, where n is the number of protons in the protein. One cycle of NOE assignment suffices when a well-defined backbone structure is computed using RDC-EXACT. In practice, it took less than one second to assign 1,783 NOE restraints from the NOE peak list picked from both the 3D ^{15}N-edited and ^{13}C-edited NOESY spectra of human ubiquitin. Right and center: The NMR structures computed from the automatically assigned NOEs. The middle panel shows the 12 best NMR structures with no NOE distance violation larger than 0.5 Å. The side-chains are blue; the backbone is magenta. The right panel is the overlay of the NMR average structure (blue) with the 1.8 Å X-ray structure (magenta) [147].

Table 17.3
NOE restraints automatically assigned by TRIANGLE

Spectrum	No. of Peaks	No. of Assigned NOEs	No. of Sequential NOEs[a]	No. of Medium[b], long-range NOEs[c]
^{15}N-NOESY	1,083	1,288	822	466
^{13}C-NOESY	420	495	228	167

(a) NOEs between residue i and $i + 1$, (b) NOEs between residue i and $i + j$ where $1 < j < 4$, and (c) NOEs between residue i and $i + j$ where $j \geq 4$. No., number.

In Wang and Donald [153], RDC-EXACT was first employed to compute an accurate backbone structure of ubiquitin using only two backbone RDCs per residue (figure 17.1L; [152, 151, 156]). Next, TRIANGLE was successfully applied to assign more than 1,700 NOEs, with better than 90% accuracy on the experimental ^{15}N- and ^{13}C-edited 3D NOESY spectra for ubiquitin [153]. The result of the automated assignment is summarized in table 17.3. Out of the 1,153 NOE peaks picked from the ^{15}N-edited NOESY spectrum, 1,083 originate from backbone amide protons, and the remaining 50 are from side-chain amide protons. Only 420 NOE peaks originating from backbone H$^\alpha$ protons could be picked from the ^{13}C-edited NOESY spectrum. TRIANGLE was able to assign 1,053 NOE peaks from the 1,083 peaks picked from ^{15}N-edited NOESY spectrum and 393 peaks from the 420 peaks picked from ^{13}C-edited NOESY spectrum. The assigned NOE distance restraints are divided into two classes: sequential NOEs and medium/long-range NOEs (table 17.3). The number of assigned NOE distance restraints is larger than the number of assigned peaks since it is possible that more than one NOE restraint could be assigned to a single peak. It took less than *one second* on a 2.4-GHz single-processor Linux workstation for TRIANGLE to compute the assignments.

To test the quality of the 1,783 NOE restraints automatically generated and assigned by TRIANGLE (table 17.3), Wang and Donald [153] then used these restraints to calculate the structure of ubiquitin, using a hybrid distance-geometry and SA protocol (XPLOR [15]). No RDC restraints were used. In the resulting ensemble of 70 structures, 163 restraints had NOE violations larger than 0.5 Å in 50 structures. However, none of the NOE violations was larger than 2.5 Å. After these 163 restraints were deleted from the NOE list, XPLOR was invoked for a second time to compute the structures using the remaining 1,620 NOE restraints. Twelve structures out of 70 total computed structures had no NOE violations larger than 0.5 Å. Thus, the NOE assignment algorithm had an accuracy of 91%, and the incorrect assignments could easily be detected and removed. The 12 best NMR structures (figure 17.5C) can be overlayed with a pairwise RMSD of 1.18 ± 0.16 Å for backbone atoms and a pairwise RMSD of 1.84 ± 0.19 Å for all heavy atoms. The accuracy of the structures computed using the automatically assigned NOEs from TRIANGLE is in the range of typical high- to medium-resolution NMR structures. The average structure (figure 17.5R) computed from the best 12 structures has a 1.43 Å backbone RMSD and 2.13 Å all heavy-atom RMSD from the 1.8 Å ubiquitin X-ray structure [147]. HANA, a successor to TRIANGLE, was tested on additional proteins, resulting in similarly good accuracies [174, 177].

RDC-EXACT and HANA were used prospectively to determine the NMR solution structure of the FF Domain 2 of human transcription elongation factor CA150 (RNA polymerase II C-terminal domain interacting protein), PDB id: 2kiq [177], and a number of algorithmic enhancements and retrospective validation tests of these software modules have been described [177].

17.5 NMR Structure Determination of Symmetric Homo-Oligomers

Symmetric homo-oligomers play pivotal roles in complex biological processes including ion transport and regulation, signal transduction, and transcriptional regulation. RDCs were recently used in studies of phospholamban, a symmetric homopentameric membrane protein that regulates the calcium levels between cytoplasm and sarcoplasmic reticulum and hence aids in muscle contraction and relaxation [111]. Ion conductance studies [75] also suggest that phospholamban might have a separate role as an ion channel. To understand the dual function of phospholamban and other such symmetric homo-oligomers, a combined experimental-computational approach, SYM-BRANE [117, 118], determined their structures using sparse intersubunit NOE distance restraints and van der Waals (vdW) packing. In this case, the RDCs were used to refine the subunit structure.

SYMBRANE is *complete* in that it tests *all* possible conformations, and it is *data-driven* in that it first tests conformations for consistency with data and only then evaluates each of the consistent conformations for vdW packing. Completeness in a structure determination method is a key requirement since it ensures that no conformation consistent with the data is missed. This avoids any bias in the search, as well as any potential for becoming trapped in local minima, problems inherent in energy minimization–based approaches. (RDC-EXACT, described above in sections 16.1–17.1, is also complete.) The data-driven nature of SYMBRANE allows one to independently quantify the amount of structural constraint provided by data alone, versus both data and packing. This avoids overreliance on subjective choices of parameters for energy minimization, and consequent false precision in determined structures. Analogously to single particle analysis in cryoelectron microscopy [107], SYMBRANE also allows determination of the oligomeric number of the complex. SYMBRANE was the first approach that determines the oligomeric number of a symmetric homo-oligomer from intersubunit NOE distance restraints. The complete ensemble of NMR structures of the human cardiac phospholamban pentamer, determined by SYMBRANE, was reported by Potluri et al. [117] and deposited in the PDB (id: 2HYN).

SYMBRANE exploits the observation that, given the structure of a subunit, the structure of a symmetric homo-oligomer is completely determined by the parameters (position and orientation) of its symmetry axis. Thus, we can formulate structure determination as a search in the space of all symmetry axis parameters, *symmetry configuration space* (SCS). SYMBRANE performs a branch-and-bound search in the SCS, which defines all possible C_n homo-oligomeric complexes for a given subunit structure. It eliminates those symmetry axes inconsistent with NOE distance restraints and then identifies conformations representing any consistent, well-packed structure.

SYMBRANE consists of two phases. The first phase performs a complete, data-driven search in SCS and returns *consistent regions* in SCS, that is, regions representing *all* conformations consistent with (i.e., *satisfying*) the data. In the second phase, all satisfying structures are evaluated for vdW packing. SYMBRANE computes a set of *well-packed satisfying (WPS) structures* that are not only consistent with data, but also have high-quality vdW packing. SYMBRANE explicitly quantifies the uncertainty in determined structures using the size of the regions in SCS and the variations in atomic coordinates in the structures. The difference in uncertainty between the satisfying structures and the WPS structures illustrates the relative precision that is possible from data alone, versus data and packing together. SYMBRANE simultaneously assigns the intermolecular NOEs during the SCS search for structure determination [118]. This assignment resolves the ambiguity as to which pairs of protons generated the observed NOE peaks, and thus should be restrained in structure determination.

For intersubunit NOEs in symmetric homo-oligomers, the ambiguity includes both the identities of the protons and the identities of the subunits to which they belong. SYMBRANE resolves both ambiguities to determine its structures, and returns all structures consistent with the available data (ambiguous or not). However, while SYMBRANE is complete, it avoids explicit enumeration of the exponential number of combinations of possible assignments. SYMBRANE geometrically prunes SCS regions and NOE assignments that are mutually inconsistent. Pruning occurs only due to *provable* inconsistency and thereby avoids the pitfall of local minima that could arise from best-first sampling-based approaches. Ultimately, SYMBRANE returns a mutually consistent set of conformations and NOE assignments. SYMBRANE can draw two types of conclusions not possible under previous methods: (a) that different assignments for an NOE would lead to different structural classes, or (b) that it is not necessary to assign an NOE since the choice of assignment would have little impact on structural precision.

In contrast to previous techniques, SYMBRANE *separately* quantifies the amount of structural constraint provided by data alone, versus data and packing (see figure 18.2d). For the human phospholamban pentamer in dodecylphosphocholine micelles, using the structure of one subunit determined from a subset of the experimental NMR data, SYMBRANE identifies a diverse set of complex structures consistent with the nine intersubunit NOE restraints [117]. The distribution of structures determined in the ensemble (PDB id: 2HYN) provides an objective characterization of structural uncertainty: Incorporating vdW packing reduced the structural diversity by 29% and average variance in backbone atomic coordinates by 44%.

By comparing data consistency and packing quality in a search using different assumptions of oligomeric number, SYMBRANE identified the C_5 pentamer as the most likely oligomeric state of phospholamban, demonstrating that it is possible to determine oligomeric number directly from NMR data. Additional tests on six other homo-oligomers, from dimer to heptamer, similarly demonstrated the power of SYMBRANE to provide unbiased determination and evaluation of homo-oligomeric complex structures [117, 118].

17.6 Applications and Connections to Other Biophysical Methods

Both NVR and SYMBRANE are structure-based (in that they exploit priors on the monomer structure) and have clear analogies to molecular replacement (MR) in X-ray crystallography and its leverage of subunit structure and non-crystallographic symmetry (NCS). At the heart of NVR is a rotation search (over 3D rotation space, $SO(3)$), and at the core of SYMBRANE is exploitation of the kinematics of C_n symmetry. It is possible to generalize and apply these algorithms to help expedite structure determination even using very different experiments (crystallography). To analyze NCS in MR-based X-ray diffraction data of biopolymers, one must "recognize" a finite subgroup of $SO(3)$ out of a large set of molecular orientations. This problem may be reduced to clustering in $SO(3)$ modulo a finite group, and solved efficiently by "factoring" into a clustering on the unit circle followed by clustering on the 2-sphere S^2, plus some group-theoretic calculations [89]. This yields a polynomial-time algorithm, CRANS, that is efficient in practice, and which enabled determination of the structure of dihydrofolate reductase-thymidylate synthase (DHFR-TS) from *Cryptosporidium hominis*, PDB id: 1QZF [110, 109, 89]. *Cryptosporidium* is an organism high on the bioterrorism list of the Centers for Disease Control, a Category B bioterrorist threat. The enzyme DHFR-TS is in the sole de novo biosynthetic pathway for the pyrimidine deoxyribonucleotide dTMP, and therefore an attractive drug target. Solving the structure of DHFR-TS from *C. hominis* enabled *species-specific drug design* [5, 116], exploiting structural differences between the human enzyme and *C. hominis* DHFR-TS.

In general, algorithms like NVR and SYMBRANE are often promiscuous, in that such tools can be applied to many problems in structural biology. The CRANS NCS method for MR [89, 110, 109] in X-ray crystallography generalized NVR rotation search algorithms that were originally proposed for a different problem, NMR assignment [82, 83, 78, 79]. Similarly, work in target selection for structural and functional genomics [84, 81, 87] has applied algorithms from NMR signal processing such as maximum entropy spectral analysis. Recent work on enzyme redesign in the nonribosomal peptide synthetase pathway [88, 136, 45, 44] (using crystallography and site-directed mutations) generalized algorithms for modeling NMR ensembles [49, 11] and rotamers for NOE assignment [153, 174]. These rotamer optimization algorithms [88, 45, 44] can, in turn, be used as a tool in structure determination as discussed below in sections 18.1.3 and 18.1.4.

18.1 Looking Under the Hood: How the Algorithms Work, and Outlook for Future Developments

In the future, one expects improvements and extensions to the algorithms described earlier. In particular, the primary goals focus on robustness and scaling. We wish the algorithms to be more robust to noise and missing data. We want the implemented systems to scale, to run successfully on more proteins, and larger proteins. A number of smaller steps and subgoals are foreseen to achieve these goals. In this section, we describe in more detail how the algorithms work so the reader can develop an appreciation both for their scope and how they may be extended and improved.

In general, but particularly for RDC-EXACT, it is useful for the algorithms and software to allow users a choice: to record either (i) one type of backbone RDC (such as NH RDCs) in two aligning media, or (ii) two types of backbone RDCs (such as NH and CH RDCs) in a single medium. This flexibility allows application to a wider range of proteins. Experimental NH RDCs in two media require only an ^{15}N-labeled sample, which is an order of magnitude cheaper to prepare than a doubly labeled ^{15}N/^{13}C sample. However, finding two aligning media for a protein is not always straightforward; in this case the algorithm can use NH and CH RDCs in a single medium since recording an extra set of RDCs in the same medium requires only slightly more spectrometer time. New methods [127] will be very useful for measuring RDCs in multiple alignment media.

18.1.1   Exact Solutions for Computing Backbone Dihedral Angles from RDCs

RDC-EXACT is an exact-solution, systematic search-based algorithm for structure determination. We now describe the outlook for such algorithms, focusing on robustness and scaling.

RDC-EXACT was developed both as a de novo structure determination algorithm and as a tool in "MR for NMR." In particular, NVR assignments are structure based. However, RDC-EXACT does not currently exploit (or require) structural priors. Thus, a naïve implementation of the MR-like pathway in figure 15.1 might ignore the structural homolog found by GD, and hence lose information. However, RDC-EXACT could use the structural homologs found by GD as a *bias* in structure determination. More specifically: RDC-EXACT is currently a de novo structure determination method, starting with resonance assignments. Those assignments could come from either NVR or conventional triple-resonance experiments. Indeed, in the case where GD detects (from unassigned NMR data) a novel fold, not in the database (figure 15.1), one can use algorithms for de novo assignment, that process either triple-resonance experiments or sequential connectivities deduced from short-range NOEs [22, 105, 149, 70, 148, 12]. However, for a non-novel fold, the structural model determined by GD can be exploited to bias the choice in RDC-EXACT of polynomial roots representing the backbone dihedral angles. Initially, these were chosen by Ramachandran fitness; we now describe an improved version of RDC-EXACT that exploits bias toward a priori structure in the regions of regular secondary structure and low dynamics. A series of filters can be employed to choose between the discrete, finite roots representing conformations. These filters (which may be viewed as a fitness function) combine the fit to the data, modeling (Ramachandran fitness), and structural homology (from GD). In the case of missing data [156], or for denatured or disordered proteins [154], one can introduce into the fitness function van der Waals or molecular mechanics energies [114, 15, 117]. This increases reliance on modeling, using an empirical energy function to compensate for missing data. Such fitness functions may combine sources of data and modeling using relative weights. The choice of these weights can be problematic [124]. By converting each term of the fitness function to a Bayesian probability (proposed by Nilges [124], and successfully employed in characterizing denatured structures [154] and automated assignments [105]) the resulting probabilities can be combined as log likelihood (the sum of the logarithms of the probabilities), without explicit weights, thereby reducing subjectivity.

When ^{13}C chemical shifts are available, TALOS [25] can be used for secondary structure determination and to provide restraint on the polynomial roots representing (ϕ, ψ) angles. Once RDC-EXACT has used sparse RDCs to determine the conformations and orientations of elements of secondary structure, these fragments are rigidly translated using sparse NOEs [152, 174] or PREs [154]. Connecting loops between these elements can then be solved (as shown in figure 17.2L and [156, 174]). *Without NMR data,* this would reduce to the kinematic loop closure problem (KLCP), which can be addressed using computer algebra or optimization techniques [74, 134, 104, 26, 27]. *With RDC data,* the polynomial system of the KLCP can be combined with the quartic and quadratic RDC polynomials (eqs. 18.6 and 18.9) to simultaneously find satisfying solutions (conformations) to the KLCP and the RDC equations. This simultaneous bicriterion optimization search requires solving higher-degree polynomials, which can be approximated using homotopy continuation or solved exactly using multivariate resultant or Gröbner basis methods, as presented in Donald et al. [37]. Since the Ramachandran plot is less informative for loops, the enhanced filters above, that incorporate packing and empirical molecular mechanics energies, are helpful outside regions of regular secondary structure. One can also employ a modified version of the robotics-based CCD algorithm [174]. One can represent the loop residues as a conformation tree (figure 16.8) [152, 88, 156, 174], and filter/search it for the loop ensemble that best fits the experimental RDCs (and any available long-range NOEs). Recent results suggest that this search can return a structural loop ensemble with good stereochemistry that simultaneously solves the KLCP consistent with the experimental NMR measurements [156, 174].

RDC-EXACT should be particularly useful for the studies of large protein complexes because backbone resonances and associated RDCs can be readily assigned and measured even in a very large protein, such as the 82 kDa malate synthase G [144, 143]. The RDC-EXACT algorithm computes backbone dihedral angles from two RDCs per residue, in constant time per residue. It is possible to derive, from the physics of RDCs, low-degree polynomials (with degree at most 4) whose solutions give the backbone (ϕ, ψ) angles.

We describe the methodology and protocol for exact solutions assuming that we are given assigned NH RDCs in two media. *These methods also hold for the case of NH and CH RDCs in one medium with slight modifications to the equations below, as shown in section 16.1 and Wang and Donald [151, 177].* For ease of exposition, we assume that the dipolar interaction constant D_{max} is equal to 1. By considering a global coordinate frame that diagonalizes the alignment tensor equation (16.1) becomes:

$$r = S_{xx}x^2 + S_{yy}y^2 + S_{zz}z^2, \tag{18.1}$$

where S_{xx}, S_{yy} and S_{zz} are the three diagonal elements of a diagonalized Saupe matrix $\mathbf{S}$ (the alignment tensor), and x, y and z are, respectively, the x, y, z-components of the unit vector $\mathbf{v}$ in a principal order frame (POF) which diagonalizes $\mathbf{S}$. Now, $\mathbf{S}$ is a 3×3 symmetric, traceless matrix with five independent elements [140, 1]. Given NH RDCs in two aligning media, the associated NH vector $\mathbf{v}$ must lie on the intersection of two conic curves [135, 159]. We now derive the

Wang-Donald equations (18.6, 18.9, and 18.12), that permit the (ϕ, ψ) backbone dihedral angles to be solved exactly and in closed form, given protein kinematics and two RDCs per reside. Very similar equations have been derived for NH and $C^\alpha H^\alpha$ RDCs measured in a single medium (see section 16.1 and [151, 177]).

Proposition 18.1 [152]   *Given the diagonal Saupe elements S_{xx} and S_{yy} for medium 1, S'_{xx} and S'_{yy} for medium 2 and a relative rotation matrix $\mathbf{R}$ between the POFs of medium 1 and 2, the square of the x-component of the unit vector $\mathbf{v}$ satisfies a univariate polynomial quartic equation.*

The following is a sketch of the proof. The methods for the computation of the seven parameters (S_{xx}, S_{yy}, S'_{xx}, S'_{yy}, and $\mathbf{R}_{12}$) and the full expressions for the polynomial coefficients and temporary variables (a_2, b_2, c_1, etc.) can be found in Wang and Donald [152].

Proof sketch   Fix a backbone NH vector $\mathbf{v}$ along the backbone and let r (eq. 18.1) and $r' = S'_{xx}x'^2 + S'_{yy}y'^2 + S'_{zz}z'^2$ be the experimental RDCs for $\mathbf{v}$ in the first and second medium, respectively. We have $(x'\ y'\ z')^T = \mathbf{R}(x\ y\ z)^T$, where x, y, z are the three components of $\mathbf{v}$ in a POF of medium 1, r' and x', y', z' are the corresponding variables for medium 2. Let $\mathbf{R} = (R_{ij})_{i,j=1,\cdots,3}$. Eliminating x', y', and z' we have

$$c_0 = a_2 x^2 + b_2 y^2 + c_1 xy + c_2 xz + c_3 yz \tag{18.2}$$

$$c_4 = a_1 x^2 + b_1 y^2 \tag{18.3}$$

where $a_2 = (S'_{xx} - S'_{zz})(R_{11}^2 - R_{13}^2) + (S'_{yy} - S'_{zz})(R_{21}^2 - R_{23}^2)$, $c_2 = 2(S'_{xx} - S'_{zz})R_{11}R_{13} + 2(S'_{yy} - S'_{zz})R_{21}R_{23}$, and a_1, b_1, b_2, c_0, $\ldots$, c_4 are similar constants; full details are given in Wang and Donald [152]. Eliminating z from equation (18.2) we obtain

$$d_8 x^4 + d_7 x^3 y + d_6 x^2 y^2 - d_5 x^2 + d_4 xy^3 - d_3 xy - d_2 y^2 + d_1 y^4 + d_0 = 0 \tag{18.4}$$

where $d_8 = a_2^2 + c_2^2$, and d_7, d_6, ..., d_0 are analogously defined; these are fully specified in [152]. (18.4) is a degree 8 polynomial in x after direct elimination of y using equation (18.3). However, it can be reduced to a quartic equation by substitution since all terms have even degree. Introducing new variables t and u such that

$$x = a \sin t, \qquad y = b \cos t, \qquad u = \cos 2t, \tag{18.5}$$

and through algebraic manipulation we finally obtain the *Wang-Donald Equation of type I for NH vectors from RDCs in two media:*

$$f_4 u^4 + f_3 u^3 + f_2 u^2 + f_1 u + f_0 = 0. \tag{18.6}$$

The full expressions for coefficients a, b, f_0, f_1, f_2, f_3, f_4 are given in Wang and Donald [152]. Since $u = 1 - 2\left(\frac{x}{a}\right)^2$ equation (18.6) is also a quartic equation in x^2.   □

Similar equations to equation (18.6) have been derived for NH and $C^\alpha H^\alpha$ RDCs measured in a single medium (see section 16.1 and [151, 177]). The y-component of $\mathbf{v}$ can be computed directly from equation (18.5). Due to twofold symmetry in the RDC equation, the number of real solutions for $\mathbf{v}$ is at most 8. We will refer to the bond vector between the N and C^{α} atoms as the NC^{α} vector. Given two unit vectors in consecutive peptide planes we can use backbone kinematics to derive quadratic equations to compute the sines and cosines of the (ϕ, ψ) angles:

Proposition 18.2 [152]　*Given the* NH *unit vectors* $\mathbf{v}_i$ *and* $\mathbf{v}_{i+1}$ *of residues i and* $i + 1$ *and the* NC^{α} *vector of residue i the sines and cosines of the intervening backbone dihedral angles* (ϕ, ψ) *satisfy the trigonometric equations* $\sin(\phi + g_1) = h_1$ *and* $\sin(\psi + g_2) = h_2$, *where* g_1 *and* h_1 *are constants depending on* $\mathbf{v}_i$ *and* $\mathbf{v}_{i+1}$, *and* g_2 *and* h_2 *depend on* $\mathbf{v}_i$, $\mathbf{v}_{i+1}$, $\sin\phi$ *and* $\cos\phi$. *Furthermore, exact solutions for* $\sin\phi$, $\cos\phi$, $\sin\psi$, *and* $\cos\psi$ *can be computed from a quadratic equation by tangent half-angle substitution.*

The following is a sketch of the proof. Full expressions for the polynomial coefficients and temporary variables $(x_1, y_1, z_1, x_2, y_2, z_2, g_1, h_1, g_2, h_2)$ introduced in the proof are given in Wang and Donald [152].

Proof sketch　Using backbone inverse kinematics, the two NH vectors $\mathbf{v}_i$ and $\mathbf{v}_{i+1}$ can be related by 8 rotation matrices between two coordinate systems in peptide planes i and $i + 1$:

$$\mathbf{v}_i = \mathbf{R}_x(\theta_7)\mathbf{R}_y(\theta_6)\mathbf{R}_x(\theta_5)\mathbf{R}_z(\psi + \pi)\mathbf{R}_x(\theta_3)\mathbf{R}_y(\phi)\mathbf{R}_y(\theta_8)\mathbf{R}_x(\theta_1)\mathbf{v}_{i+1}. \tag{18.7}$$

The definitions of the coordinate systems, the expressions for the rotation matrices $\mathbf{R}_x, \mathbf{R}_y$ and $\mathbf{R}_z$ and the definitions of the six backbone angles $(\theta_1, \theta_3, \theta_5, \theta_6, \theta_7,$ and $\theta_8)$ are given in Wang and Donald [152]. The backbone (ϕ, ψ) angles are defined according to the standard convention. Given the values of these six angles $\mathbf{R}_l = \mathbf{R}_x(\theta_7)\mathbf{R}_y(\theta_6)\mathbf{R}_x(\theta_5)$ and $\mathbf{R}_r = \mathbf{R}_y(\theta_8)\mathbf{R}_x(\theta_1)$ are two 3×3 *constant* matrices. Define two new vectors $\mathbf{w}_1 = (x_1, y_1, z_1) = \mathbf{R}_l^{-1}\mathbf{v}_i$ and $\mathbf{w}_2 = (x_2, y_2, z_2) = \mathbf{R}_r\mathbf{v}_{i+1}$ to obtain

$$x_1 = -(\cos\phi\cos\psi + \sin\theta_3\sin\phi\sin\psi)x_2 - \cos\theta_3\sin\psi\ y_2 + (\cos\psi\sin\phi - \cos\phi\sin\theta_3\sin\psi)\ z_2$$

$$y_1 = (\cos\phi\sin\psi - \sin\theta_3\sin\phi\cos\psi)x_2 - \cos\theta_3\cos\psi\ y_2 - (\sin\phi\sin\psi + \cos\phi\sin\theta_3\cos\psi)\ z_2$$

$$z_1 = \cos\theta_3\sin\phi\ x_2 - \sin\theta_3\ y_2 + \cos\theta_3\cos\phi\ z_2. \tag{18.8}$$

By equation (18.8) we can then obtain the *Wang-Donald Equation of type II for the ϕ dihedral angle from RDCs in two media:*

$$\sin(\phi + g_1) = h_1, \tag{18.9}$$

where

$$h_1 = \frac{z_1 + y_2\sin\theta_3}{\sqrt{(x_2\cos\theta_3)^2 + (z_2\cos\theta_3)^2}}, \tag{18.10}$$

and g_1 is a similar constant; see Wang and Donald [152] for details. The values of $\sin\phi$ and $\cos\phi$ can be computed from a quadratic equation by the substitution

$$w = \tan \frac{\phi}{2}, \sin \phi = \frac{2w}{1 + w^2}, \cos \phi = \frac{1 - w^2}{1 + w^2} \qquad (18.11)$$

Substituting the computed $\sin \phi$ and $\cos \phi$ into equation (18.8), we can obtain the analogous *equation of type II for the ψ dihedral angle from RDCs in two media* [152]; it is another simple trigonometric equation:

$$\sin (\psi + g_2) = h_2. \qquad (18.12)$$

Hence, $\sin \psi$ and $\cos \psi$ can be computed similarly from a quadratic equation where both g_2 and $h_2 \leq 1$ are computed from $y_1, x_2, y_2, z_2, \theta_3, \sin \phi$, and $\cos \phi$. $\qquad \square$

Similar equations to equations (18.9–18.12) have been derived for NH and $C^\alpha H^\alpha$ RDCs measured in a single medium (see section 16.1 and [151, 177]).

The structures computed by RDC-EXACT may be slightly different from structures computed using traditional protocols (section 17.1), which use different data, more data, and rely more on modeling. In contrast to, for example, simulated annealing approaches, RDC-EXACT is built on the exact solutions for computing backbone (ϕ, ψ) angles from RDC data and systematic search. RDC-EXACT is guaranteed to compute the global minimum (the best fit, or maximum likelihood solution to the data) and hence is *provable* in terms of completeness, correctness, and complexity. We caution that in the context of RDC-EXACT, the term "provably correct" refers to the computational correctness of the algorithm. It is not meant to imply that the RDC-EXACT structures computed using the preceding modeling assumptions (i.e., standard protein covalent geometries, pruning of (ϕ, ψ) solutions using the Ramachandran plot and steric clashes) will always compute biologically correct structures. More precisely, the proofs guarantee accuracy (of the computed structures) and optimality (of the maximum likelihood structure) up to, but only up to, the accuracy of the modeling assumptions. One can relax the modeling assumptions and improve their accuracy, as outlined earlier. However, a helpful first step has been to develop an algorithm that is provably correct up to the limitations of the initial model. It is unusual for protein structure determination algorithms to have guarantees of computing the optimal structure, in polynomial time. RDC-EXACT has both these desirable properties [156, 177], and hence promises significant advantages as a method to compute structures accurately using only very sparse restraints.

For smaller proteins it may be possible to collect side-chain RDCs, in which case RDC-EXACT can be extended to compute side-chain conformations. One can extend the algorithms to compute complete protein structures (including side-chains), since exact equations analogous to equations (18.6 and 18.9) can be derived *mutatis mutandis* to compute the side-chain dihedral angles $\chi_1, \chi_2, \ldots$ from side-chain RDCs. In this case, the average or Ramachandran angles used as root-filters in Wang and Donald [152], may be replaced with modal side-chain rotamer library angles $\chi_{a,1}, \chi_{a,2}, \ldots$. One can also expedite protein structure determination using pseudocontact shift restraints [72] or carbonyl chemical shift anisotropy upon alignment [167], since similar equations can be derived for computing the corresponding vector orientations and dihedral angles.

18.1.2 Nuclear Vector Replacement and Fold Recognition Using Unassigned RDCs

NVR-like algorithms for automated assignment form a cornerstone for molecular-replacement (MR) type methods for NMR starting with only unassigned data. Since GD can be viewed as the application (filtering) of NVR modules to a database (as opposed to a single structure), improvements to NVR will, *mutatis mutandis*, improve the accuracy and performance of GD. Note that unlike MR (in crystallography), NVR/GD is sufficiently efficient computationally that it can be applied to thousands of template structures in a short time.

NVR uses two classes of constraints: *geometric* and *probabilistic* (see figure 17.3). The H^N-^{15}N RDC, H-D exchange and ^{15}N-NOESY each provide independent geometric constraints on assignment. A sparse number of d_{NN}s are extracted from the unassigned NOESY after the diagonal peaks of the NOESY are cross-referenced to the peaks in the HSQC. These d_{NN}s provide distance restraints for assignments. In general, there are only a small number of unambiguous d_{NN}s that can be obtained from an unassigned 3D ^{15}N-edited NOESY. The amide exchange information probabilistically identifies the peaks in the HSQC corresponding to non-hydrogen-bonded, solvent-accessible backbone amide protons. RDCs provide probabilistic constraints on each backbone amide-bond vector's orientation in the POF. Finally, chemical shift prediction is employed [79] to compute a probabilistic constraint on assignment. NVR exploits the geometric and probabilistic constraints by combining them within an Expectation/Maximization framework.

NVR can be improved by exploiting different, but still efficient experiments. Currently, NVR uses amide exchange experiments to distinguish surface from buried residues [79]. This "inside/outside" information is correlated with the structural model, and increases assignment accuracy. When these experiments may be impractical for larger proteins, one can, in principle, employ paramagnetic quenching [172] or "water HSQC" experiments [41, 51], which identify fast-exchanging or solvent-exposed backbone protons. Since this information is complementary to the slow-exchange protons identified by H-D exchange, it likely could be integrated into NVR by complementing the corresponding edge weights in a bipartite graph, which we shall discuss later. Similarly, it is probable that assignment accuracy will be improved by incorporating CH RDCs into NVR and GD, since in helices the C^α-H^α bond vector orientations are less degenerate than NH vectors, thereby both improving tensor accuracy in the bootstrapping phase and increasing disambiguation of helical RDCs in the assignment phase. Four-dimensional ^{15}N-,^{15}N- and ^{13}C-,^{15}N-edited NOESY experiments [21] from singly and doubly labeled samples, respectively, will reduce the ambiguity in both spin systems and assignments, and facilitate matching of NOE restraints to structural models. While the acquisition time for 4D NOESY is longer, the direct polar Fourier transformation holds great potential to speed up data collection significantly, by producing quantitatively accurate spectra from radial and concentric sampling [23, 13].

The ultimate goal is to make NVR useful in an analogous manner to molecular replacement (MR) in crystallography. To do this NVR has been extended to operate on more distant structural homologs [8]. The relative sensitivity of RDCs to structural noise is evident in equation (16.1), from the quadratic dependence on the bond vector **v** versus the linear dependence on the alignment tensor **S**. In the NVR scoring function, the M step of the EM algorithm uses edge weights to compute the joint probability of the resonance assignments in an *ensemble* of bipartite graphs. This score is a maximum likelihood estimator of assignment accuracy, and as parameters and inputs to the algorithm (such as the structural model and alignment tensor) are varied, its log likelihood may be optimized. As cross-validation, an alternative scoring function, proposed in Langmead and Donald [79], employs the *mean tensor consistency* computed from an ensemble of subsets of assignments against the structural model. In principle, both methods allow a *parametric family* of structural models to be explored about the putative homolog, enabling NVR to bootstrap assignments analogously to the role of MR, starting with a structural homolog in crystallography. Indeed, traditional (X-ray) MR had previously been extended to solve phase problems from more distant models by using normal mode analysis (NMA) [138, 137]. Analogously, one can perform a geometric exploration of conformation space to generate an ensemble of neighboring structures to the initial homolog using NMA [8].

NMA can be performed as described traditionally [58, 17] or with a Gaussian network model [141]. It would also be valuable to compare and geometrically simulate diffusive motions to sample the internal mobility of the protein using alternative techniques such as FRODA, FIRST, and ROCK [65, 173, 161], that are based on graph rigidity theory. The structures that maximize the likelihood of assignment accuracy can then be used for NVR [8]. An allied idea was used earlier by Baker and coworkers for decoy detection in the context of pruning structure predictions by NMR data [97]. NVR-NMA ensembles [8] are different in that they systematically represent the deformation space for equilibrium protein motions, rather than a family that mixes correct and incorrect structures. It appears that as in Suhre and Sanejouand [138], the NMA-like approaches above can adequately sample the necessary convolution of structural variation, noise, and dynamics around the target structure [8].

We now describe a number of key technical and algorithmic features of NVR. NVR uses a variation of the EM algorithm [79], which is a statistical method for computing the maximum likelihood estimates of parameters for a generative model. EM has been a popular technique in a number of different fields, including machine learning and image understanding. It has been applied to bipartite matching problems in computer vision [30]. Bipartite matching is extremely sensitive to noise in the edge weights. There is evidence that sophisticated algorithms such as EM or NVR must be employed if bipartite matching is to be used as a subroutine on data that is sparse or noisy [79, 8]. In the EM framework there are both observed and hidden (i.e., unobserved) random variables. In the context of structure-based assignment, the observed variables are the chemical shifts, unassigned d_{NN}s, amide exchange rates, RDCs, and the 3D structure of the target protein.

Let X be the set of observed variables. The hidden variables $Y = Y_G \cup Y_S$ are the true (i.e., correct) resonance assignments Y_G, and Y_S, the correct, or "true" alignment tensors. Of course, the values of the hidden variables are unknown. Specifically, Y_G is the set of edge weights of a bipartite graph, $G = \{K, R, K \times R\}$, where K is the set of peaks in the HSQC and R is the set of residues in the protein. The weights Y_G represent *correct* assignments, and therefore encode a perfect matching in G. Hence, for each peak $k \in K$ (respectively, residue $r \in R$), exactly one edge weight from k (respectively r) is 1 and the rest are 0.

The probabilities on all variables in Y are parameterized by the "model," which is the set Θ of all assignments made so far by the algorithm. Initially, Θ is empty. As EM makes more assignments, Θ grows, and both the probabilities on the edge weights Y_G and the probabilities on the alignment tensor values Y_S will change. The goal of the EM algorithm is to estimate Y accurately to discover the correct edge weights Y_G, thereby computing the correct assignments. The EM algorithm has two steps; the Expectation (E) step and the Maximization (M) step. The E step computes the expectation $E(\Theta \cup \Theta'|\Theta) = E(\log \mathbf{P}(X, Y|\Theta \cup \Theta'))$. Here, Θ' is a nonempty set of candidate new assignments that is disjoint from Θ. The M step computes the maximum likelihood new assignments $\Theta^* = \underset{\Theta'}{\operatorname{argmax}} E(\Theta \cup \Theta'|\Theta)$. Then the master list of assignments is updated, $\Theta \leftarrow \Theta \cup \Theta^*$.

The alignment tensors are recomputed at the end of each iteration, using all the assignments in Θ. Thus, the tensor estimates will be continually refined during the run of the algorithm. The algorithm terminates when each peak has been assigned.

Care must be taken to implement the probabilistic EM framework efficiently (for details see [79]). In brief, individual bipartite graphs are constructed for each of the 7 sources of data (see figure 17.3). NVR operates on bipartite graphs between peaks and residues. The edge weights from each peak to all residues form a probability distribution. The probabilities are derived from (1) "inside/outside" experiments (H-D exchange, paramagnetic quenching), (2) d_{NN}s, (3) chemical shift predictions based on average chemical shifts from the BMRB [132], (4) chemical shift predictions made by the program SHIFTS [169], (5) chemical shift predictions made by the program SHIFTX [101], and (6–7) constraints from RDCs in two media.

Data (1–7) above are converted into constraints on assignment. The difference between the experimentally determined chemical shifts and the set of predicted chemical shifts are converted into assignment probabilities. Let $\mathbf{S}_1$ and $\mathbf{S}_2$ be the alignment tensors computed from *unassigned* data as described in section 17.2 and other studies [79, 82, 83, 78]. The amide bond vectors from the structural model are used to back-compute a set of RDCs using equation (16.1). The difference between each back-computed RDC and each experimentally recorded RDC is converted into a probability on assignment. Let D_m be the set of observed RDCs in medium m, and F_m be the set of back-computed RDCs using the model and $\mathbf{S}_m$. Two bipartite graphs M_1 and M_2 are constructed on the peaks in K and residues in R. The edge weights are computed as probabilities as follows: $w(k, r) = \mathbf{P}(k \mapsto r|S_m) = g(k, r)$ where $k \in K$ and $r \in R$. Here, $g(k, r) = \mathcal{N}(d_m(k) - b_m(r), \sigma_m)$ where $d_m(k) \in D_m, b_m(r) \in F_m$. The probabilities are computed using a 1-dimensional Gaussian distribution $\mathcal{N}(x - \mu, \sigma) = \frac{1}{\sigma\sqrt{2\pi}} \exp\left(-\frac{(x-\mu)^2}{2\sigma^2}\right)$, with mean $d_m(k) - b_m(r)$ and standard deviation σ_m. Langmead et al. [79] used $\sigma = L/8$ Hz in their experiments (section 17.2), where L is the range of the RDCs in that medium (the maximum-valued RDC minus the minimum valued RDC). If an RDC is missing in medium i for a peak k, then we set the weight $w(k, r) = 1/n_0$ in graph M_i, for each residue r of the n_0 remaining (i.e., *unassigned*) residues. The details of how the chemical shifts, d_{NN}s, and amide exchange rates are converted into constraints on assignment are analogous (see [79]).

NVR uses a unique voting scheme for combining multiple sources of information [8, 79]. For each possible combination of data (1–7) above, a *combined graph* is constructed whose edge weights are the joint probabilities of the edges in the single-spectra graphs. NVR then constructs a new bipartite graph V, called the *expectation graph*, whose edge weights are initialized to 0. For each combined graph c, we compute a maximum bipartite matching. Let $H_c \subset E$ be the matching computed on c. For each edge $e \in H_c$ NVR increments the weight on the same edge in the expectation graph, V, by 1. This is done for all combined graphs. Hence, each bipartite matching "votes" for a set of edges. Thus, edge weights in V record the number of times a particular edge was part of a maximum bipartite matching. Note that the edge weights are probabilities in the bipartite graphs. Thus, a bipartite matching gives a maximum likelihood solution on that graph, which in turn maximizes the expected log-likelihood of the average edge weight.

Let w_0 be the largest edge weight in V. The M step is computed, and assignments are made by $\Theta \leftarrow \Theta \cup w^{-1}(w_0)$. As previously stated, each of the constituent votes used to construct V is a maximum likelihood solution. Hence, it maximizes the expected edge weight in the matching. Therefore, in the bipartite graph V, those edges with maximum votes have the highest expected values over all combinations of the data, thus correctly computing the maximum likelihood new assignments Θ^* as the argmax in the M-Step described earlier.

When an assignment is made, the associated nodes $k \in K$ and $r \in R$ are removed from B_{NOE}, the NOE graph. The geometric NOE restraints are applied to prune matching. The three graphs for chemical shift prediction are synchronized with B_{NOE} so that all graphs have the same set of zero-weight edges. Each graph is then renormalized so that the edge weights from each peak form a probability distribution. This completes a single iteration of the EM algorithm.

Recently, a new version of the GD module of NVR, called HD, improved the accuracy of NVR/GD [80]. In particular, HD eliminates the false-positives evident in figure 17.5, even against a larger database of 4,500 folds (versus only 2,456 in GD [78]; see figure 18.1 and [80]). Furthermore, it has been found that combining HD with NMA ensembles (described above) can boost the assignment accuracy using distant structural homologs (3-7 Å backbone RMSD) by up to 22% [8].

The advantage of NVR over maximum bipartite matching lies in its iterative nature. NVR takes a conservative approach, making only likely assignments, given the current information. After making these assignments, the edge weights between the remaining unassigned peaks and residues are updated. Suppose that, during the ith iteration of the algorithm, peak k is assigned to residue r. The edge weight between peak k and residue r is then set to 1, indicating the certainty of that assignment. The edge weights form a probability distribution. Accordingly, the edge weight between peak k and any other residue is set to 0. Similarly, the weights on the edges from

(a)

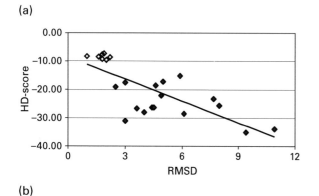

(b)

(a)	Homolog[b]	Sequence Identity (%)[c]	RMSD[d] (Å)	HD Score[e]
(i)	1H8C	15.7	1.9	-7.65
	1RFA	15.9	2.2	-8.69
	1VCB:B	11.8	1	-8.31
	1EF1:A [4-84]	10	1.6	-8.50
(ii)	1DK8:A	28.7	1.8	-7.18
(iii)	1JML:A	12.8	1.8	-9.27
	1HEZ:E	12.7	2.0	-9.65

(c)

(a)	Homolog[b]	Sequence Identity (%)[c]	RMSD[d] (Å)	HD Score[e]
(i)	1C9F:A	12.1	3	-17.52
	1XGM:A	6.1	4.6	-18.52
	1ESR:A	0	5.9	-15.21
(ii)	1VLK	8.2	3.6	-26.65
	1I4Y:D	4.8	4	-28.06
	1SWG:A	0	4.5	-26.38
	1B33:D	5.8	4.9	-22.08
	1CFC	5.1	6.1	-28.48
	1J95:A	0	7.7	-23.37
	1IDR:B	5.2	8	-25.61
	1E8E:A	7.1	9.4	-35.18
	1IDR:A	5.2	10.9	-33.90
(iii)	1EX4:B	8.7	2.5	-19.04
	1EXQ:B	6.2	3	-31.04
	1MPG:A	7.3	4.4	-26.31
	1DH3:A	0	5	-17.27

Figure 18.1

Improved fold detection from unassigned NMR data [80]: (a) score computed by HD versus backbone RMSD (Å). The line is a least-squares fit to the data. The correlation coefficient is -0.75. Open diamonds are homologous structures with low sequence similarity, detected by HD score ≥ -10 (table b). Solid diamonds are structures classified as nonhomologs by HD score < -10 (table c). Tables b-c: (a) 3 test proteins: (i) ubiquitin, (ii) GαIP, (iii) SPG. [b]PDB/chain Id for the structures detected by HD. [c]Sequence identity of the 3 test proteins[(a)] vs. the primary sequence of the structure.[b] [d]Backbone RMSD between the structures[b] and the native structures of the test proteins.[(a)] [e]Score computed by the algorithm HD. Higher HD scores (closer to 0) indicate closer structural similarity. Table (B) does not include those structures detected by HD that have > 30% sequence identity. Copyright 2004 IEEE.

any *other* peak to r are immediately set to 0. The (nonzero) edge weights from each remaining unassigned peak are renormalized prior to the next iteration. Thus, a peak whose assignment may be ambiguous in iteration i may become unambiguous in iteration $i + 1$.

Somewhat surprisingly, results of perturbation studies [79, pp. 120–130] suggest that NVR is not very sensitive to the quality of the initial tensor estimates, because the additional, non-RDC lines of evidence (chemical shift prediction, amide exchange, d_{NN}'s) can overcome these inaccuracies. The NVR voting algorithm (see above and [79, 8]) used to integrate different lines of evidence is essentially a means to increase the signal-to-noise ratio. Here, the signal is the computed likelihood of the assignment between a peak and the (correct) residue. The noise is the uncertainty in the data, where the probability mass is distributed among multiple residues. Each line of evidence (i.e., experiment) has noise, but the noise tends to be random and thus cancels when the lines of evidence are combined. Conversely, the signals embedded in each line of evidence tend to reinforce each other, resulting in (relatively) unambiguous assignments. Hence, even if the two initial tensor estimates are poor, it is unlikely that they can conspire (by voting together) to force an incorrect assignment. More generally, given the NVR voting scheme (above, and [79, 8]), any pair of lines of evidence is unlikely to outvote the majority. Finally, the iterative update (described above) by the alignment tensors at each cycle of the EM, allows the tensor accuracy to improve as more assignments are made [79].

18.1.3 Automated NOE Assignment

TRIANGLE is discussed for illustrative purposes; the interested reader is referred to HANA [174] for a more sophisticated algorithm with the same flavor. The TRIANGLE [153] and HANA [174] algorithms for automated NOE assignment use a rotamer library ensemble and RDCs. Both algorithms employ a rotamer-library–derived ensemble of intra- and interresidue vectors between the backbone atoms and side-chain protons to reduce the ambiguity in the assignment of NOE restraints involving side-chain protons, especially aliphatic protons. For example, consider the putative assignment of an NOE to an (H^N, H^δ) (respectively, (H^α, H^δ)) pair of protons. The *triangle relationship* (see figure 17.5L) defines a decision procedure to filter NOE assignments by *fusing* information from RDCs ($\mathbf{v}_{N\beta}$), rotamer modeling ($\mathbf{v}_{\beta\delta}$), and NOEs ($|\mathbf{v}_{N\delta}|$). Specifically,

Proposition 18.3 [153]   *Given the vector $\mathbf{v}_{\beta\delta}$ from a C^β atom to a side-chain proton S (e.g., H^δ), and the vector $\mathbf{v}_{N\beta}$ from the backbone H^N to the C^β (respectively, the vector $\mathbf{v}_{\alpha\beta}$ from the H^α atom to C^β), the vector $\mathbf{v}_{NS}$ from the H^N to S (respectively, the vector $\mathbf{v}_{\alpha S}$ from the H^α to S) can be computed by: $\mathbf{v}_{NS} = \mathbf{v}_{N\beta} + \mathbf{v}_{\beta\delta}$, $\mathbf{v}_{\alpha S} = \mathbf{v}_{\alpha\beta} + \mathbf{v}_{\beta\delta}$.*

$\mathbf{v}_{N\beta}$, and $\mathbf{v}_{\alpha\beta}$ are computed directly from the backbone structure determined by RDC-EXACT and can be inter- or intraresidue. The vectors $\mathbf{v}_{NS}$, $\mathbf{v}_{N\beta}$, and $\mathbf{v}_{\beta\delta}$ form three sides of a triangle (figure 17.5), where $\mathbf{v}_{N\beta}$ is computed from the backbone structure and $\mathbf{v}_{\beta\delta}$ lies in a finite set of vectors, mined from the PDB rotamers. Hence, for each side-chain proton, $\mathbf{v}_{\beta\delta}$ lies in a finite set of vectors. Consequently, there is a finite set V_N (respectively V_α) of vectors containing $\mathbf{v}_{NS}$ (respectively $\mathbf{v}_{\alpha S}$). The triangle relationship (Proposition 18.1.3) is then used to filter the assignments of NOE peaks involving side-chain protons. The algorithm first estimates the NOE distances, d_N and d_α, using the NOE intensities from, respectively, the experimental NOE peaks picked from the 3D ^{15}N- and ^{13}C-NOESY spectra. TRIANGLE tests d_N and d_α to see whether they satisfy the constraints: $\min |\mathbf{v}_{NS}| - \varepsilon_N \le d_N \le \max |\mathbf{v}_{NS}| + \varepsilon_N$, $\min |\mathbf{v}_{\alpha S}| - \varepsilon_\alpha \le d_\alpha \le \min |\mathbf{v}_{\alpha S}| + \varepsilon_\alpha$, where min and max for $\mathbf{v}_{NS}$ (respectively, $\mathbf{v}_{\alpha S}$) are both computed over $\mathbf{v}_{NS} \in V_N$ (respectively, $\mathbf{v}_{\alpha S} \in V_\alpha$). ε_N and ε_α are the NOE distance error bounds for the corresponding 3D ^{15}N- (respectively, ^{13}C-) edited NOESY peaks.

Tests of TRIANGLE and HANA are described in section 17.4 and other studies [153, 174, 177]. In the future, it will be important to quantify any possible bias introduced by the rotamer library, using both rotamer subsampling and rotamer energy minimization (generalizing prior work on rotamer optimization [88, 45, 44]), and comparison to conventional assignments. For example, one can use a subsampled rotamer library that is more than 6,000 times finer than the library in Wang and Donald [153]. In preliminary tests on ubiquitin NOE spectra, the finer library resulted in modest improvements in assignment accuracy (about 5%) at a considerable computational cost (2 hours versus 1 second). This suggests that the optimal number of rotamers should lie in between.

Further study is required of the effect of spin diffusion on cross-peak intensity and the NOE distance $|\mathbf{v}_{N\delta}|$ in the triangle relationship (figure 17.5L) for assignment pruning. An extension of TRIANGLE has been used with RDC-EXACT to assign the NOEs and compute structures, including side-chains [174]. It will be valuable to extend TRIANGLE to account for the various labeling patterns necessary in large proteins to solve the structures of such large systems. These labeling strategies can sometimes vitiate the HCCH-TOCSY [108] for side-chain resonance assignments. One can extend TRIANGLE to assign NOEs involving these ambiguous proton resonances, by replacing their side-chain assignment with a *set* of possible labels (analogous to [1]). The labels, which represent competing assignments for the indirect ^{1}H dimension, can be pruned using Proposition 18.1.3. The resulting NOESY assignments from TRIANGLE would provide side-chain resonance assignments as a byproduct. This extension to TRIANGLE would not only reduce the dependence of NOE assignment algorithms upon the (less sensitive) TOCSY experiments, but also increase their tolerance of incomplete side-chain resonance assignments.

Side-chain/side-chain NOEs may be assigned by "lifting" TRIANGLE's rotamer-based ensemble of internuclear vectors (via Cartesian product). 4D NOESY experiments can expedite TRIANGLE's computation of these assignments. A discussion of the information content of 4D NOESY versus selective labeling is therefore warranted. TRIANGLE and its successor, HANA [174], employ a "symmetry check" [60] to increase assignment confidence using "back" NOEs. Consider a putative assignment for an NOE, ^{15}N-H^N ↔ ^{1}H, where $\omega(^1$H) is assigned to ^{1}H$^\delta$ of F16. If the assignment is correct, then we should see a symmetric NOE in the ^{13}C-NOESY: H^N ↔ ^{1}H$^\delta$-^{13}C$^\delta$ of F16. Finding this peak requires searching a box in the 3D ^{13}C-NOESY, about the (H^N, ^{1}H, ^{13}C$^\delta$) shifts. Since the "back" NOE might be missing and there might be other peaks in this box, the symmetry check must be treated inferentially, as changing the Bayesian probability of the assignment. Using a 4D NOESY reduces this ambiguity, and in principle the symmetric interaction will always be observed. As a pruning condition, back NOEs and 4D NOEs complement the Triangle Relationship (Proposition 18.1.3) by integrating the geometric and spectral constraint of the additional observed ^{13}C$^\delta$. However, under some isotopic labeling schemes, 4D NOEs to protons bonded to ^{12}C will not be observed, and the previously described set-of-labels strategy must be employed. Finally, TRIANGLE and HANA should be helpful to assist with proposed experimental methods for 24–65 kDa proteins that currently use manual 4D NOE assignment (e.g., [170]), which could then be automated, saving time and reducing subjectivity.

18.1.4   NMR Structure Determination of Symmetric Oligomers

Many experimental challenges exist for determining the structure of membrane protein complexes. Current algorithms in this area, including SYMBRANE (section 17.5) do not address all of these problems, and make a number of assumptions. Most important, it can sometimes be assumed that the overall fold of a subunit can be obtained from RDCs and other NMR measurements in the holo state, and that intersubunit NOEs can be distinguished from intrasubunit NOEs using iso-tope filtering, as was done for phospholamban [111, 117]. In particular, the automated ambiguous intersubunit NOE assignment in SYMBRANE relies on these assumptions [118].

One wishes to extend and relax these assumptions; although a long-term goal is to develop algorithms for general membrane proteins, this chapter focuses next on the challenges of sym-metric oligomers, which include many biomedically important systems. Moreover, RDCs can play a crucial role for such complexes, not only in determining the conformation and orienta-tion of the individual subunits, but also in computing the symmetry axis. The overall scheme of SYMBRANE is shown in figure 18.2 (plate 22). In contrast, traditional protocols [102] for structure determination of protein complexes from NMR data use simulated annealing (SA) and molec-ular dynamics (MD). However, the SA/MD mechanism is not complete and could get trapped in local minima. Furthermore, the precision of the determined structure is strongly affected by the temperature. SYMBRANE avoids temperature dependence and local minima problems, and does not suffer from "false precision" in characterizing the diversity of determined structures. SYMBRANE's search in the symmetry configuration space (SCS) takes advantage of the "closed-ring" constraint of a symmetric homo-oligomer. AMBIPACK [150] used a branch-and-bound algorithm to compute rigid body transformations satisfying potentially ambiguous intersubunit distance restraints. In contrast, SYMBRANE uses the oligomeric number to enforce an a priori symmetry constraint. In this sense it is analogous to the manner in which noncrystallographic symmetry is handled in molecular replacement for X-ray crystallography [89]. By formulating the structure determination problem in the SCS rather than in the space of atom positions, SYM-BRANE is able to exploit directly the kinematics of the "closed-ring" constraint, and thereby derive an *analytical* bound for pruning, which is tighter and more accurate than previous randomized numerical techniques [150]. SYMBRANE's guarantees of completeness should be especially impor-tant for membrane proteins (where sparse data is a particularly important problem), but should also prove useful for water-soluble oligomers.

SYMBRANE simultaneously assigns NOEs and determines structure (by identifying regions in SCS), and is guaranteed to find all consistent assignments and structures [118]. Let R be a set of (possibly ambiguous) NOEs. Let r_{kl} represent the lth possible assignment of the kth NOE r_k of R, and let $s_{kl} \subset S^2 \times \mathbb{R}^2$ (see figure 18.2a) be the region of the SCS in which distance restraint r_{kl} is satisfied. An *assignment* specifies the subunit and the protons involved in the NOE. A determined structure must satisfy all ambiguous NOEs; it satisfies each by satisfying one or more of its possible assignments. In the SCS, this translates into finding the region

$$p(R) = (s_{11} \cup s_{12} \cup s_{13} \cup \ldots) \cap (s_{21} \cup s_{22} \cup \ldots) \cap \ldots = \bigcap_{k=1}^{\|R\|} \left(\bigcup_{l=1}^{t_k} s_{kl} \right), \tag{18.13}$$

where t_k is the number of possible assignments for r_k, the kth NOE.

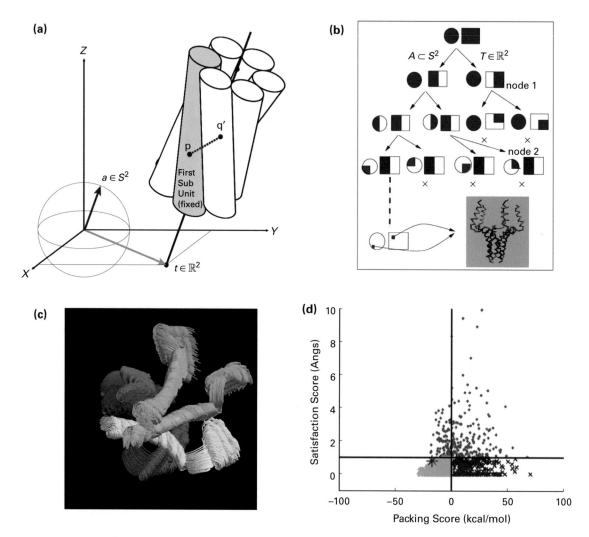

Figure 18.2 (plate 22)

SYMBRANE Algorithm. (a) Given the subunit structure, the relative position $\mathbf{t} \in \mathbb{R}^2$ and orientation $\mathbf{a} \in S^2$ of the symmetry axis uniquely determines structure. Structure determination is a search problem in 4-dimensional *symmetry configuration space (SCS)*, $S^2 \times \mathbb{R}^2$. An NOE is shown between protons $\mathbf{p}$ and $\mathbf{q}'$ [117]. (b) The branch-and-bound algorithm proceeds as a tree search in SCS. The 4D SCS is represented as two 2D regions, a sphere representing the orientation space S^2 and a square representing the translation space $\mathbb{R}^2$. The dark shaded regions at each node of the tree represent the region in SCS being explored ($A \times T \subset S^2 \times \mathbb{R}^2$). Ultimately (bottom left of the tree), the branch-and-bound search returns regions in 4D space representative of structures that possibly satisfy all the restraints. At each node, we test satisfaction of each restraint of form $\mathbf{pq}' \leq d$ by testing intersection between the ball of radius d centered at $\mathbf{p}$ and the convex hull bounding possible positions of $\mathbf{q}'$. If there exists an intersection between the ball and the convex hull for each restraint, further branching is done (node 1); otherwise, the entire node and its subtree are pruned (node 2) [117]. (c) Representative results: complete ensemble of NMR structures of the unphosphorylated human phospholamban pentamer, PDB id: 2HYN. (d) Phospholamban restraint satisfaction score vs. packing score for all structures. The vertical and horizontal lines indicate the cutoffs for WPS structures: 1 Å for the satisfaction score and 0 kcal/mol for the packing score. The green stars and the blue crosses indicate the set of satisfying structures. The magenta points indicate the set of nonsatisfying structures that have been pruned. The green stars indicate the set of WPS structures and the red star indicates the reference structure [117].

Using equation (18.13), the algorithm computes *redundant* and *inconsistent* NOEs and assignments, via unions and intersections of geometric sets in SCS. Let $Q \subset R$ be a set of NOEs. Let r_k be an (arbitrary) ambiguous NOE (not in Q) with t_k assignments. We compute that NOE assignment r_{kl} is *redundant with respect to Q* if and only if $p(Q) \subset s_{kl}$. Similarly, we compute that r_{kl} is *inconsistent with respect to Q* if and only if $p(Q) \cap s_{kl} = \emptyset$, where $\emptyset$ is the empty set. We can extend these algorithms from assignments to NOEs as follows. Let $S_k = \bigcup_{l=1}^{t_k} s_{kl}$ be the region of the SCS that satisfies at least one assignment of r_k. We then compute that r_k is *redundant with respect to Q* if and only if $p(Q) \subset S_k$, and compute that r_k is *inconsistent with respect to Q* if and only if $p(Q) \cap S_k = \emptyset$. SYMBRANE eliminates inconsistent NOEs and assignments, and is able to detect redundant ones. SYMBRANE's approach to identifying the mutually consistent, complete set of NOE assignments and SCS regions is summarized in figure 18.3 and described in detail in Potluri et al. [118].

When RDCs can be obtained in combination with NOEs, structure determination of symmetric homooligomers reduces to an inverse kinematics problem that can be solved *exactly* (analogously to the procedures in section 17.1), without resorting to the branch-and-bound search used in studies by Potluri et al. [117, 118]. It is likely that ambiguous NOE assignment [118] will also be improved since the ensemble of exact solutions will be smaller, increasing the structural precision that geometrically prunes the assignments. SYMBRANE currently allows for limited uncertainty in the subunit structure: SYMBRANE computes satisfying structures using rigid monomers, but energy scores allowing side-chain minimization [117]. This allows for some side-chain flexibility, but is subject to local minima and is not systematic. Instead, one could use a rotamer library to systematically sample different distinct conformational side-chain states, exploiting recent advances in rotamer optimization [88, 45, 44], particularly in the presence of provable rotamer energy minimization [45, 44]. Considering rotamers increases the complexity of the hierarchical subdivision scheme, but techniques such as dead-end elimination (DEE) [34, 85, 48, 115, 44] and minimized-DEE [45, 44] can be used to prune rotamers and compute solutions more quickly. It may also be possible to derive exact or analytic solutions for NOE assignment when RDCs are available. While more difficult, backbone flexibility could also be incorporated in a DEE-like search [42, 43]. Allowing greater uncertainty in the subunit structure would make SYMBRANE more robust and eventually useful for multimers with a more intricate folding topology.

Two novel features of SYMBRANE are of particular note: (1) the ability to assign intersubunit NOEs in a complete search that avoids exhaustive, exponential enumeration [118]; and (2) the ability to determine the oligomeric number using NMR data alone [117]. Specifically:

1. Errors in resolving intersubunit NOE ambiguities have led to incorrect NMR structures [20]. Beyond the importance of this particular problem in NMR data analysis, SYMBRANE represents a departure from the stochastic methods frequently employed by the NMR community. A corollary is that such stochastic methods, now routinely employed, should be reconsidered in light of their inability to ensure identification of the unique or globally optimal structural models consistent with a set of NMR observations. SYMBRANE can provide more confident answers to these problems.

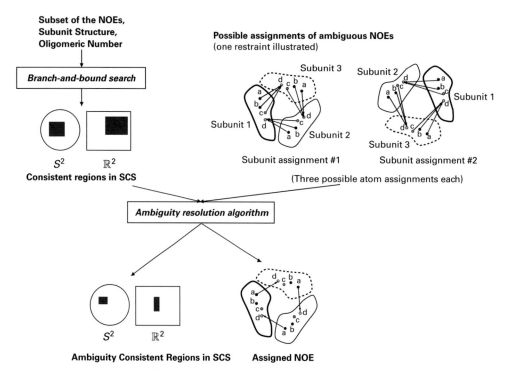

Figure 18.3

SYMBRANE strategy to resolve ambiguous NOEs in structure determination of symmetric homo-oligomers. Top: Different possible assignments for one ambiguous NOE are illustrated for a trimer. The NOE could be ordered between subunits as 1-2-3 or 1-3-2, and has chemical shift degeneracy between protons *a*, *b*, and *c*. The former is called "subunit ambiguity" and the latter "atom ambiguity." *Phase 1*: Consistent regions in the SCS are obtained using a branch-and-bound algorithm [117], taking a subset of available NOEs (typically atom-unambiguous NOEs, if any) as input. The 4-dimensional SCS is cartoon-represented as in figure 18.2b. *Phase 2:* SYMBRANE uses the consistent regions output from the branch-and-bound search plus possible assignments of the ambiguous NOEs, to determine a mutually consistent, complete set of NOE assignments and ambiguity consistent regions (ACR) in SCS. On p. 177: Pseudocode for SYMBRANE's ambiguity resolution algorithm. Each NOE's score is the average of the scores of all of its possible assignments (the line marked (*), bottom). This score provides an ordering criterion for NOE assignment. Potluri et al. [118] showed that the order does not affect the completeness, results, or accuracy, but can improve the efficiency.

2. In general, the *oligomeric state* of the complex includes the *symmetry type* (C_n, D_n, etc.) as well as the *oligomeric number* ($n = 2, 3, 4, \ldots$). In addition to assigning the intermolecular NOEs and determining the complex structure, SYMBRANE tests whether the available data suffices to determine the oligomeric number for C_n symmetry. For each possible symmetry type and oligomeric number, SYMBRANE can determine a set Y of WPS structures. For some oligomeric states, $Y = \emptyset$: we can place higher confidence on the oligomeric state that (a) *has some* WPS structures (i.e., $Y \neq \emptyset$), and (b) affords better vdW packing. Thus, we can determine the oligomeric state using the NMR data and vdW packing. SYMBRANE provides for an independent verification of the oligomeric state (which is typically determined using experiments such as chemical cross-linking

Input:
R : Set of NOEs
$U \subset R$: Subset of the NOEs (typically atom-unambiguous NOEs)
C : Set of cells from branch-and-bound with U
$N : R \to \mathbb{Z}_{n-1} \times A$: Possible assignments for each NOE
    due to subunit ambiguity (subunit offset 1 to $n - 1$)
    and atom ambiguity (proton in the set A).

Output:
$C' \subset C$: Remaining cells
$N' \subset N$: Remaining assignments

Algorithm:
initialize $R' \leftarrow R - U$, $C' \leftarrow C$
while $R' \neq \emptyset$
    // Determine the next NOE to assign according to our criterion
    $S \leftarrow$ representative structures for C'
    $q \leftarrow \underset{r \in R'\ a \in N(r)}{\arg\max\ \text{average}} |\{s \in S \mid r \text{ under assignment } a \text{ is violated in } s\}|$ (*)
    $R' \leftarrow R' - \{q\}$

    // Remove inconsistent cells
    $C' \leftarrow \{c \in C' \mid \exists a \in N(q) \text{ s.t. } c \text{ is consistent with } q \text{ under assignment } a\}$
end while

// Remove inconsistent assignments
$N' \leftarrow \{(r, a) \mid r \in R, a \in N(r), \exists c \in C' \text{ s.t. } c \text{ is consistent with } r \text{ under assignment } a\}$

return N', C'

Figure 18.3
(Continued)

followed by SDS-PAGE, or by equilibrium sedimentation). The results from SYMBRANE show how novel algorithms can draw conclusions from NMR data that were previously possible only using additional biophysical experiments. Such algorithms can reduce experimental time and provide cross-validation of structure and assembly.

References

[1] A. B. Eiso, D. J. R. Pugh, R. Kaptein, R. Boelens, and A. M. J. J. Bonvin. Direct use of unassigned resonances in NMR structure calculations with proxy residues. *Journal of the American Chemical Society*, 128;23(2006):7566–7571.

[2] H. M. Al-Hashimi, P. J. Bolon, and J. H. Prestegard. Molecular symmetry as an aid to geometry determination in ligand protein complexes. *Journal of Magnetic Resonance*, 142(2000):153–158.

[3] H. M. Al-Hashimi, A. Gorin, A. Majumdar, Y. Gosser, and D. J. Patel. Towards structural genomics of RNA: Rapid NMR resonance assignment and simultaneous RNA tertiary structure determination using residual dipolar couplings. *Journal of Molecular Biology*, 318(2002):637–649.

[4] H. M. Al-Hashimi and D. J. Patel. Residual dipolar couplings: Synergy between NMR and structural genomics. *Journal of Biomolecular NMR* 22;1(2002):1–8.

[5] A. C. Anderson. Two crystal structures of dihydrofolate reductase-thymidylate synthase from Cryptosporidium hominis reveal protein-ligand interactions including a structural basis for observed antifolate resistance. *Acta Crystallographica, Section F, Structural Biology and Crystallization Communications* 61;3(2005):258–262.

[6] M. Andrec, P. Du, and R. M. Levy. Protein backbone structure determination using only residual dipolar couplings from one ordering medium. *Journal of Biomolecular NMR* 21;4(2001):335–347.

[7] A. Annila, H. Aitio, E. Thulin, and T. Drakenberg. Recognition of protein folds via dipolar couplings. *Journal of Biomolecular NMR* 14(1999):223–230.

[8] S. Apaydin, V. Conitzer, and B. R. Donald. Structure-based protein NMR assignments using native structural ensembles. *Journal of Biomolecular NMR* 40(2008):263–276.

[9] P. J. Artymiuk, C. C. F. Blake, D. W. Rice, and K. S. Wilson. The structures of the monoclinic and orthorhombic forms of hen egg-white lysozyme at 6 angstroms resolution. *Acta Crystallographica, Section B Biology Crystallographica* 38(1982):778.

[10] C. R. Babu, P. F. Flynn, and A. J. Wand. Validation of protein structure from preparations of encapsulated proteins dissolved in low viscosity fluids. *Journal of the American Chemical Society*, 123(2001):2691.

[11] C. Bailey-Kellogg, J. J. Kelley III, R. Lilien, and B. R. Donald. Physical geometric algorithms for structural molecular biology; in *the Special Session on Computational Biology & Chemistry, Proceedings of the IEEE International Conference on Robotics and Automation (ICRA-2001)*, May 2001, pp. 940–947.

[12] C. Bailey-Kellogg, A. Widge, J. J. Kelley III, M. J. Berardi, J. H. Bushweller, and B. R. Donald. The NOESY Jigsaw: Automated protein secondary structure and main-chain assignment from sparse, unassigned NMR data. *Journal of Computational Biology* 3-4;7(2000):537–558.

[13] B. E. Coggins and P. Zhou. Sampling of the NMR time domain along concentric rings. *Journal of Magnetic Resonance* 184;2(2007):207–221.

[14] B. Berger, J. Kleinberg, and F. T. Leighton. Reconstructing a three-dimensional model with arbitrary errors. *Journal of the ACM*, 46;2(1999):212–235.

[15] A. T. Brünger. *XPLOR: A System for X-ray Crystallography and NMR*. New Haven, CT: Yale University Press, 1993.

[16] J. Cavanaugh, W. J. Fairbrother, A. G. Palmer III, and N. J. Skelton. *Protein NMR Spectroscopy: Principles and Practice*. New York: Academic Press, 1995.

[17] C. N. Cavasotto, J. A. Kovacs, and R. A. Abagyan. Representing receptor flexibility in ligand docking through relevant normal modes. *Journal of the American Chemical Society* 127;26(2005):9632–9640.

[18] Y. Chen, J. Reizer, M. H. Saier Jr., W. J. Fairbrother, and P. E. Wright. Mapping of the binding interfaces of the proteins of the bacterial phosphotransferase system, HPr and IIAglc. *Biochemistry* 32;1(1993):32–37.

[19] J. J Chou, S. Li, and A. Bax. Study of conformational rearrangement and refinement of structural homology models by the use of heteronuclear dipolar couplings. *Journal of Biomolecular NMR* 18(2000):217–227.

[20] G. M. Clore, J. G. Omichinski, K. Sakaguchi, N. Zambrano, H. Sakamoto, E. Appella, and A. M. Gronenborn. Interhelical angles in the solution structure of the oligomerization domain of p53: Correction. *Science* 267;5203(1995): 1515–1516.

[21] B. E. Coggins, R. A. Venters, and P. Zhou. Filtered backprojection for the reconstruction of a high-resolution (4,2)D CH_3-NH NOESY spectrum on a 29 kDa protein. *Journal of the American Chemical Society* 127;33(2005):11562–11563.

[22] B. E. Coggins and P. Zhou. PACES: Protein sequential assignment by computer-assisted exhaustive search. *Journal of Biomolecular NMR* 26;2(2003):93–111.

[23] B. E. Coggins and P. Zhou. Polar Fourier transforms of radially sampled NMR data. *Journal of Magnetic Resonance* 182;1(2006):84–95.

[24] F. Cordier, M. Rogowski, S. Grzesiek, and A. Bax. Observation of through-hydrogen-bond $^{2h}J_{HC'}$ in a perdeuterated protein. *Journal of Magnetic Resonance* 40;2(1999):510–512.

[25] G. Cornilescu, F. Delaglio, and A. Bax. Protein backbone angle restraints from searching a database for chemical shift and sequence homology. *Journal of Biomolecular NMR* 13(1999):289–302.

[26] E. A. Coutsias, C. Seok, M. P. Jacobson, and K. A. Dill. A kinematic view of loop closure. *Journal of Computational Chemistry* 25;4(2004):510–528.

[27] E. A. Coutsias, C. Seok, M. J. Wester, and K. A. Dill. Resultants and loop closure. *International Journal of Quantum Chemistry* 106;1(2006):176–189.

[28] G. Crippen. Chemical distance geometry: Current realization and future projection. *Journal of Mathematical Chemistry* 6(1991):307–324.

[29] G. M. Crippen and T. F. Havel. *Distance Geometry and Molecular Conformations*. New York: John Wiley, 1988.

[30] A. D. J. Cross and E. R. Hancock. Graph matching with a dual-step EM algorithm. *IEEE Transactions on Pattern Analysis and Machine Intelligence*, 20;11(1998):1236–1253.

[31] F. del Rio-Portílla, V. Blechta, and R. Freeman. Measurement of poorly-resolved splittings by J-doubling in the frequency domain. *Journal of Magnetic Resonance* 111a(1994):132–135.

[32] F. Delaglio, G. Kontaxis, and A. Bax. Protein structure determination using molecular fragment replacement and NMR dipolar couplings. *Journal of the American Chemical Society*, 122;9(2000):2142–2143.

[33] A. Dempster, N. Laird, and D. Rubin. Maximum likelihood from incomplete data via the EM algorithm. *Journal of the Royal Statistical Society, Series B*, 39;1(1977):1–38.

[34] J. Desmet, M. DeMaeyer, B. Hazes, and I. Lasters. The dead-end elimination theorem and its use in protein side-chain positioning. *Nature*, 356(1992):539–542.

[35] R. Diamond. Real-space refinement of the structure of hen egg-white lysozyme. *Journal of Molecular Biology*, 82(1974):371–391.

[36] B. R. Donald. Plenary lecture: Algorithmic challenges in structural molecular biology and proteomics, in *Proceedings of the Sixth International Workshop on the Algorithmic Foundations of Robotics (WAFR)*. Utrecht/Zeist, The Netherlands: July 2004, pp. 1–10. University of Utrecht. in *Algorithmic Foundations of Robotics VI*, Springer Tracts in Advanced Robotics (Vol. 17), M. Erdmann, M. Overmars, D. Hsu, and A. F. van der Stappen (eds.). Berlin: Springer, 2005, pp. 1–10.

[37] B. R. Donald, D. Kapur, and J. Mundy. *Symbolic and Numerical Computation for Artificial Intelligence*. London: Academic Press, Harcourt Jovanovich, 1992.

[38] A. C. Fowler, F. Tian, H. M. Al-Hashimi, and J. H. Prestegard. Rapid determination of protein folds using residual dipolar couplings. *Journal of Molecular Biology* 304;3(2000):447–460.

[39] T. Gallagher, P. Alexander, P. Bryan, and G. L. Gilliland. Two crystal structures of the B1 immunoglobulin-binding domain of streptococcal protein G and comparison with NMR. *Biochemistry*, 33(1994):4721–4729.

[40] K. H. Gardner and L. E. Kay. Production and incorporation of ^{15}N, ^{13}C, ^{2}H (^{1}H-δ1 methyl) isoleucine into proteins for multidimensional NMR studies. *Journal of the American Chemical Society* 119;32(1997):7599–7600.

[41] G. Gemmecker, W. Jahnke, and H. Kessler. Measurement of fast proton exchange rates in isotopically labelled compounds. *Journal of American Chemical Society* 115;24(1993):11620–11621.

[42] I. Georgiev and B. R. Donald. Dead-end elimination with backbone flexibility. *Bioinformatics*, 23(13), 2007. Special issue on papers from the International Conference on Intelligent Systems for Molecular Biology (ISMB 2007), Vienna, Austria: July 21-25, 2007.

[43] I. Georgiev, D. Keedy, J. Richardson, D. Richardson, and B. R. Donald. Algorithm for backrub motions in protein design. *Bioinformatics*, 22 (Jul 2008):e174–183. Special issue on papers from International Conference on Intelligent Systems for Molecular Biology (ISMB 2008), Toronto, CA, July 2008.

[44] I. Georgiev, R. Lilien, and B. R. Donald. Improved pruning algorithms and divide-and-conquer strategies for dead-end elimination, with application to protein design. *Bioinformatics* 22;14(2006):e174–183. Special issue on papers from the International Conference on Intelligent Systems for Molecular Biology (ISMB 2006), Fortaleza, Brazil.

[45] I. Georgiev, R. Lilien, and B. R. Donald. The minimized dead-end elimination criterion and its application to protein redesign in a hybrid scoring and search algorithm for computing partition functions over molecular ensembles. *Journal of Computational Chemistry* 29;10(2008):1527–1542.

[46] A. W. Giesen, S. W. Homans, and J. M. Brown. Determination of protein global folds using backbone residual dipolar coupling and long-range NOE restraints. *Journal of Biomolecular NMR* 25(2003):63–71.

[47] E. Girard, L. Chantalat, J. Vicat, and R. Kahn. Gd-HPDO3A, a complex to obtain high-phasing-power heavy atom derivatives for SAD and MAD experiments: Results with tetragonal hen egg-white lysozyme. *Acta Crystallographica, Section D, Biological Crystallography* 58(2001):1–9.

[48] R. Goldstein. Efficient rotamer elimination applied to protein side-chains and related spin glasses. *Biophysical Journal* 66(1994):1335–1340.

[49] M. J. Gorczynski, J. Grembecka, Y. Zhou, Y. Kong, L. Roudaiya, M. G. Douvas, et al. Allosteric inhibition of the protein-protein interaction between the leukemia-associated proteins RUNX1 and CBFβ. *Chemistry & Biology* 14;10(2007).

[50] A. M. Gronenborn, D. R. Filpula, N. Z. Essig, A. Achari, M. Whitlow, P. T. Wingfield, and G. M. Clore. A novel, highly stable fold of the immunoglobulin binding domain of streptococcal protein G. *Science* 253(1991):657.

[51] S Grzesiek and A Bax. Measurement of amide proton exchange rates and NOEs with water in 13C/15N-enriched calcineurin B. *Journal of Biomolecular NMR*, 3(6)(1993)627–638.

[52] P. Güntert, C. Mumenthaler, and K. Wüthrich. Torsion angle dynamics for NMR structure calculation with the new program DYANA. *Journal of Molecular Biology* 273(1997):283–298.

[53] Peter Guntert. Automated NMR structure calculation with CYANA. *Methods in Molecular Biology* 278(2004): 353–378.

[54] P. J. Hajduk, R. P. Meadows, and S. W. Fesik. Drug design: Discovering high-affinity ligands for proteins. *Science* 278(1997):497–499.

[55] B. Hendrickson. Conditions for unique graph realizations. *SIAM Journal on Computing* 21(1992):65–84.

[56] B. Hendrickson. The molecule problem: Exploiting structures in global optimization. *SIAM Journal on Optimization* 5(1995):835–857.

[57] T. Herrmann, P. Güntert, and K. Wüthrich. Protein NMR structure determination with automated NOE assignment using the new software CANDID and the torsion angle dynamics algorithm DYANA. *Journal of Molecular Biology*, 319(2002):209–227.

[58] K. Hinsen. Normal mode theory and harmonic potential approximations. Course and Lecture Notes. Available at http://dirac.cnrs-orleans.fr/~hinsen/, Centre de Biophysique Moléculaire (CNRS), Orleans, France, 2000.

[59] Y. J. Huang, G. V. Swapna, P. K. Rajan, H. Ke, B. Xia, K. Shukla, et al. Solution NMR structure of ribosome-binding factor a (RbfA), a cold shock adaptation protein from *Escherichia coli*. *Journal of Molecular Biology*, 327(2003):521–536.

[60] Y.J. Huang, R. Tejero, R. Powers, and G.T. Montelione. A topology-constrained distance network algorithm for protein structure determination from NOESY data. *Proteins* 62;3(2006):587–603.

[61] J.-C. Hus, D. Marion, and M. Blackledge. *De novo* determination of protein structure by NMR using orientational and long-range order restraints. *Journal of Molecular Biology*, 298;5(2000):927–936.

[62] J. C. Hus, D. Marion, and M. Blackledge. Determination of protein backbone using only residual dipolar couplings. *Journal of the American Chemical Society* 123(2001):1541–1542.

[63] J.C. Hus, J. Prompers, and R. Brüschweiler. Assignment strategy for proteins of known structure. *Journal of Magnetic Resonance* 157(2002):119–125.

[64] J. Korukottu, M. Bayrhuber, P. Montaville, V. Vijayan, Y. S. Jung, S. Becker, and M. Zweckstetter. Fast high-resolution protein structure determination by using unassigned NMR data. *Angewandte Chemie (International Edition in English)*, 46(2007):1176–1179.

[65] D. J. Jacobs, A. J. Rader, L. A. Kuhn, and M. F. Thorpe. Protein flexibility predictions using graph theory. *Proteins* 44;2(2001):150–165.

[66] E. C. Johnson, G. A. Lazar, J. R. Desjarlais, and T. M. Handel. Solution structure and dynamics of a designed hydrophobic core variant of ubiquitin. *Structure Folding Design*, 7(1999):967–976.

[67] Y.S. Jung and M. Zweckstetter. Backbone assignment of proteins with known structure using residual dipolar couplings. *Journal of Biomolecular NMR* 30;1(2004):25–35.

[68] K. Juszewski, C. D. S. Schwieters, D. S. Garrett, R. A. Byrd, N. Tjandra, and G. M. Clore. Completely automated, highly error-tolerant macromolecular structure determination from multidimensional nuclear Overhauser enhancement spectra and chemical shift assignments. *Journal of the American Chemical Society* 126(2004):6258–6273.

[69] K. Ruan, K. B. Briggman, and J. R. Tolman. De novo determination of internuclear vector orientations from residual dipolar couplings measured in three independent alignment media. *Journal of Biomolecular NMR* 41;2(2008):61–76.

[70] H. Kamisetty, C. Bailey-Kellogg, and G. Pandurangan. An efficient randomized algorithm for contact-based NMR backbone resonance assignment. *Bioinformatics* 22;2(2006):172–180.

[71] L. E. Kay. Protein dynamics from NMR. *Nature Structural Biology* 5(1998):513–517.

[72] M. D. Kemple, B. D. Ray, K. B. Lipkowitz, F. G. Prendergast, and B. D. Rao. The use of lanthanides for solution structure determination of biomolecules by NMR: Evaluation of the methodology with EDTA derivatives as model systems. *Journal of the American Chemical Society* 110;25(1988):8275–8287.

[73] D. E. Kim, D. Chivian, and D. Baker. Protein structure prediction and analysis using the Robetta server. *Nucleic Acids Research* 32(Web Server issue)(2004):526–531.

[74] R. Kolodny, L. Guibas, M. Levitt, and P. Koehl. Inverse kinematics in biology: The protein loop closure problem. *The International Journal of Robotics Research* 24(2005):151–163.

[75] R. J. Kovacs, M. T. Nelson, H. K. Simmerman, and L. R. Jones. Phospholamban forms Ca^{2+}-selective channels in lipid bilayers. *The Journal of Biological Chemistry* 263(1988):18364–18368.

[76] H. W. Kuhn. Hungarian method for the assignment problem. *Naval Research Logistics Quarterly* 2(1955):83–97.

[77] I. V. Kurinov and R. W. Harrison. The influence of temperature on lysozyme crystals—structure and dynamics of protein and water. *Acta Crystallographica, Section D, Biological Crystallography* 51(1995):98–109.

[78] C. Langmead and B. R. Donald. 3D structural homology detection via unassigned residual dipolar couplings, in *Proceedings of the IEEE Computer Society Bioinformatics Conference (CSB)*, pp. 209–217, Stanford, August 2003. PMID: 16452795.

[79] C. Langmead and B. R. Donald. An expectation/maximization nuclear vector replacement algorithm for automated NMR resonance assignments. *Journal of Biomolecular NMR* 29;2(2004):111–138.

[80] C. Langmead and B. R. Donald. High-throughput 3D structural homology detection via NMR resonance assignment, in *Proceedings of the IEEE Computational Systems Bioinformatics Conference (CSB)*, pp. 278–289, Stanford, CA, August 2004. PMID: 16448021.

[81] C. Langmead, C. R. McClung, and B. R. Donald. A maximum entropy algorithm for rhythmic analysis of genome-wide expression patterns, in *Proceedings of the IEEE Computer Society Bioinformatics Conference (IEEE CSB)*, pp. 237–245, August 2002. PMID: 15838140.

[82] C. Langmead, A. Yan, R. Lilien, L. Wang, and B. R. Donald. A polynomial-time nuclear vector replacement algorithm for automated NMR resonance assignments; in *Proceedings of The Seventh Annual International Conference on Research in Computational Molecular Biology (RECOMB)*, pp. 176–187, Berlin, Germany, April 2003. ACM Press.

[83] C. Langmead, A. Yan, R. Lilien, L. Wang, and B. R. Donald. A polynomial-time nuclear vector replacement algorithm for automated NMR resonance assignments. *Journal of Computational Biology* 11;2–3(2004):277–298.

[84] C. Langmead, A. Yan, C. R. McClung, and B. R. Donald. Phase-independent rhythmic analysis of genome-wide expression patterns. *Journal of Computational Biology* 10;3–4(2003):521–536.

[85] I. Lasters and J. Desmet. The fuzzy-end elimination theorem: Correctly implementing the side chain placement algorithm based on the dead-end elimination theorem. *Protein Engineering* 6(1993):717–722.

[86] M. Li, H. P. Phatnani, Z. Guan, H. Sage, A. L. Greenleaf, and P. Zhou. Solution structure of the Set2-Rpb1 interacting domain of human Set2 and its interaction with the hyperphosphorylated C-terminal domain of Rpb1. *Proceedings of the National Academy of Sciences (U S A)* 102;49(2005):17636–17641.

[87] R. Lilien, H. Farid, and B. R. Donald. Probabilistic disease classification of expression-dependent proteomic data from mass spectrometry of human serum. *Journal of Computational Biology* 10;6(2003):925–946.

[88] R. Lilien, B. Stevens, A. Anderson, and B. R. Donald. A novel ensemble-based scoring and search algorithm for protein redesign, and its application to modify the substrate specificity of the Gramicidin Synthetase A phenylalanine adenylation enzyme. *Journal of Computational Biology*, 12;6–7(2005):740–761.

[89] R. H. Lilien, C. Bailey-Kellogg, A. C. Anderson, and B. R. Donald. A subgroup algorithm to identify cross-rotation peaks consistent with non-crystallographic symmetry. *Acta Crystallographica Section D, Biological Crystallography* 60;6(2004):1057–1067.

[90] K. Lim, A. Nadarajah, E. L. Forsythe, and M. L. Pusey. Locations of bromide ions in tetragonal lysozyme crystals. *Acta Crystallographica, Section D, Biological Crystallography* 54(1998):899–904.

[91] B Lopez-Mendez and P Guntert. Automated protein structure determination from NMR spectra. *Journal of the American Chemical Society*, 128;40(2006):13112–13122.

[92] J. A. Losonczi, M. Andrec, M. W. Fischer, and J. H. Prestegard. Order matrix analysis of residual dipolar couplings using singular value decomposition. *Journal of Magnetic Resonance* 138;2(1999):334–342.

[93] M. Bryson, F. Tian, J. H. Prestegard, and H. Valafar. REDCRAFT: A tool for simultaneous characterization of protein backbone structure and motion from RDC data. *Journal of Magnetic Resonance* 191;2(2008):322–334.

[94] A. Marin, T. E. Malliavin, P. Nicolas, and M. A. Delsuc. From NMR chemical shifts to amino acid types: Investigation of the predictive power carried by nuclei. *Journal of Biomolecular NMR* 30;1(2004):47–60.

[95] J. Meiler, N. Blomberg, M. Nilges, and C. Griesinger. A new approach for applying residual dipolar couplings as restraints in structure elucidation. *Journal of Biomolecular NMR* 16(2000):245–252.

[96] J. Meiler, W. Peti, and C. Griesinger. DipoCoup: A versatile program for 3D-structure homology comparison based on residual dipolar couplings and pseudocontact shifts. *Journal of Biomolecular NMR* 17(2000):283–294.

[97] J. Meiler and D. Baker. Rapid protein fold determination using unassigned NMR data. *Proceedings of the National Academy of Sciences (U S A)* 100;26(2003):15404–15409.

[98] J. Meiler and D. Baker. The fumarate sensor DcuS: Progress in rapid protein fold elucidation by combining protein structure prediction methods with NMR spectroscopy. *Journal of Magnetic Resonance* 173;2(2005):310–316.

[99] R. Mettu, R. Lilien, and B. R. Donald. High-throughput inference of protein-protein interfaces from unassigned NMR data. *Bioinformatics* 21;Suppl. 1(2005):i292–i301.

[100] G A Mueller, W. Y. Choy, D. Yang, J. D. Forman-Kay, R. A. Venters, and L. E. Kay. Global folds of proteins with low densities of NOEs using residual dipolar couplings: Application to the 370-residue maltodextrin-binding protein. *Journal of Molecular Biology* 300;1(2000):197–212.

[101] S Neal, A. M. Nip, H. Zhang, and D. S. Wishart. Rapid and accurate calculation of protein ^{1}H, ^{13}C and ^{15}N chemical shifts. *Journal of Biomolecular NMR* 26(2003):215–240.

[102] M. Nilges. A calculation strategy for the structure determination of symmetric dimers by 1H NMR. *Proteins* 17;3(1993):297–309.

[103] M. Nilges, M. Macias, S. Odonoghue, and H. Oschkinat. Automated NOESY interpretation with ambiguous distance restraints: The refined NMR solution structure of the pleckstrin homology domain from β-spectrin. *Journal of Molecular Biology* 269(1997):408–422.

[104] K. Noonan, D. O'Brien, and J. Snoeyink. Probik: protein backbone motion by inverse kinematics. *International Journal of Robotics Research* 24;11(2005):971–982.

[105] O. Vitek, C. Bailey-Kellogg, B. Craig, and J. Vitek. Inferential backbone assignment for sparse data. *Journal of Biomolecular NMR* 35;3(2006):187–208.

[106] H. Oki, Y. Matsuura, H. Komatsu, and A. A. Chernov. Refined structure of orthorhombic lysozyme crystallized at high temperature: Correlation between morphology and intermolecular contacts. *Acta Crystallographica, Section D, Biological, Crystallography* 55(1999):114.

[107] A.L. Okorokov, M.B. Sherman, C. Plisson, V. Grinkevich, K. Sigmundsson, G. Selivanova, et al. The structure of p53 tumour suppressor protein reveals the basis for its functional plasticity. *EMBO Journal* 25;21(2006):5191–5200.

[108] E. T. Olejniczak, R. X. Xu, and S. W. Fesik. A 4D HCCH-TOCSY experiment for assigning the side-chain ^{1}H and ^{13}C resonances of proteins. *Journal of Biomolecular NMR* 2;6(1992):655–659.

[109] R. O'Neil, R. Lilien, B. R. Donald, R. Stroud, and A. Anderson. The crystal structure of dihydrofolate reductase-thymidylate synthase from *Cryptosporidium hominis* reveals a novel architecture for the bifunctional enzyme. *The Journal of Eukaryotic Microbiology* 50;6(2003):555–556.

[110] R. O'Neil, R. Lilien, B. R. Donald, R. Stroud, and A. Anderson. Phylogenetic classification of protozoa based on the structure of the linker domain in the bifunctional enzyme, dihydrofolate reductase-thymidylate synthase. *Journal of Biological Chemistry* 278;52(2003):52980–52987.

[111] K. Oxenoid and J. J. Chou. The structure of phospholamban pentamer reveals a channel-like architecture in membranes. *Proceedings of the National Academy of Sciences (U S A)* 102(2005):10870–10875.

[112] A. G. Palmer. Probing Molecular Motion By NMR. *Current Opinion in Structural Biology* 7(1997):732–737.

[113] A. G. Palmer, J. Williams, and A. McDermott. Nuclear magnetic resonance studies of biopolymer dynamics. *Journal of Physical Chemistry* 100(1996):13293–13310.

[114] D. A. Pearlman, D. A. Case, J. W. Caldwell, W. S. Ross, T. E. Cheatham, S. DeBolt, et al. AMBER, a package of computer programs for applying molecular mechanics, normal mode analysis, molecular dynamics and free energy calculations to simulate the structures and energies of molecules. *Computer Physics Communications* 91(1995):1–41.

[115] N. Pierce, J. Spriet, J. Desmet, and S. Mayo. Conformational splitting: A more powerful criterion for dead-end elimination. *Journal of Computational Chemistry* 21(2000):999–1009.

[116] V. M. Popov, D. C. M. Chan, Y. A. Fillingham, W. A. Yee, D. L. Wright, and A. C. Anderson. Analysis of complexes of inhibitors with Cryptosporidium hominis DHFR leads to a new trimethoprim derivative. *Bioorganict Medicinal Chemistry Letters* 16;16(2006):4366–4370.

[117] S. Potluri, A. Yan, , J. J. Chou, B. R. Donald, and C. Bailey-Kellogg. Structure determination of symmetric homo-oligomers by a complete search of symmetry configuration space using NMR restraints and van der Waals packing. *Proteins: Structure, Function and Bioinformatics* 65;1(2006):203–219.

[118] S. Potluri, A. Yan, B. R. Donald, and C. Bailey-Kellogg. A complete algorithm to resolve ambiguity for inter-subunit NOE assignment in structure determination of symmetric homo-oligomers. *Protein Science* 16;1(2006).

[119] J. H. Prestegard. New techniques in structural NMR—anisotropic interactions. *Nature Structural Biology*, (1998):517–522.

[120] J. H. Prestegard, C. M. Bougault, and A. I. Kishore. Residual dipolar couplings in structure determination of biomolecules. *Chemical Reviews* 104;8(2004):3519–3540.

[121] Y. Qu, J. T. Guo, V. Olman, and Y. Xu. Protein fold recognition through application of residual dipolar coupling data. *Pacific Symposium on Biocomputing*, (2004):459–470.

[122] R. Bertram, J. R. Quine, M. S. Chapman, and T. A. Cross. Atomic refinement using orientational restraints from solid-state NMR. *Journal of Magnetic Resonance* 147(2000):9–16.

[123] R. Ramage, J. Green, T. W. Muir, O. M. Ogunjobi, S. Love, and K. Shaw. Synthetic, structural and biological studies of the ubiquitin system: The total chemical synthesis of ubiquitin. *Journal of Biochemistry* 299(1994):151–158.

[124] W. Rieping, M. Habeck, and M. Nilges. Inferential structure determination. *Science* 209;5732(2005):303–306.

[125] C. A. Rohl and D. Baker. De Novo determination of protein backbone structure from residual dipolar couplings using Rosetta. *Journal of the American Chemical Society* 124;11(2002):2723–2729.

[126] C. A. Rohl. Protein structure estimation from minimal restraints using Rosetta. *Methods in Enzymology* 394(2005):244–260.

[127] K. Ruan and J. R. Tolman. Composite alignment media for the measurement of independent sets of NMR residual dipolar couplings. *Journal of the American Chemical Society* 127;43(2005):15032–15033.

[128] S. Bansal, X. Miao, M.W. Adams, J. H. Prestegard, and H. Valafar. Rapid classification of protein structure models using unassigned backbone RDCs and probability density profile analysis (PDPA). *Journal of Magnetic Resonance* 192;1(2008):60–68.

[129] S. Rumpel, S. Becker, and M. Zweckstetter. High-resolution structure determination of the CylR2 homodimer using paramagnetic relaxation enhancement and structure-based prediction of molecular alignment. *Journal of Biomolecular NMR* 40;1(2008):1–13.

[130] A. Saupe. Recent results in the field of liquid crystals. *Angewandte Chemie* 7(1968):97–112.

[131] J. B. Saxe. Embeddability of weighted graphs in k-space is strongly NP-hard, in *Proceedings of the 17th Allerton Conference on Communications, Control, and Computing*, (1979):480–489.

[132] B. R. Seavey, E. A. Farr, W. M. Westler, and J. L. Markley. A relational database for sequence-specific protein NMR data. *Journal of Biomolecular NMR* 1(1991):217–236.

[133] S. B. Shuker, P. J. Hajduk, R. P. Meadows, and S. W. Fesik. Discovering high-affinity ligands for proteins: SAR by NMR. *Science* 274(1996):1531–1534.

[134] R. Singh and B. Berger. ChainTweak: Sampling from the neighbourhood of a protein conformation. *Pacific Symposium on Biocomputing 2005* (2005):54–65.

[135] N. R. Skrynnikov and L. E. Kay. Assessment of molecular structure using frame-independent orientational restraints derived from residual dipolar couplings. *Journal of Biomolecular NMR* 18;3(2000):239–252.

[136] B. Stevens, R. Lilien, I. Georgiev, B. R. Donald, and A. Anderson. Redesigning the PheA domain of Gramicidin Synthetase leads to a new understanding of the enzyme's mechanism and selectivity. *Biochemistry* 45;51(2006):15495–15504.

[137] K. Suhre and Y.H. Sanejouand. ElNemo: A normal mode web server for protein movement analysis and the generation of templates for molecular replacement. *Nucleic Acids Research* 32(Web Server issue)(2004):610–614.

[138] K. Suhre and Y. H. Sanejouand. On the potential of normal-mode analysis for solving difficult molecular-replacement problems. *Acta Crystallographica, Section D, Biological Crystallography* 60;Pt 4(2004):796–799.

[139] F. Tian, H. Valafar, and J. H. Prestegard. A dipolar coupling based strategy for simultaneous resonance assignment and structure determination of protein backbones. *Journal of the American Chemical Society* 123;47(2001):11791–11796.

[140] N. Tjandra and A. Bax. Direct measurement of distances and angles in biomolecules by NMR in a dilute liquid crystalline medium. *Science* 278(1997):1111–1114.

[141] D. Tobi and I. Bahar. Structural changes involved in protein binding correlate with intrinsic motions of proteins in the unbound state. *Proceedings of the National Academy of Sciences (U S A)* 102;52(2005):18908–18913.

[142] J. R. Tolman, J. M. Flanagan, M. A. Kennedy, and J. H. Prestegard. Nuclear magnetic dipole interactions in field-oriented proteins: Information for structure determination in solution. *Proceedings of the National Academy of Sciences USA* 92(1995):9279–9283.

[143] V. Tugarinov, W. Y Choy, V. Y. Orekhov, and L. E. Kay. Solution NMR-derived global fold of a monomeric 82-kDa enzyme. *Proceedings of the National Academy of Sciences of the United States* 102;3(2005):622–627.

[144] V. Tugarinov, R. Muhandiram, A. Ayed, and L. E. Kay. Four-dimensional NMR spectroscopy of a 723-residue protein: Chemical shift assignments and secondary structure of malate synthase G. *Journal of the American Chemical Society* 124;34(2002):10025–10035.

[145] H. Valafar and J. H. Prestegard. Rapid classification of a protein fold family using a statistical analysis of dipolar couplings. *Bioinformatics* 19;12(2003):1549–1555.

[146] M. C. Vaney, S. Maignan, M. Ries-Kautt, and A. Ducruix. High-resolution structure (1.33 Angstrom) of a HEW lysozyme tetragonal crystal grown in the APCF apparatus. Data and structural comparison with a crystal grown under microgravity from SpaceHab-01 mission. *Acta Crystallographica, Section D, Biological Crystallography* 52(1996):505–517.

[147] S. Vijay-Kumar, C. E. Bugg, and W. J. Cook. Structure of ubiquitin refined at 1.8 Å resolution. *Journal of Molecular Biology* 194(1987):531–544.

[148] O. Vitek, C. Bailey-Kellogg, B. Craig, P. Kuliniewicz, and J. Vitek. Reconsidering complete search algorithms for protein backbone NMR assignment. *Bioinformatics* 21;Suppl 2(2005):ii230–ii236.

[149] O. Vitek, J. Vitek, B. Craig, and C. Bailey-Kellogg. Model-based assignment and inference of protein backbone Nuclear Magnetic Resonances. *Statistical Applications in Genetics and Molecular Biology* 3;1(2004).

[150] C. E. Wang, T. Lozano-Pérez, and B. Tidor. AMBIPACK: A systematic algorithm for packing of macromolecular structures with ambiguous distance constraints. *Proteins: Structure, Function, and Genetics* 32(1998):26–42.

[151] L. Wang and B. R. Donald. Analysis of a systematic search-based algorithm for determining protein backbone structure from a minimal number of residual dipolar couplings, in *Proceedings of the IEEE Computational Systems Bioinformatics Conference (CSB)*, pp. 319–330, Stanford, CA, August 2004. PMID: 16448025.

[152] L. Wang and B. R. Donald. Exact solutions for internuclear vectors and backbone dihedral angles from NH residual dipolar couplings in two media, and their application in a systematic search algorithm for determining protein backbone structure. *Journal of Biomolecular NMR* 29;3(2004):223–242.

[153] L. Wang and B. R. Donald. An efficient and accurate algorithm for assigning nuclear Overhauser effect restraints using a rotamer library ensemble and residual dipolar couplings, in *Proceedings of the IEEE Computational Systems Bioinformatics Conference (CSB)*, pp. 189–202, Stanford, CA, August 2005. PMID: 16447976.

[154] L. Wang and B. R. Donald. A data-driven, systematic search algorithm for structure determination of denatured or disordered proteins, in *Proceedings of the LSS Computational Systems Bioinformatics Conference (CSB)*, pp. 68–78, Stanford, CA, August 2006. PMID: 17369626.

[155] L. Wang, R. Mettu, and B. R. Donald. An algebraic geometry approach to protein backbone structure determination from NMR data, in *Proceedings of the IEEE Computational Systems Bioinformatics Conference (CSB)*, pp. 235–246, Stanford, CA, August 2005. PMID: 16447981.

[156] L. Wang, R. Mettu, and B. R. Donald. A polynomial-time algorithm for *de novo* protein backbone structure determination from NMR data. *Journal of Computational Biology* 13;7(2006):1276–1288.

[157] Y. X. Wang, J. Jacob, F. Cordier, P. Wingfield, S. J. Stahl, S. Lee-Huang, et al. Measurement of 3hJNC' connectivities across hydrogen bonds in a 30 kDa protein. *Journal of Biomolecular NMR* 14;2(1999):181–184.

[158] Y. X. Wang, J. L. Marquardt, P. Wingfield, S. J. Stahl, S. Lee-Huang, D. Torchia, and A. Bax. Simultaneous measurement of ^{1}H-^{15}N, ^{1}H-^{13}C', and ^{15}N-^{13}C' dipolar couplings in a perdeuterated 30 kda protein dissolved in a dilute liquid crystalline phase. *Journal of the American Chemical Society* 120;29(1998):7385–7386.

[159] W. J. Wedemeyer, C. A. Rohl, and H. A. Scheraga. Exact solutions for chemical bond orientations from residual dipolar couplings. *Journal of Biomolecular NMR* 22(2002):137–151.

[160] H. Weidong and L. Wang. Residual dipolar couplings: Measurements and applications to biomolecular studies. *Annual Reports on NMR Spectroscopy* 58(2006):231–303.

[161] S. Wells, S. Menor, B. Hespenheide, and M. F. Thorpe. Constrained geometric simulation of diffusive motion in proteins. *Physical Biology* 2;4(2005):127–136.

[162] Wikipedia. NP-completeness. *Wikipedia, the Free Encyclopedia*, September 2008. Available at http://en.wikipedia .org/wiki/NP-complete.

[163] D. S. Wishart and B. D. Sykes. The ^{13}C chemical shift index: A simple method for the identification of protein secondary structure using ^{13}C chemical shift data. *Journal of Biomolecular NMR* 4(1994):171–180.

[164] D. S. Wishart, B. D. Sykes, and F. M. Richards. Relationship between nuclear magnetic resonance chemical shift and protein secondary structure. *Journal of Molecular Biology* 222;2(1991):311–333.

[165] D. S. Wishart, B. D. Sykes, and F. M. Richards. The ^{13}C chemical shift index: A fast and simple method for the identification of protein secondary structure using ^{13}C chemical shift data. *Biochemistry* 31;6(1992):1647–1651.

[166] K. Wuthrich. Protein recognition by NMR. *Nature Structural Biology* 7;3(2000):188–189.

[167] W. Y. Choy, M. Tollinger, G. A. Mueller, and L. E. Kay. Direct structure refinement of high molecular weight proteins against residual dipolar couplings and carbonyl chemical shift changes upon alignment: An application to maltose binding protein. *Journal of Biomolecular NMR* 21;1(2001):31–40.

[168] X. Wang, S. Bansal, M. Jiang, and J. H. Prestegard. RDC-assisted modeling of symmetric protein homo-oligomers. *Protein Science*; 17;5(2008):899–907.

[169] X. P Xu and D. A. Case. Automated prediction of ^{15}N, ^{13}C'alpha', ^{13}C'beta' and ^{13}C' chemical shifts in proteins using a density functional database. *Journal of Biomolecular NMR* 21(2001):321–333.

[170] Y. Xu, Y. Zheng, J. S. Fan, and D. Yang. A new strategy for structure determination of large proteins in solution without deuteration. *Nature Methods* 3;11(2006):931–937.

[171] A. Yan, C. Langmead, and B. R. Donald. A probability-based similarity measure for Saupe alignment tensors with applications to residual dipolar couplings in NMR structural biology. *International Journal of Robotics Research* 24;2–3(2005):165–182. (Special Issue on Robotics Techniques Applied to Computational Biology.)

[172] J. Zamoon, A. Mascioni, D. D. Thomas, and G. Veglia. NMR solution structure and topological orientation of monomeric phospholamban in dodecylphosphocholine micelles. *Biophysical Journal* 85;4(2003):2589–2598.

[173] M. I. Zavodszky, M. Lei, M. F. Thorpe, A. R. Day and L. A. Kuhn. Modeling correlated main-chain motions in proteins for flexible molecular recognition. *Proteins* 57;2(2004):243–261.

[174] J. Zeng, C. Tripathy, P. Zhou, and B. R. Donald. A Hausdorff-based NOE assignment algorithm using protein backbone determined from residual dipolar couplings and rotamer patterns, in *Proceedings of the LSS Computational Systems Bioinformatics Conference (CSB)*, pp. 169–181, Stanford, CA, August 2008. PMID. 19122773.

[175] M. Zweckstetter. Determination of molecular alignment tensors without backbone resonance assignment: Aid to rapid analysis of protein-protein interactions. *Journal of Biomolecular NMR* 27;1(2003):41–56.

[176] M. Zweckstetter and A. Bax. Single-step determination of protein substructures using dipolar couplings: Aid to structural genomics. *Journal of the American Chemical Society* 123;38(2001):9490–9491.

[177] J. Zeng, J. Boyles, C. Tripathy, L. Wang, A. Yan, P. Zhou, B.R. Donald. High-resolution protein structure determination starting with a global fold calculated from exact solutions to the RDC equations. *Journal of Biomolecular NMR* 45;3(2009):265–81.

19 Proteomic Disease Classification Algorithm

In this chapter, we study an algorithm [1] to automatically classify the status of a patient to be healthy or to have a particular disease, based on the mass spectrum (MS) data of his or her blood serum. We have been studying proteins in this book using atomic-resolution biophysical methods such as NMR and X-ray crystallography. MS, in contrast, allows us to assay and compare the masses of many proteins at the same time.

19.1 Proteomic Disease Classification

Mass spectrometry (MS) provides ultrahigh-resolution mass information. A mass spectrum consists of a set of m/z ("mass-to-charge") values and corresponding relative intensities that are a function of all ionized molecules present with that m/z ratio; m is mass and z is charge; m/z is plotted on the horizontal axis, and the corresponding relative intensity is plotted on the vertical axis of a typical mass spectrum. *Mass spectrometry classification algorithms (MSCAs)* are designed to discriminate one condition from another by comparing their mass spectra. For example, an MSCA can tell whether or not a patient has a particular disease from MS data of his peripheral blood.

19.1.1 Methods

Spectral space Each spectrum is represented as a point $(p_1, \ldots, p_r)$ in an r-dimensional space, where r is the number of m/z values for which relative intensities p_i are recorded per spectrum.

Training set A set of data-class pairs (r, c), where r is the data and c is the class that r belongs to. For example, r is the mass spectrum of a patient, and $c \in \{diseased, healthy\}$ indicating the health status of the patient. In this study, *diseased* can only mean one particular disease, for example, ovarian cancer. An extension was considered in Lilien et al. [1] to multiple classes (more than two).

Classifier A classifier is a mapping from data to classes. Usually, the classifier is first trained on a training set, then, for any data (points in the spectral space), it outputs a class for the data (for example, diseased or healthy).

19.1.2   Q5: An MSCA Algorithm

An MSCA called Q5 was introduced in Lilien et al. [1]. In the training step, it first reduces the dimension of the data in the training set by *PCA*. The dimension is reduced to the *intrinsic dimensionality* of the problem, which is the number of samples (in this case, patient blood samples) minus the number of classes. Then, a hyperplane H is computed by *linear discriminant analysis (LDA)* so that the projections of data in the training set onto H best discriminate the two classes, in terms of the ratio of the *in-class variance* to the *between-class variance*.

In the testing step, for a point p in the spectral space, Q5 first reduces the dimension of p by projecting it to the same subspace computed in the training step. Then, the reduced-dimension data is further projected to H. Finally, a *probabilistic classifier* is applied to compute which class p belongs to.

More precisely, Q5 works as follows:

Training Stage

Input   Data spectrum represented as column vectors. $\mathbf{x} \in X$ (healthy) and $\mathbf{y} \in Y$ (disease). Let $n_x = |X|, n_y = |Y|, n = n_x + n_y$.

PCA

- Compute *all-class-mean* $\mu' = \dfrac{1}{n}(\sum_{\mathbf{x} \in X} x + \sum_{\mathbf{y} \in Y} y)$.
- Let X' and Y' be the zero-mean matrices obtained by subtracting μ' from X and Y, respectively. Let $\mathbf{P} = [X'Y']$ be the $r \times n$ matrix.
- Let $\mathbf{C} = \mathbf{PP}^T$. Compute the *eigenvalues* and normalized *eigenvectors* of $\mathbf{C}$, denoted by λ_i and v_i ($i = 1, \ldots, w$), respectively. That is, for any $i = 1, \ldots, w$, $\mathbf{C}v_i = \lambda_i v_i$.
- Sort $\lambda_1, \ldots, \lambda_w$ and discard the eigenvectors corresponding to small eigenvalues. Let the remaining eigenvectors be $\mathbf{V}$.
- Reduce the dimension of $\mathbf{P}$ by $\mathbf{V}$. For any $\mathbf{x}' \in X'$, let $\mathbf{x}_p = \mathbf{Vx}'$; for any $\mathbf{y}' \in Y'$, let $\mathbf{y}_p = \mathbf{Vy}'$.

LDA

- Compute the *within-class means*. Let $\mu_x = \dfrac{1}{n_x} \sum_{\mathbf{x}_p \in X_p} \mathbf{x}_p$ and $\mu_y = \dfrac{1}{n_y} \sum_{\mathbf{y}_p \in Y_p} \mathbf{y}_p$.
- Compute the *within-class scatter matrix* $\mathbf{S_w}$ by

$$\mathbf{S_w} = \mathbf{M_x M_x}^T + \mathbf{M_y M_y}^T,$$

where the columns of $\mathbf{M_x}$ are $\mathbf{x}_p - \mu_x$; the columns of $\mathbf{M_y}$ are $\mathbf{y}_p - \mu_y$.
- Compute the *between-class scatter matrix* $\mathbf{S_b}$ by

$$\mathbf{S_b} = n_x(\mu_x - \mu)(\mu_x - \mu)^T + n_y(\mu_y - \mu)(\mu_y - \mu)^T.$$

• Let $\mathbf{e}$ be the normalized *generalized eigenvector* of $\mathbf{S_b}$ and $\mathbf{S_w}$ corresponding to the largest *generalized eigenvalue* λ. That is, $\mathbf{S_b}\mathbf{e} = \lambda\mathbf{S_w}\mathbf{e}$.

Let p_1 (respectively p_2) be the mean of the $(\mathbf{x}_p)^T\mathbf{e}$ (respectively $(\mathbf{y}_p)^T\mathbf{e}$).

Output $\mathbf{V}$, $\mathbf{e}$, p_1, and p_2.

Testing or Prediction Stage

Input z in the spectral space, and a threshold $t \in [0.5, 1]$.

Let $z_d = (\mathbf{V}(z - \mu'))^T\mathbf{e}$.

Probabilistic Classifier

• Let q be the midpoint of p_1 and p_2. Choose σ such that $\exp[-(q - p_1)^2/\sigma^2] = \dfrac{1}{2}$.
• Compute $Pr(z_d \in C_i) = \exp[-(z_d - p_i)^2/\sigma^2]$ $(i = 1, 2)$. If $Pr(z_d \in C_i) > t$ for some i, then output z_d is in class C_i. Otherwise output *ambiguous*.

19.2   Results and Discussion

Q5 was demonstrated on several MS data sets for patients with ovarian and prostate cancer. In all cases, excellent accuracy was obtained in disease classification. In Lilien et al. [1], an extension was described to identify biomarkers as well. In addition to impressive experimental accuracy, Q5 has guarantees, as an algorithm, of efficiency, soundness, and completeness. Such algorithms can probe the network of proteins that interact in cancer processes, and may help translate MCSA diagnosis toward the clinic.

   This chapter shows how geometric techniques from statistical estimation and machine learning can be used to develop an algorithm for cancer proteomics, which uses data from a mass spectrometer to distinguish between healthy and diseased blood in humans. While geometry pervades the algorithms in this book, Lilien et al. [1] can be viewed as an investigation of geometry at a larger scale—instead of considering a *single* protein (or a small number of proteins) and their interactions, this chapter discussed algorithms for classifying and characterizing the geometry of the *oncoproteome* (i.e., the space of "all" expressed cancer proteins) as projected onto the mass-to-charge ratios of their proteolytic digest.

Reference

[1] R. Lilien, H. Farid and B. Donald. *Journal of Computational Biology*. 10;6(2003):925–946.

20 Protein Flexibility: Introduction to Inverse Kinematics and Loop Closure Problem

This chapter describes the loop closure problem and inverse kinematics, and their applications in two computational tools for modeling protein backbones: Probik [1] and ChainTweak [2]. We will spend several chapters on the topic of protein flexibility, and computational algorithms to handle it.

20.1 Loop Closure Problem and Exact Inverse Kinematics

20.1.1 Protein Backbone Representations

There are two main models to represent a protein backbone: the *all-atom Cartesian coordinates–based model* and the *dihedral angles–based model*. Probik applies a Cartesian coordinate–based model, while ChainTweak uses a dihedral angle–based model.

In the Cartesian-based model, an energy minimization step is usually required to compute the correct bond lengths or angles of the protein structure. Due to the efficiency and convergence issues, it is not suitable for larger perturbations of the structure. Although a MD-like approach can be applied to relax this limitation, it incurs an expensive computational cost. Alternatively, the dihedral angle–based model has no restriction on small perturbation size, since no energy minimization step is required. However, a dihedral angle change at one residue position will influence the other positions of all subsequent residues.

20.1.2 Loop Closure Problem

Definition: *The n-atom Loop Closure Problem*   Given the fixed positions and orientations of the first and last bonds in an n-atom chain, how can one find the possible positions of the internal $n - 4$ atoms that satisfy the desired constraints on bond lengths, bond angles, and peptide dihedrals?

It has been shown that the loop closure problem with 6 unknown dihedral angles—that is, 6 degrees of freedom (DOFs)—has at most 16 possible solutions, whereas the number of possible solutions is (generically) infinite for the problem with more than 6 DOFs [3].

20.1.3 Denavit-Hartenberg Local Frames

Consider two bonds, b_{i-1} and b_i, as shown in figure 20.1. The *Denavit-Hartenberg (DH)* local frame on bond b_i is defined as $F_i = \{x_i, y_i, z_i, u_i\}$, where x_i, y_i, z_i are three frame axes, and especially z_i is in the same direction as b_{i-1}, and u_i is the origin on bond b_i and closest to bond b_{i-1}. To consider the rotational and translational relationship between DH local frames F_{i-1} and F_i, we define the following geometric parameters [1]:

d_{i-1}: distance along bond b_{i-1} from u_{i-1} to the closest point to u_i;

θ_i: dihedral angle measured from x_i to x_{i+1} about z_i;

a_i: offset distance from u_{i+1} to bond b_i;

α_i: angle from z_i to z_{i+1}.

Let v_{i-1} denote the representation of a vector in frame F_{i-1}, and let v_i denote the representation of the same vector in frame F_i. Then the relationship between v_i and v_{i-1} is formulated as

$$v_{i-1} = A_i v_i,$$

where $A_i = T_z(d) \cdot R_z(\theta) \cdot R_x(\alpha) \cdot T_x(a)$, and $R_x(\omega)$ denotes the 4×4 homogeneous rotation matrix along the x-axis by angle ω, and $T_z(d)$ denotes the 4×4 homogeneous translation along the z-axis by d distance.

Let F_1 and F_{hand} denote the DH frames on the anchor and target atoms (respectively) in an n-bond chain $b_1, \ldots, b_n$. The terminology "hand" comes from thinking of the kinematic chain as a robot arm. Let v' be the representation of a vector with respect to frame F_{hand}, and let v be the representation of the same vector in frame F_1. Then we have $v = A_1 A_2 \cdots A_n v'$, where A_i denotes the transformation matrix from frame F_i to F_{i-1}.

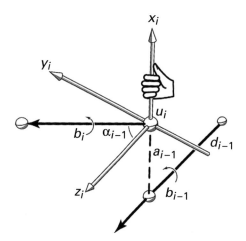

Figure 20.1
Denavit-Hartenberg local frame [1].

20.2   Probik

20.2.1   Overview

Probik models the range of possible local motions of the protein backbone by varying the geometric parameters, such as bond angles, lengths, and dihedrals. It computes derivatives of each atom's position based on an inverse kinematics solver.

20.2.2   Algorithm Description

The following describes the four steps of the Probik algorithm:

1. *Calculate wiggled loops*   Compute the geometric representation and parameters of the 6-DOF loop in the inverse kinematics chain, based on the structural information from the PDB file. At the same time, we generate a list of possible values for each variable parameter, based on the actual value, the ideal value, and the standard deviation of this parameter.

2. *Get exact solutions by using inverse kinematics solver*   Use the inverse kinematics solver in Manocha and Canny [3] to compute all possible conformations for small variations of one variable parameter at a time.

3. *Classify families of poses*   Organize the solution poses and determine the range of allowed motions from the variations of each variable parameter.

4. *Estimate the derivatives*   Estimate the derivatives of motion for each atom in the 6-DOF chain. A quadratic polynomial curve, representing a geometric constraint of each atom, is computed to fit the solution poses.

20.2.3   Exploring Control Parameters Based on Principal Component Analysis

Probik uses principal component analysis (PCA) to find a small number of combinations of parameters, acting as "control handles," that can cover the same space of possible backbone conformations obtained by varying one bond length or angle at a time. The reduction in the number of controls decreases the complexity of backbone manipulation and design.

Probik first generates a large number of perturbed loops by wiggling bond angles and peptide dihedral angles in a normal distribution. Then PCA is applied to compute a low-dimensional approximation of the loop solution space.

20.3   ChainTweak

20.3.1   Overview

Given a base conformation, ChainTweak generates a set of backbone conformations within a user-defined neighborhood of this base. Each new conformation has the first and last residues in their same position as in the base, whereas some dihedral angles in between the first and last residues have changed.

ChainTweak uses a sliding window approach to compute the dihedral angles of some atoms within $0 - 2$ Å while keeping other atoms fixed. In each window, some perturbations are generated to satisfy the tripeptide loop closure problem.

20.3.2 Algorithm Description

Input   A backbone conformation with fixed bond lengths, bond angles, and dihedral angles.

Output   A set of conformations whose RMSDs roughly follow a user-defined distribution.

Preprocessing Step   Produce a set of conformations by randomly sampling dihedral angles at the end-residues (i.e., dihedral angles near the terminal regions of the chain).

A subroutine based on a sliding window approach, SLIDEWIN, is called iteratively by Chain-Tweak until a certain large deviation from the starting conformation is achieved. Each iteration of SLIDEWIN moves each residue by around $0.5 - 1.5$ Å. The output conformation in each iteration is used as the input of the next iteration. Between two iterations, some conformations may be pruned out based on their RMSD from the original conformation and user-defined parameters.

The SLIDEWIN Subroutine   Given an input backbone conformation, a window of 3 residues, that is, 9 points, is chosen. Then the residues within the window, that is, with 3 fixed points on both ends, form a 6-DOF loop closure problem, which can be solved by robotics inverse kinematics algorithms, such as that in Manocha and Canny [3]. There are at most 16 possible solutions for each fixed window. A solution within the acceptable regions of the Ramachandran plot is randomly selected as output of this subroutine.

Table 20.1
How Probik and ChainTweak Work

How They Work	Probik	ChainTweak
Restrict output to realistic conformations	Must satisfy the 6-DOF loop closure problem	Must satisfy the 6-DOF loop closure problem
Loop size	3 or 4 residues	3 residues
Input backbone size	Unlimited Iterates through all 6-DOF loops in the desired region	Unlimited "Sliding window" used to iterate through desired region
Backbone representation	Cartesian coordinates for atoms and derived joints*	Phi and psi dihedral angles
Cover all of the conformational space	Not addressed	"Suggested" to be true, by work on polygonal chains

Credit: Bruce Donald and John MacMaster.

Table 20.2
What Probik and ChainTweak do

Probik	ChainTweak
"Modeling local protein backbone motion"	"Sampling from the neighborhood of a given base conformation"
"Local" = allowed by varying only one peptide dihedral, or bond angle or length within about one standard deviation	"Neighborhood = within 4 Angstroms RMSD of the given base conformation
"Provides the biochemist with the range of motions possible for a fragment of the protein backbone"	"Generates a set of backbone conformations"

Credit: Bruce Donald and John MacMaster. All quotes are from the papers [1, 2].

Table 20.3
Why the results of Probik and ChainTweak are useful

Probik	ChainTweak
Reduce clashes in X-ray crystallographic refinement	Can improve efficiency of loop modeling programs*
Enable backbone movement in protein design	Can generate ensembles to be used to generate decoys*
Estimate the derivative of motion for each atom and model each atom's motion	Being independent of an energy function, can be used in energy function design
PCA-based control handles reduce the complexity of protein backbone manipulation*	Can be used in homology modeling to pick the best energy function*
PCA-based control handles for loops in alpha-helices are applicable in any alpha-helix*	

Credit: Bruce Donald and John MacMaster.

20.4 Comparisons Between Probik and ChainTweak

Tables 20.1, 20.2, 20.3 compare Probik and ChainTweak.

Exercise For each of the "starred" claims (*) in table 20.3, how would you argue these claims are true? For each point, construct a detailed theoretical or empirical demonstration.

References

[1] K. Noonan, D. O'Brien, and J. Snoeyink. Probik: Protein backbone motion by inverse kinematics. *The International Journal of Robotics Research* 24;11(2005): 971–982.

[2] R. Singh, B. Berger. ChainTweak: Sampling from the neighbourhood of a protein conformation. *Pacific Symposium on Biocomputing* (2005): 54–65.

[3] D. Manocha and J. Canny. Efficient inverse kinematics of general 6R maipulators. *IEEE Transactions on Robotics and Automation* 10;5(1994): 648–657.

[4] Available at http://www4.cs.umanitoba.ca/~jacky/Teaching/Courses/74.795-Humanoid-Robotics/ReadingList/chap3-forward-kinematics.pdf.

21   Normal Mode Analysis (NMA) and Rigidity Theory

This chapter introduces normal mode analysis [5], and then presents the FIRST algorithm [9] for evaluating the protein flexibility.

21.1   Normal Mode Analysis

21.1.1   Introduction

Normal mode analysis (NMA) is a standard technique for studying the dynamics of biological systems. It provides an analytical description of the dynamics in a macromolecular system near a minimum by using a harmonic approximation of the potential. NMA can be applied in the analysis of protein flexibility. It has been shown that over half of the 3,800 Yale database of protein movements can be modeled in a structure using at most two low-frequency normal modes [11].

As shown in figure 21.1, there are two kinds of harmonic approximations of the potential: the local minimum approximation and the effective harmonic (global) approximation. These two approximations correspond to different structural points of view. The effective harmonic approximation overlooks the protein structure (around $0.1 \sim 10\,nm$ scale), and thus describes a stable conformation in the overall minimum of potential. In contrast, the local minimum approximation looks closer at the protein structure (around $0.001 \sim 0.1$ nm), and describes a given conformational substate.

The starting point of NMA is finding a particular stable conformation that represents a minimum of the potential energy surface, which can be computed by an energy minimization algorithm. After that, we can construct a *harmonic approximation* of the potential well around this conformation. A harmonic potential well is defined by

$$U(r) = 0.5(r - R) \cdot K(R) \cdot (r - R), \qquad (21.1)$$

where R denotes a $3N$-dimensional[1] vector describing the stable conformation, r denotes the current conformation vector, and K denotes a symmetric and positive semidefinite matrix that describes the shape of the potential well. The matrix K can be obtained by the second derivative

---

[1] N is the total number of atoms.

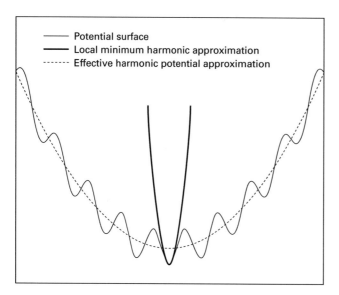

Figure 21.1
Two kinds of harmonic approximations (in one-dimensional view) [8].

of the potential:

$$K_{ij} = \left[\frac{\partial^2 U}{\partial r_i \partial r_j} \right]_{r=R_{\min}} ,$$
(21.2)

where $R_{\min}$ denotes the local minimum.

Figure 21.2 shows the basic idea of normal modes, where r_1, r_2 denote two coordinates of the system, and e_1, e_2 denote the normal modes of the harmonic potential well. The motions in the normal mode directions are independent, whereas the motions in the coordinate directions are not. The normal mode vectors can be obtained from the eigenvectors e_i of the matrix K:

$$K \cdot e_i = \lambda_i e_i, \quad i = 1, \dots, 3N.$$
(21.3)

The associated eigenvalues λ_i represents the curve shape of the potential along the normal mode directions.

Based on the independence of the normal modes, we can simplify the harmonic potential function as follows:

$$U(c) = \frac{1}{2} c \cdot \Lambda \cdot c,$$
(21.4)

where $c_i = (r - R) \cdot e_i$, and $r = R_{\min} + \sum_{i=1}^{3N} c_i e_i$, and Λ is a diagonal matrix consisting of the eigenvalues λ_i.

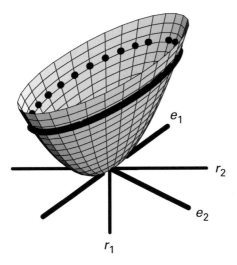

e_1

r_2

e_2

r_1

Figure 21.2
A two-dimensional harmonic potential well [5].

21.1.2 Different Normal Modes

There are four methods for NMA: *vibrational NMA, Brownian NMA, coarse-grained mode*, and *essential dynamics*. Both vibrational NMA and Brownian NMA are derived from a standard all-atom potential by energy minimization, and thus require minimization and Hessian computation. The differences are that vibrational NMA has a short timescale (less than the residence time in a local minimum), and thus is appropriate for studying fast motions, while Brownian NMA has a timescale larger than the residence time in a local minimum, and thus is appropriate for studying slow motions. The coarse-grained mode is used for a given structure, and does not require the expensive minimization and Hessian computation, but instead fits an energy function to the given structure using a spring model. Its timescale is much larger than the residence time in a local minimum, and it is suitable for studying slow, diffusive motions. Essential dynamics is used to find the coordinates that contribute significantly to the fluctuations from a given set of structures that reflect the flexibility of the molecule. It has a timescale much larger than the residence time in a minimum, but cannot capture the fine-level intricacies of the motion. Table 21.1 compares of these four modes.

We recommend the tutorial articles [5, 1] for a more detailed discussion of NMA. A great deal of work has been done on developing and applying NMA methods to analyze biological macromolecular structures, including proteins and ribonucleic acids. For more on these NMA methods and applications see studies listed in the References [2, 3, 4, 6, 7, 10, 12, 15, 16, 17, 18, 19, 20, 21, 22, 23].

Table 21.1
Comparison of normal modes

	Amplitude	Timescale	Starting Structure	Practical
Vibrational	Small	Short	By minimization	N
Brownian	Large	Long	By minimization	Y/N
Coarse-grained	Large	Long	Given	Y
Essential	Large	Long	Given	N

21.2 Protein Flexibility Predictions Using Graph Theory

21.2.1 Overview of FIRST

Given a set of distance constraints, including covalent bonds, hydrogen bonds, and salt bridges, Floppy Inclusion and Rigid Substructure Topography (FIRST) [9] provides the precise mechanical properties of a protein structure. FIRST can compute the rigid regions, and other regions that can move collectively or independently.

Table 21.2 compares FIRST and NMA.

The mathematical basis of the graph theoretic FIRST approach has its roots in Laman's theorem and a family of algorithms generally known as *rigidity percolation*, including the "Pebble Game." We introduce these concepts in the following sections. They are sometimes taught as an allied part of *distance geometry* (cf. chapters 31, 32, and 36). Most distance geometry algorithms are introduced with the goal of determining a *unique* or *maximum likelihood* macromolecular structure. In contrast, in this chapter we discuss rigidity theory from the contrapositive point of view, to calculate *flexible* regions of a protein. In this light, it is natural to compare and contrast with other methods for modeling protein dynamics, such as NMA. We explore these ideas next.

21.2.2 Rigidity Theory

A protein structure can be formulated into a constraint network, where nodes represent the atoms and edges represent the distance constraints. A constraint is *redundant* if its breakage does not change the flexibility of the network, otherwise it is considered *independent*. The following procedure can be used to test whether a constraint is redundant or independent:

1. Construct a constraint network representing the structure of the given protein.

2. Replace each distance constraint by a spring, and construct a dynamical matrix for the spring network. Then calculate all normal mode frequencies and count the number of zero eigenvalues.

3. Add the test constraint to the network, and count the number of zero eigenvalues. If the number of zero eigenvalues remains constant, then the constraint is redundant; otherwise, it is independent.

Table 21.2
Comparison of FIRST and NMA

	All-Atom?	Speed?	Given Starting Pt.	Freq. Spectrum?	Way of Generating New Conformations?	Flexibility or Mobility Index
FIRST	Y	Y	Y	Low-frequency motion	Y(with ROCK or FRODA)	Y(incorrect for rigid regions flanked by flexible hinges)
NMA	Y / N (N for coarse-grained)	Y (N for VNMA, Brownian, ED)	Y	All frequencies	Y	Y

This algorithm runs in $O(n^2 \cdot n^3) = O(n^5)$ time, where n is the number of atoms in the network. Laman's theorem is used to calculate the generic rigidity of a 2D framework:

Laman's Theorem [13, 14] *Let G be a graph having exactly $2n - 3$ edges, where n is the number of vertices in G. Then G is* generically rigid *in $\mathbb{R}^2$ if, and only if, $e' \leq 2n' - 3$ for every subgraph of G having n' vertices and e' edges.*

Applying Laman's theorem directly requires exponential time complexity. However, applying it recursively, as shown in the Pebble Game algorithm [9], takes only $O(n^2)$ time, where n is the number of the atoms in the framework.

Notes Laman's theorem cannot be simply generalized to a 3D framework. So far, it can be generalized only in some special class of 3D frameworks. For more details, please refer to the generic rigidity section in [9].

21.2.3 Pebble Game Analysis

In the 3D Pebble Game algorithm, each node or atom has three pebbles that represent the 3 degrees of freedom of the atom. Each independent edge uses up 1 degree of freedom from one of its incident nodes. We can rearrange pebbles throughout the network while obeying the following pebble covering rule: once an edge has been covered by a pebble, it should be kept covered by one of the pebbles from its incident nodes. Figure 21.3 shows a pebble covering in a network.

The 3D Pebble Game algorithm runs in the following recursive way. Each distance constraint is added at a time, and for each new independent distance constraint, pebbles are rearranged to test whether this constraint is independent. The testing is done by checking whether a pebble can be moved from one of the incident nodes to this new constraint. If a pebble can be collected at both ends, then it is independent; otherwise, it is redundant. This process is repeated until all distance constraints have been placed and tested. From the pebble covering rule, the free pebbles are restricted to certain regions during all rearrangements. The total number of free pebbles on the

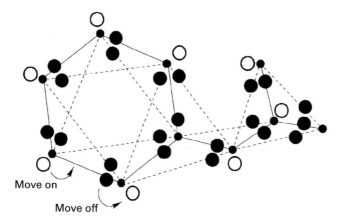

Figure 21.3
A pebble covering. Free pebbles are placed directly on vertices and denote degrees of freedom (o). Pebbles covering
a bond are placed directly on the edges of the graph and represent distance constraints (•). The pebble covering is not
unique, because pebbles can be rearranged according to a few simple rules; see [9]. An example is shown of how an
elementary pebble exchange works (two arrows). A free pebble can be moved onto a covered edge (adjacent to a vertex),
provided that a corresponding pebble, presently covering the edge and associated with the neighboring vertex, is moved
off [9].

nodes in a framework represents the number of degrees of freedom associated to this framework.
Thus, based on the location and number of free pebbles throughout the framework, we can determine the rigid, overconstrained, or underconstrained regions.

References

[1] I. Bahar and A. J. Rader. Coarse-grained normal mode analysis in structural biology. *Current Opinion in Structural
Biology* 15;5(2005 Oct):586–592.

[2] B. Brooks and M. Karplus. Harmonic dynamics of proteins: Normal modes and fluctuations in bovine pancreatic
trypsin inhibitor. *Proceedings of the National Academy of Sciences (U S A)* 80;21(1983 Nov):6571-6575.

[3] B. Brooks and M. Karplus. Normal modes for specific motions of macromolecules: application to the hinge-bending
mode of lysozyme. *Proceedings of the National Academy of Sciences U S A*. 82;15(1985 Aug):4995-4999.

[4] A. Ghysels, V. Van Speybroeck, E. Pauwels, S. Catak, B. R. Brooks, D. Van Neck, and M. Waroquier. Compara-
tive study of various normal mode analysis techniques based on partial Hessians. *Journal of Computational Chemistry*
31;5(2010 Apr 15):994–1007.

[5] K. Hinsen. Normal Mode Theory and Harmonic Potential Approximation, 2000. Available at: http://dirac.cnrs-orleans
.fr/~hinsen/.

[6] K. Hinsen. Analysis of domain motions by approximate normal mode calculations. *Proteins* 33;3(1998 Nov 15):
417–429.

[7] K. Hinsen, A. Thomas and M. J. Field. Analysis of domain motions in large proteins. *Proteins* 34;3(1999 Feb 15):
369–382.

[8] K. Hinsen. Normal mode analysis: Applications. EMBO practical course on biomolecular simulation. European
Molecular Biology Laboratory (EMBL), Meyerhofstr. 1, D-69117 Heidelberg, Germany, 2000. Accessed 2005–2006 at
http://dirac.cnrs-orleans.fr/plone/Members/hinsen/courses-and-lecture-notes/embo-course-on-biomolecular-simulations/
embo-course- 2000/embo_2000_normal_modes_practical.pdf/view.

[9] D. J. Jacobs, A. J. Rader, L. A. Kuhn, and M. F. Thorpe. Protein flexibility predictions using graph theory. *Proteins: Structure, Function, and Genetics* 44(2004):150–165.

[10] M. K. Kim, R. L. Jernigan, and G. S. Chirikjian. An elastic network model of HK97 capsid maturation. *Journal of Structural Biology* 143;2(2003 Aug):107–117.

[11] W. G. Krebs, V. Alexandrov, C. Wilson, N. Echols, H. Yu, and M. Gerstein. Normal mode analysis of macromolecular motions in a database framework: Developing mode concentration as a useful classifying statistic. *Proteins: Structure, Function, and Genetics* 48;4(2002): 682–695.

[12] O. Kurkcuoglu, Z. Kurkcuoglu, P. Doruker, and R. L. Jernigan. Collective dynamics of the ribosomal tunnel revealed by elastic network modeling. *Proteins* 75;4(2009 Jun):837–845.

[13] G. Laman. On graphs and rigidity of plane skeletal structures. *Journal of Engineering Mathematics* 4(1970): 331–340.

[14] Eric W. Weisstein. Laman's Theorem. *From MathWorld—A Wolfram Web Resource*. Available at: http://math world.wolfram.com/LamansTheorem.html.

[15] G. Li and Q. Cui. A coarse-grained normal mode approach for macromolecules: An efficient implementation and application to Ca(2+)-ATPase. *Biophysical Journal* 83;5(2002 Nov):2457-2474.

[16] A. D. Schuyler and G. S. Chirikjian. Normal mode analysis of proteins: A comparison of rigid cluster modes with C(alpha) coarse graining. *Journal of Molecular Graphics Modeling* 22;3(2004 Jan):183–193.

[17] A. D. Schuyler, R. L. Jernigan, P. K. Qasba, B. Ramakrishnan, and G. S. Chirikjian. Iterative cluster-NMA: A tool for generating conformational transitions in proteins. *Proteins* 74;3(2009 Feb 15):760–776.

[18] F. Tama and Y. H. Sanejouand. Conformational change of proteins arising from normal mode calculations. *Protein Engineering* 14;1(2001 Jan):1–6.

[19] F. Tama, O. Miyashita, and C. L. Brooks, 3rd. Flexible multi-scale fitting of atomic structures into low-resolution electron density maps with elastic network normal mode analysis. *Journal of Molecular Biology* 337;4(2004 Apr 2): 985–999.

[20] F. Tama, F. X. Gadea, O. Marques, and Y. H. Sanejouand. Building-block approach for determining low-frequency normal modes of macromolecules. *Proteins* 41;1(2000 Oct 1):1–7.

[21] A. Van Wynsberghe, G. Li, and Q. Cui. Normal-mode analysis suggests protein flexibility modulation throughout RNA polymerase's functional cycle. *Biochemistry* 43;41(2004 Oct 19):13083–13096.

[22] W. Zheng, B. R. Brooks, and D. Thirumalai. Allosteric transitions in the chaperonin GroEL are captured by a dominant normal mode that is most robust to sequence variations. *Biophysical Journal* 93;7(2007 Oct 1):2289–2299. (Epub June 8, 2007.)

[23] W. Zheng, B. R. Brooks, and D. Thirumalai. Allosteric transitions in biological nanomachines are described by robust normal modes of elastic networks. *Current Protein and Peptide Science* 10;2(2009 Apr):128–132.

22   ROCK and FRODA for Protein Flexibility

This chapter describes two examples of using the algorithmic approach FIRST to explore protein flexibility: ROCK [1, 2] and FRODA [3]. FIRST is introduced in chapter 21.

22.1   The ROCK Algorithm

22.1.1   Terminology

Ring   A closed loop of bonds in which any two atoms are connected by two distinct paths.

Interlocked rings   Two rings that share common bond(s).

Ring cluster   A collection of rings that are interlocked with one another.

Side branches   All atoms that do not belong to any ring cluster.

Figure 22.1 shows examples of ring, interlocked ring, ring cluster, and side branch.

22.1.2   Overview

Rigidity Optimized Conformational Kinetics (ROCK) is an algorithm that generates conformations for a protein that satisfy the flexibility constraints. ROCK explores flexible regions identified by FIRST by a random-walk sampling of the rotatable bonds.

The following shows the steps of the ROCK algorithm:

1. Get the flexibility information from FIRST (chapter 21).

2. Generate conformations for ring clusters. Reject unclosed conformations.

3. Attach and perturb side branches. Reject conformations with steric clash or bad bond length/angles.

4. Reject conformations with bad stereochemistry (analogous to the Ramachandran plot for proteins).

5. Reject conformations with missing hydrophobic interactions.

The following sections present the details of these steps.

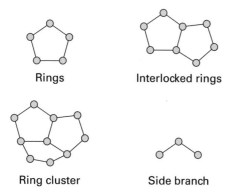

Figure 22.1
Examples of ring, interlocked ring, ring cluster, and side branch.

22.1.3 Conformation Sampling in Single-Ring Closure

Although it is easy to generate conformations for side branches, it is not trivial to generate conformations for ring clusters, since the dihedral angles in a ring are all correlated.

An N-fold ring closes if the following equations are satisfied:

$$P_0 + T_0 R_1 P_1 + T_0 R_1 T_1 R_2 P_2 + \cdots + T_0 R_1 \cdots T_{N-2} R_{N-1} P_{N-1} = 0 \tag{22.1}$$

$$T_0 R_1 T_1 R_2 \cdots T_{N-2} R_{N-1} T_{N-1} R_N = I, \tag{22.2}$$

where P_0 is the distance vector, T_i is the rotational matrix along the bond angle, and R_i is the rotational matrix along the dihedral angle.

The total DOFs of an N-fold ring are $3N - N - N - 6 = N - 6$. If $N - 6$ dihedral angles in an N-fold ring are fixed, the remaining six dihedral angles can be obtained by solving the ring closure equations. Conformational sampling in an N-fold ring is performed by altering $N - 6$ dihedral angles randomly and then computing the remaining six unknown dihedral angles by solving the ring closure equations.

22.1.4 Conformation Sampling in Multiple-Ring Closure

First, a fictitious closure potential $\mathcal{F}$ is defined by

$$\mathcal{F} = [P_0 + T_0 R_1 P_1 + T_0 R_1 T_1 R_2 P_2 + \cdots + T_0 R_1 \cdots T_{N-2} R_{N-1} P_{N-1}]^2$$

$$+ \sum_{i,j=3}^{3} [T_0 R_1 T_1 R_2 \cdots T_{N-2} R_{N-1} T_{N-1} R_N - I]_{ij}^2.$$

To close all rings, we need to minimize $\mathcal{A} = \sum_{\text{all ring}} \mathcal{F}$, the total fictitious potential of all rings in the ring cluster. In a closed ring cluster, the total fictitious potential is minimized to zero.

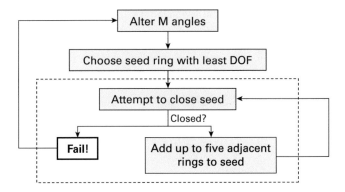

Figure 22.2
The iterative approach for minimizing the total fictitious ring closure potential.

Suppose that we have a set of M dihedral angles in an N-atom ring cluster, then computing the minimum value of the total fictitious potential takes $O((N - M)^3) = O(N^3)$ time, which is not feasible for large N. An improved approach is to iteratively close rings, that is, to close one ring in the ring cluster at each iteration. Let n denote the total number of rings. Since the average number of variables per ring is N/n, the total computational cost is $O((N/n)^3 \cdot n) = O(N^3/n^2)$, which is $O(N)$ when $n = \Theta(N)$. In practice, however, it is impossible to close all the rings in an interlocked ring cluster iteratively, since closing some rings unavoidably opens other rings. ROCK applies an iterative scheme to solve the ring closure equations of all rings in a ring cluster, as shown in figure 22.2.

22.1.5 Conformation Sampling in Side Branches
After new conformations of the ring clusters have been generated, side branches are set with correct bond lengths and angles. The atomic positions in side branches are first randomly perturbed, and then relaxed such that (1) bond lengths and angles of side branch atoms are maintained; (2) there are no van der Waals overlaps between side branches themselves, nor between side branches and ring clusters; and (3) chirality at side branches is maintained. A nonlinear optimization toolbox (DONNLP2) is used to find the optimal solutions of side branches while satisfying these constraints.

22.1.6 Hydrophobic Interactions and Ramachandran Checks
Once each new conformation is generated, the distances between pairs of carbon atoms need to be checked to determine whether they are in the hydrophobic interaction list. The new conformation is accepted if the distances are within the allowed hydrophobic interaction distance; otherwise it is rejected.

The (ϕ, ψ) dihedral angles are also checked against the Ramachndran plot. Conformations with bad stereochemistry are rejected.

22.2   Application of ROCK in Flexible Docking

The following steps show how to apply ROCK in protein-ligand flexible docking:

1. Analyze rigidity of protein and ligand by using FIRST.
2. Apply ROCK to generate conformations for protein and ligand.
3. Apply SLIDE to dock all possible pairs of conformations.

SLIDE is a modeling tool for protein-ligand interactions based on steric complementarity, hydrophobic, and hydrogen-bonding interactions.

22.3   FRODA

22.3.1   Overview

Framework rigidity optimized dynamic algorithm (FRODA) is designed for exploring the internal mobility of proteins. In FRODA, the rigid regions of the protein are first identified by using FIRST, and replaced by ghost templates that guide the movements of atoms. The conformational space is explored using random moves. The covalent, hydrophobic, and hydrogen bond constraints, and the avoidance of van der Waals overlaps are all considered.

22.3.2   The FRODA Algorithm

In FRODA, a ghost template represents a rigid body that consists of a set of mutually rigid atoms. Atoms are bound to the vertices of the ghost templates. The following iterative process is used to enforce the bond length/angle constraints: The ghost templates are fit to the atom positions by using a least-squares approach, and then atoms are fit to the vertices to which they are bound. During geometric simulation of the protein motions, we vary some dihedral angles while keeping the bond length and angle constraints unchanged. For intuition, imagine there are only two ghost templates. The preceding iterative process is quite similar to the *expectation/maximization* approach in linear regression, and will typically converge to a new conformer. For example, in an E/M linear regression problem, we are given a set of points in an Euclidean space, and try to find two optimal lines that best fit these points. A difference between the linear regression and the iterative process here may be that, not only the lines (corresponding to the ghost templates) but also the points (corresponding to atoms) are allowed to move, which in fact speeds up the convergence. Figure 22.3 (plate 23) shows a simulation example from an ethane molecule's motion.

The steric clash problem is handled by adding a small bias to the atomic motion during the fitting process between atoms and templates, as shown in figure 22.4 (plate 24). A similar method

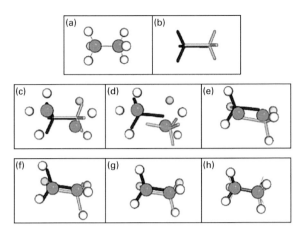

Figure 22.3 (plate 23)
The motion of an ethane molecule as determined by geometric simulation: (a) Initial atomic positions; (b) ghost templates; (c) random atomic displacement; (d) fitting of ghost templates to atoms; (e) refitting of atoms to ghost templates; (f) and (g) further iterations of (d) and (e); (h) until a new conformer is found [3].

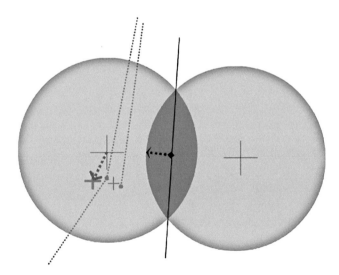

Figure 22.4 (plate 24)
The method to handle steric overlap. When atoms come into contact, the overlap influences the movement of the atoms during the relaxation. The atom on the left is tethered to sites (blue dots) in two ghost templates (shown as blue dotted lines). The center of the atom, shown as a large black cross, would move to the small light blue cross in order to match the midpoint of its two associated ghost template sites. However, the steric overlap (black dashed arrows) causes it to move further away from the touching atom, so that its new position is at the larger blue cross, and the resultant motion is shown by the dashed red arrow [3].

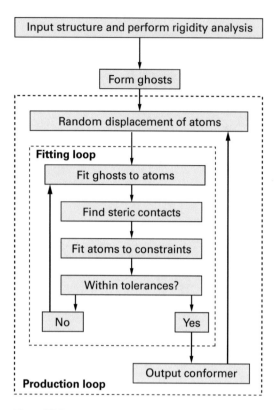

Figure 22.5
The program flow of the FRODA algorithm. An initial rigidity analysis leads to the creation of the ghost templates. Each iteration of the FRODA loop involves an initial random displacement of the atoms followed by multiple iterations of the fitting loop, in which the geometric and steric constraints are enforced [3].

is also applied to ensure hydrophobic relationships. The Ramachandran plot is used to check and reject conformers with bad (ϕ, ψ) dihedral angles.

Figure 22.5 shows the flow of the FRODA algorithm, and figure 22.6 (plate 25) illustrates the exploration of the conformation space.

22.3.3 Comparisons Between ROCK and FRODA
FRODA is much faster (about 100 to 1,000 times) than ROCK, since ROCK spends a lot of time in rejecting possible conformers, while FRODA does not solve expensive closure equations or perform nonlinear optimization.

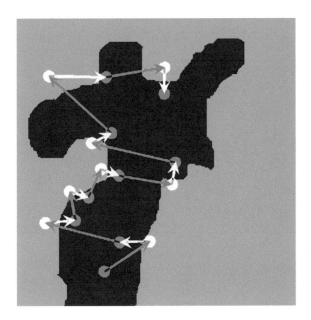

Figure 22.6 (plate 25)
Exploration of the conformation space. Here we show a sketch of a two-dimensional slice through the $3N$-dimensional phase space of the protein, divided into allowed (blue) and forbidden (red) zones. Successive random moves (green arrows) followed by enforcement of the constraints (yellow arrows) produces a set of conformers (green circles) that explore the allowed region [3].

References

[1] M. Lei, M. I. Zavodszky, L. A. Kuhn, and M. F. Thorpe. Sampling protein conformations and pathways. *Journal of Computational Chemistry* 25(2004):1133–1148.

[2] M. I. Zavodszky, M. Lei, M. F. Thorpe, A. R. Day, and L. A. Kuhn. Modeling correlated main-chain motions in proteins for flexible molecular recognition. *Proteins: Structure, Function, and Genetics* 44(2004):150–165.

[3] S. Wells, S. Menor, B. Hespenheide, and M. F. Thorpe. Constrained geometric simulation of diffusive motion in proteins. *Physical Biology* 2, (2005):S127–S136.

23 Applications of NMA to Protein-Protein and Ligand-Protein Binding

This chapter describes the applications of NMA to protein-protein interactions [1] and protein-ligand binding [2].

23.1 Structure Changes for Protein Binding in the Unbound State

23.1.1 Classical Models for Protein-Protein Interactions

There are three main models for describing the mechanisms of protein interactions: the "lock-and-key" model, the "induced fit" model, and the preexisting equilibrium model, as shown in figure 23.1.

The *lock-and-key* model emphasizes the importance of shape complementarity between two rigid structures. It cannot be used to explain the conformational changes before and after binding. The *induced fit* model considers the conformation change due to nonrigid protein interactions. A geometric fit is ensured only after the structural rearrangement of the base protein. In the *preexisting equilibrium model*, an ensemble of closely related conformations in equilibrium is used to represent the native state of a protein, and for different substrates, there are different suitable conformations to be used for binding. The preexisting equilibrium model can be combined with the induced fit model to explain protein binding.

23.1.2 Gaussian Network Model (GNM)

In a GNM, each residue is represented by a node centered at its C^α atom position. Any two nodes within the distance r_c are connected by a spring of force constant γ. Thus, we have Gaussian fluctuations for each node and within interresidue distance. The topology of an N-node network is represented by a *Kirchhoff matrix* Γ, that is, for any given i, $1 \leq i \leq N$, we have

$$\Gamma_{ij} = \begin{cases} -1, & \text{if } i \neq j \text{ and } R_{ij} \leq r_c \\ 0, & \text{if } i \neq j \text{ and } R_{ij} > r_c \\ -\sum_{i \neq j} \Gamma_{ij}, & \text{if } i = j \ . \end{cases} \tag{23.1}$$

(a)

(b)

(c)

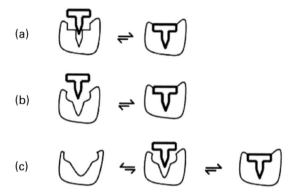

Figure 23.1
Models for protein-protein interactions: (a) Lock-and-key mechanism; (b) induced fit; (c) preexisting equilibrium followed by induced fit [1]. Copyright 2005 National Academy of Science.

The motions along different modes are computed by eigenvalue decomposition $\Gamma = U \Lambda U^{-1}$, where U denotes the orthogonal matrix of eigenvectors, and Λ denotes the diagonal matrix of the eigenvalues λ_k, $1 \leq k \leq N$. The eigenvalues represent the frequencies of the nonzero GNM modes, and the ith element $(u_k)_i$ along the kth eigenvector direction represents the fluctuation of residue i from its equilibrium position in the kth principal axis direction.

The cross-correlation between fluctuations of residues i and j is

$$\langle \Delta \vec{R}_i \cdot \Delta \vec{R}_j \rangle = \frac{3k_B T}{\gamma} [\Gamma^{-1}]_{ij},$$

where $[\Gamma^{-1}]_{ij}$ denotes the inverse of the ijth element in Γ. The mean-square fluctuation is obtained by replacing j by i.

The limitation of GNM is that it assumes fluctuations are isotropic, and thus only predicts the motions in their amplitudes, not their directions. On the other hand, the anisotropic network model (ANM), detailed in the next subsection, can provide the directions of collective motions.

23.1.3 Anisotropic Network Model (ANM)

An ANM is a normal mode analysis, where the Hessian matrix H is based on a harmonic potential:

$$V_{ANM} = \frac{\gamma}{2} \sum_{i,j}^{N} (R_{ij} - R_{ij}^0) \Gamma_{ij},$$

where R_{ij}^0 and R_{ij} represent the original and deformed distances between residues i and j correspondingly. By eigenvalue decomposition of H, we have $3N - 6$ eigenvectors $(\vec{u}_k^{ANM})^T =$

$[u_k^{X1}, u_k^{Y1}, u_k^{Z1}, \cdots, u_k^{ZN}]$, where $u_k^{Xi}, u_k^{Yi}, u_k^{Zi}$ represent the motions of residue i in the x, y, z directions along the kth mode.

The ANM modes can be mapped to GNM ones by comparing the squares of fluctuations between the resulting modes in two models.

23.2 Receptor Flexibility Representation Through Relevant Normal Modes

23.2.1 Methodology Overview

In Cavasotto et al. [2], a low-frequency normal-mode-based algorithm is presented to generate *multiple receptor conformations (MRCs)* that incorporate the receptor flexibility in ligand docking and virtual screening. First, we have the following observations:

1. The first low-frequency model is not always the best candidate for docking.

2. The conformation space grows rapidly with the number of normal modes considered; so does the time complexity.

3. Irrelevant modes add noise to energy calculation and minimization.

Based on these observations, the notion of *measure of relevance* of normal modes on selected regions is used to limit the number of modes for generating MRCs.

The following are the four main steps of the methodology:

1. Determine the relevant normal modes that represent binding pocket flexibility.

2. Generate a de novo ensemble of MRCs through perturbing the structure along the relevant modes.

3. Optimize side-chain conformations by complexing the receptor conformational ensemble with known binders. Minimize the global energy by using a flexible ligand/flexible side-chain approach.

4. Screen the generated MRCs by using *receptor ensemble docking (RED)*. Combine the results and search for improvements.

The next sections give the details of these four steps.

23.2.2 Determination of the Relevant Normal Mode

First, a coarse-grained NMA method using C^α atoms is applied. Second, two adjacent regions along the C^α chain are defined, as shown in figure 23.2, where region A surrounds the ends of the loop, and region B includes the central atoms of the loop. The relevance of mode n on the loop is defined by

$$\rho(n) = \frac{\|d_n\|_{2,A}}{\|d\|_{2,A}} + \frac{\|d - d_n\|_{2,B}}{\|d\|_{2,B}} + \frac{\|m_n\|_{2,B}}{\|m\|_{2,B}} + \frac{\|m - m_n\|_{2,B}}{\|m\|_{2,B}}, \tag{23.2}$$

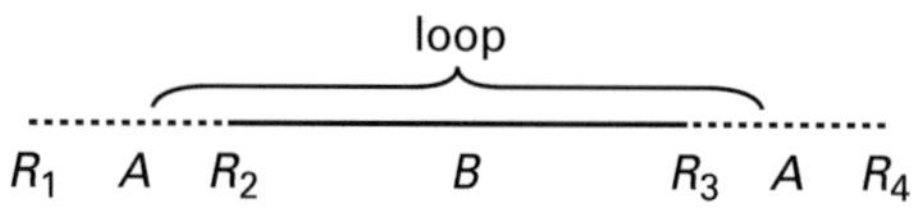

loop

R_1 A R_2 B R_3 A R_4

Figure 23.2
Notation in the definition of the *measure of relevance* [2]. Reprinted with permission. Copyright 2005 American Chemical Society.

where

$$m^2 = \sum_{n=7}^{3N}\left(\frac{\|u^n\|}{\omega_n}\right)^2 \equiv \sum_{n=7}^{3N} m_n^2, \ \ d^2 = \sum_{n=7}^{3N}\left(\frac{\|S_{u^n}\|}{\lambda_n}\right)^2, \ \text{and} \ \ \|d\|_{2,A} = \left(\sum_{j\in A} d(j)^2\right)^{1/2}.$$

23.2.3   Generation of MRCs

Given a set of selected relevant modes, an ensemble of MRCs is generated in the following way:

1. Normalize the selected relevant modes.

2. Form a linear combination of normalized relevant modes.

3. Deform the original structure by displacing all atoms in each residue by the linear combination of relevant modes.

4. Generalize an ensemble of conformations by tethering and energy-minimizing the original structure.

5. Cluster the structures within the ensemble, and then reduce the number of conformations.

23.2.4   Side-Chain Optimization

The global energy is evaluated and minimized by

1. random conformation change according to a predefined probability distribution;

2. local energy minimization of differentiable terms followed by global energy reevaluation including nondifferentiable terms; and

3. acceptance or rejection on the basis of the Metropolis criterion.

There are three atom layers for side-chain optimization: Zone 1, the ligand plus side-chains within 5.5 Å of the ligand-binding pocket; Zone 2, side-chains within 4.5 Å of Zone 1; Zone 3, the rest of the receptor.

23.2.5   Small-Scale Virtual Screening Using RED

Virtual screening is a strategy using computational analysis to select a subset of compounds appropriate for a given receptor. The following pairs of receptor structures were used for a small-scale virtual screening: 1FMO and receptor A, 1JLU and receptor B, where receptor structures A and B were extracted from complexes A and B. A 1,000-compound library consisting of randomly selected molecules seeded with known cAPK binders was docked and scored to the MRCs using RED. A merging-shrinking procedure was used to sort and merge the screening results, and the best rank for each compound was kept.

References

[1] D. Tobi and I. Bahar. Structural changes involved in protein binding correlate with intrinsic motions of proteins in the unbound state. *PNAS* 102;52(2005):18908–18913.

[2] C. N. Cavasotto, J. A. Kovacs, and R. A. Abagyan. Representing receptor flexibility in ligand docking through relevant normal modes. *Journal of the American Chemistry Society* 127;26(2005 Jul 6):9632–9640.

24 Modeling Equilibrium Fluctuations in Proteins

In this lecture, we introduce the *fragment ensemble method* to capture the mobility of a protein fragment, such as a missing loop, and its extension into the *protein ensemble method* to characterize the mobility of an entire protein at equilibrium [1, 2].

24.1 Missing Loops and Protein Flexibility

Proteins are flexible macromolecules. Under equilibrium conditions, a protein may populate a large ensemble of different structures that could be essential for their biological functions. In particular, loop fragments are sometimes highly flexible even in generally stable proteins, and therefore they cannot always be characterized by X-ray crystallography. In such cases, partially-resolved protein structures are reported with loop fragments missing.

Methods for loop modeling can be broadly classified into two categories. *Database methods* search for candidate loops that satisfy geometric constraints and constraints on length in homologous proteins available in structural databases, such as Protein Data Bank (PDB) and loop libraries. The limitation of this approach is the loop diversity, which was addressed recently using a divide-and-conquer approach or by assembling long missing loops from small fragments sampled from a loop library. Loops of size up to 15 residues can be modeled by database methods. *Ab initio methods* may sample from a discretized solution space (e.g., (ϕ, ψ) map) and then refine through molecular dynamics simulations, Monte Carlo searches with simulated annealing, genetic algorithms, dynamic programming, bond scaling with relaxation or multicopy searches, or adapt efficient robotics-inspired sampling algorithms to model loops of arbitrary length. Robotics-inspired ab initio methods employ a probabilistic sampling framework: Loop conformations are first sampled ignoring the constraints and later these constraints are enforced through gradient descent, or the satisfaction of constraints is integrated in the sampling process. In the latter case, a loop conformation that satisfies the constraints on its termini is found by solving an inverse kinematics (IK) problem.

24.2 Materials and Methods

24.2.1 Fragment Ensemble Method (FEM)

Given an incomplete protein structure and the amino acid sequence of the missing fragment, FEM can generate an ensemble of physical conformations that fit with the given protein structure. The FEM procedure can be described as follows (figure 24.1, plate 26):

Step 0: Obtain initial conformations   The unknown fragments (i.e., a missing loop) are initially obtained from a sequence-homologous protein structure from PDB, and the missing atom information is obtained from PSFGEN package. The fragments generally do not fit with the rest of the protein, owing to their nontrivial dependence on the amino acid sequence and the environment provided by the rest of the protein.

Step 1. Backbone geometric exploration   An idealized geometry model is used here, and the bond lengths and bond angles are kept fixed to their equilibrium values. The only degrees of freedom (DOFs) employed at this stage are the (ϕ, ψ) backbone dihedral angles starting at residue $n_1 + 1$ and ending at residue $n_2 - 1$, where the missing fragment is between the residues n_1 and n_2. A set of initial conformations for the backbone of the fragment is generated by sampling the values of the DOFs uniformly at random from $[-\pi, \pi]$, and each of these generated fragments was closed by using the cyclic coordinate descent (CCD) algorithm that finds the minimum distance between the residue n_2 of the fragment and its target pose in the given protein structure by solving the following IK problem: "Given the positions of the backbone atoms of the stationary anchor n_2, assign values to the DOFs of the kinematic chain modeling the fragment so that the backbone atoms of the mobile anchor n_2 assume their target positions in the stationary anchor." Note that before applying the CCD algorithm, the mobile anchor n_1 of the fragment is attached

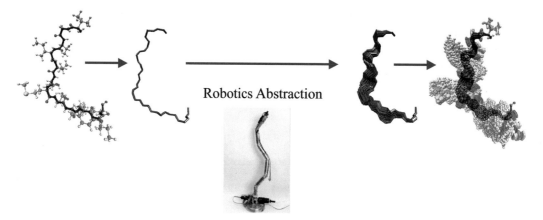

Figure 24.1 (plate 26)
Overview of fragment ensemble method (FEM) (1) Backbone geometric exploration: ensemble of backbones that fit the missing loop; (2) side-chain exploration for a fixed backbone: loop in all-atom detail and without steric clash; (3) all-atom energy refinement: energy-minimized ensemble of loops. Credit: Lydia Kavraki.

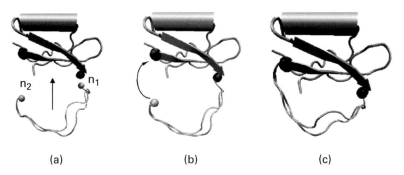

Figure 24.2
(a) The mobile anchors in two different polypeptide chains for the CI2 VAL53-ASP64 loop fragment, drawn in gray, are not attached to the stationary anchors drawn in black. (b) Mobile anchor n_1 is attached to its corresponding stationary anchor through rigid body transformations. (c) Rotations of the dihedral bonds of the fragment steer the other mobile anchor n_2 toward its target pose in the stationary anchor [1].

to the corresponding stationary anchor by rigid body transformations. Refer to figure 24.2. In the backbone geometric exploration step, a fragment is said to be *collision-free* if its energy is below a maximum energy value E_{max} (i.e., 5,000 kcal/mol).

Step 2. Side-chain exploration for a fixed backbone   In this step, we sample the values of the side-chain dihedrals uniformly at random in the interval $[\pi, -\pi]$ until an all-atom fragment conformation is found such that the corresponding protein conformation C is collision-free.

Step 3. Energetic refinement of a modeled fragment   The *closure-constrained backbone re-finement* and the *closure-constrained conjugate gradient descent* strategies are performed in this step to refine each completed protein conformation C generated from the last step. The first strategy is to minimize the energy of a completed protein conformation while keeping the fragment anchored and modifying the dihedrals during minimization, and the latter method relaxes the idealized geometry model and allows more mobility of the fragment's atoms for stable interactions with the rest of the protein. Since both of the aforementioned methods can be trapped in local minima, an interleaving minimization procedure used the two preceding methods in interleaving mode in the hope of discovering a better energy minimum (which can still be a local minimum, if we are unlucky).

FEM yields many all-atom closed fragment conformations of low energy, and the work [1] indicates that the fragment conformations can be generated independently from one another and their computations are easily distributed. Also, it is possible to employ a Boltzmann probability-based statistical mechanics model to weight the probability of a local fluctuation.

24.2.2   Protein Ensemble Method (PEM)
PEM, which employs FEM as a component to capture equilibrium fluctuations over an entire protein, has the following three steps.

Step 1. Define consecutive overlapping fragments The polypeptide chain of the protein is divided into a set of consecutive fragments $\mathcal{F}$ with significant overlap to characterize the flexibility of the entire protein. In Shehu et al. [1], the fragment size (or *window size*) is chosen to be 30 residues long and the overlap between two consecutive fragments is 25 residues.

Step 2. Obtain fragment ensembles For each fragment $F \in \mathcal{F}$, an ensemble of physical fragment conformations are obtained using FEM.

Step 3. Combine fragment ensembles The fluctuations of the overlapping fragments are combined to obtain equilibrium fluctuations of the entire chain. Any measurable quantity X_i for residue i can be obtained as a weighted average of the values measured for the residue i over all the overlapping fragments in which residue i is present. For example, if the window size and the overlap are set to be 30 and 25, respectively, the residue 19 is then contained in fragments [1, 30], [5, 35], [10, 40], and [15, 35]. Here the square brackets denote fragments, e.g., [1, 30] is residues [1 − 30]. Therefore, any measurable quantity X_i, for $i = 19$ can be obtained independently over these four fragment ensembles $\langle X_{19} \rangle_{[1,30]}$, $\langle X_{19} \rangle_{[5,35]}$, $\langle X_{19} \rangle_{[10,40]}$, and $\langle X_{19} \rangle_{[15,45]}$. Finally, the average value $\langle X_i \rangle$, is defined by averaging over all the different fragment ensembles embracing residue i, $\{[n_1, n_2] \mid i \in [n_1, n_2]\}$ as

$$\langle X_i \rangle = \sum_{\{[n_1,n_2]|i\in[n_1,n_2]\}} \frac{\langle X_i \rangle w(i, [n_1, n_2])}{\mathbf{N}},$$

in which $\mathbf{N} = \sum_{\{[n_1,n_2]|i\in[n_1,n_2]\}} w(i, [n_1, n_2])$ is the normalization factor and $w(i, [n_1, n_2])$ is the weight function used to downplay the finite-size effects introduced by the finite length of each fragment. Furthermore, the contribution of the terminal residues of each fragment and a few neighboring residues to the total average needs to be either discarded or strongly reduced, because they are attached to the reference protein structure through the CCD algorithm, and hence are artificially restricted.

Measuring Robustness to Different Approximations Following is a list of approximations to measure their effects on the equilibrium mobility captured:

1. The order of update of the DOFs in the CCD algorithm. (N terminus to C terminus or random order).

2. Inaccuracy of the force field employed: CHARMM versus AMBER.

3. Finite-size effects because of the definition of the fragments and inherent to the CCD algorithm.

4. Energy minimization of obtained conformations (can be trapped in a local minimum).

5. Statistical weighting scheme and the weight functions used.

The errors from these approximations compared to the error bars for ensemble averages of NMR data such as residual dipolar couplings and order parameters indicate that these approximations do not significantly affect the equilibrium mobility of the proteins employed (refer to figure 24.5).

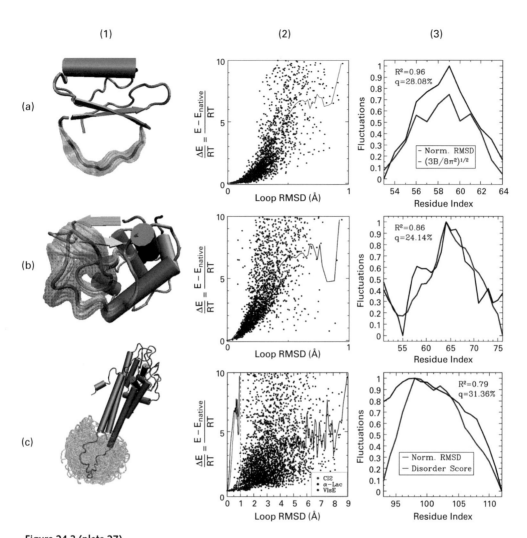

Figure 24.3 (plate 27)
(a1, b1, c1) 5,000 transparent loop conformations versus opaque reference structure (equilibriated native structure for CI2 and α-Lac and lowest energy structure for VlsE). (a2, b2, c2) Energy landscapes associated with generated ensembles are shown by plotting the energetic difference versus the RMSD of each conformation relative to a reference structure. Energy landscapes are shown only for conformations with energy less than 10 RT units away from the reference structure. An average energy profile is computed by distributing conformations in bins every 0.001 Å away from the reference structure and measuring the energy of each bin as an average over its conformations. Average energy profiles obtained for CI2 and α-Lac are very steep compared to the flat average energy profile of VlsE. (a3, b3, c3) Obtained fluctuations versus B factor-derived fluctuations for the CI2 loop, fluctuations in the literature for the α-Lac loop, and disorder scores for the VlsE loop [1].

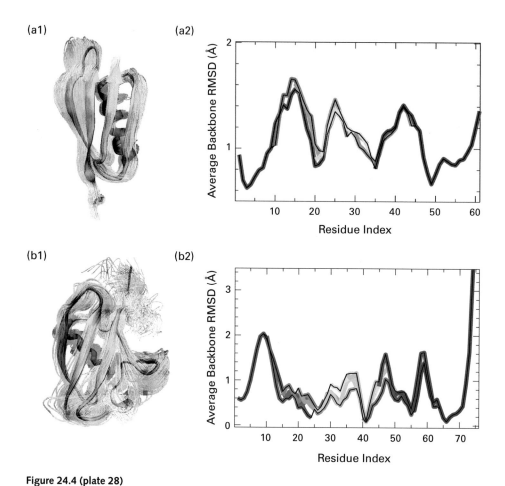

Figure 24.4 (plate 28)

(a1 and b1) Computed native ensembles for protein G and ubiquitin, respectively. (a2 and b2) Average RMSD per residue obtained by combining the local fluctuations of all the different regions. Results for different regions are shown in different colors, from red to blue as a window of 30 residues slides from the N- to the C-terminus of the protein. The black lines mark the highest and lowest RMSD values recorded from all the different windows including each given residue, and provide an estimate for the uncertainty of the procedure. Two consecutive 30-residue windows have an overlap of 25 residues. The results corresponding to the first and last 5 residues of each fragment are discarded since they are biased by the finite size of the window [1].

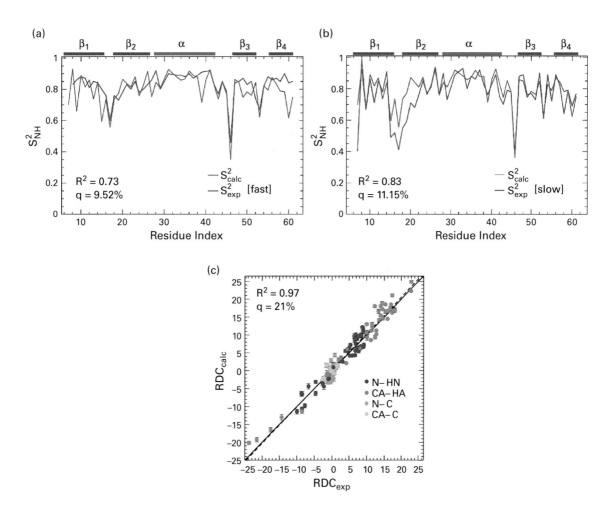

Figure 24.5 (plate 29)
Comparison of NMR relaxation data measured versus the generated ensemble for protein G. (a) Comparison of S^2 backbone (amide) order parameters calculated over the ensemble (S^2_{calc}) with fast (S^2_{NH}) data obtained from NMR relaxation measurements (S^2_{exp}). (b) Comparison of S^2 backbone (amide) order parameters measured over the ensemble (S^2_{calc}) with slow (S^2_{NH}) data obtained from NMR relaxation measurements (S^2_{exp}). (c) Comparison of residual dipolar coupling (RDC) parameters as obtained in the ensemble (RDC$_{calc}$, on the y-axis), and from NMR relaxation measurements (RDC$_{exp}$, on the x-axis). Results for different bond types are shown in different colors. (a–c) The dashed black line indicates the linear least-squares regression fit on the two sets of data, while the continuous line represents the identity diagonal [1].

Therefore, it is reasonable to combine the results obtained in different fragments to generate global fluctuations over an entire protein.

24.3   Results

To validate the accuracy of FEM, as shown in figure 24.3 (plate 27), the work [1] reproduces some native loops in stable proteins, such as CI2 (PDB ID: 1COA, 12-residue loop between VAL53 and ASP64), and α-Lac (PDB ID: 1HML, the 26-residue loop between LYS51 and THR76).

Figure 24.4 presents the results of the application of PEM to characterize the equilibrium ensemble of two proteins: streptococcal protein G (PDB ID: 1IGD) and human ubiquitin (PDB ID: 1UBQ). Figures 24.4 (a1) and (b1) qualitatively show the structural variability of the generated structures for protein G and ubiquitin. The consistency of fluctuations measured over ensembles of neighboring overlapping fragments can be seen in figure 24.4 (a2) and (b2), where the average RMSD of each residue measured over ensembles of the fragments that encompass that residue is plotted. See plate 28.

Order parameters—experimentally available S^2 data for protein G—are derived from ^{15}N NMR relaxation experiments and capture the fast dynamics of this protein in the picosecond to nanosecond timescale, which we refer to as *fast S^2*. Figure 24.5a shows the agreement between the ensemble-calculated backbone (amide) S^2 and the fast S^2 data of protein G. The agreement is better for α-helix and the β_2- and β_3-helix loops (residues 22–48), indicating that most of the mobility captured by the ensemble for these residues happens on the picosecond to nanosecond timescale. Figure 24.5b shows a better agreement between the S^2 data measured over the ensemble and the slow S^2 data reported in literature. In figure 24.5c, it is shown that the RDCs measured over the ensemble and those experimentally measured in bicelle medium agree with a Pearson correlation of 97% and q-factor of 21%.

Exercise   In order to compare RDCs from the computationally-predicted ensemble versus experimental RDCs, how should the alignment tensor(s) be computed? And how should the back-calculated RDCs be weighted in calculating the ensemble average?

References

[1] A. Shehu, C. Clementi, and L. E. Kavraki. Modeling protein conformational ensembles: from missing loops to equilibrium fluctuations. *Proteins: Structure, Function, and Bioinformatics* 65;1(2006 Oct 1):164–179.

[2] Available at: http://www.cs.rice.edu/CS/Robotics/bioinformatics/EquilibriumFluctuations.html.

25 Generalized Belief Propagation, Free Energy Approximations, and Protein Design

In this chapter we introduce graphical models and belief propagation algorithms. Then we discuss two applications. First, we describe a generalized belief propagation algorithm for approximating the free energy of a protein structure. Second, we describe how belief propagation algorithms can be useful in protein design.

Graphical models are a powerful, and relatively new algorithmic framework, which shows significant promise for structural biology.

25.1 Free Energy

A general definition for the *free energy* of a system is "the amount of energy which can be converted into work." Although there are several types of free energy, the most widely used is the *Gibbs free energy*, which can be defined as the amount of thermodynamic energy that can be converted into work at constant temperature and pressure. Formally, we write

$$G = H - T \cdot S = (E + P \cdot V) - T \cdot S,$$

where G = Gibbs free energy, H = enthalpy (the heat content of the system), S = entropy (a measure of the degree of randomness of the system), E = internal energy, T = absolute temperature, P = pressure, and V = volume.

The change in the Gibbs free energy of a system is $\Delta G = (\Delta E + P \cdot \Delta V) - T \cdot \Delta S$, and since the change in volume is small for nearly all biochemical reactions, we can write $\Delta G = \Delta E - T \cdot \Delta S$.

Free energy functions have been successfully used in protein structure prediction, fold recognition, homology modeling, and protein design; for example, see chapters 11 and 12. However, most free energy functions only model the internal energy E using inter- and intramolecular interactions terms (van der Waals, electrostatic, solvent, etc.) [11, 13, 8, 2]. The entropy S is usually ignored because it is harder to compute, since it involves summing over an exponential number of terms. Another approach is to compute the free energy using statistical potentials derived from known protein structures (e.g., from PDB). Such methods have the advantage that the derived potentials encode both the entropy S and the internal energy E. But there are several disadvantages, the most important being the fact that the observed interactions are usually not independent [14]. Kamisetty et al. [15] use a generalized belief propagation algorithm to compute an approximation of the free energy function that includes the entropy term.

25.2 Graphical Models

Graphical models are graphs that represent the dependencies among discrete random variables. The variables are usually represented as nodes in the graph and the dependencies as edges. Examples of graphical models include Bayesian networks (figure 25.1a), Markov random fields (figure 25.1b), and factor graphs (figure 25.1c). We use capital letters for the random variables $(X_1, X_2, \ldots, X_n)$ and small letters for the discrete values that the random variables can take $(x_1, x_2, \ldots, x_n)$.

Let $p(x_1, x_2, \ldots, x_n)$ be the joint probability of all the variables in the model. The *marginal probability* of each variable can be written as

$$p(x_i) = \sum_{x \setminus x_i} p(x) = \sum_{\{x_1, \ldots, x_n\} \setminus \{x_i\}} p(x_1, x_2, \ldots, x_n).$$

Computing these marginal probabilities is usually computationally expensive because it involves a sum over an exponential number of terms. Belief propagation algorithms try to overcome this problem by approximating the marginals $p(x_i)$ with the *beliefs* $b(x_i)$.

25.2.1 Bayesian Networks
In a Bayesian network (BN), the dependencies (conditional probabilities) are represented by arrows. Each variable is independent of all the nondescendants, given its parent. For example,

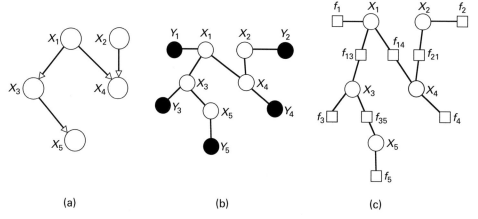

(a)   (b)   (c)

Figure 25.1
(a) Bayesian network. (b) Pairwise Markov random field. (c) Factor graph.

for the BN in figure 25.1a we have $p(x_4|x_1, x_2, x_3, x_5) = p(x_4|x_1, x_2)$. The joint probability can be written as

$$p(x_1, \ldots, x_n) = \prod_{i=1}^{n} p(x_i \mid Parents(x_i)).$$

25.2.2 Pairwise Markov Random Fields

A Markov random field (MRF) contains two types of random variables: *hidden* $(X_1, X_2, \ldots)$ and *observed* $(Y_1, Y_2, \ldots)$. The dependencies among variables are represented using *compatibility functions* (also known as *potentials*): $\phi_i(x_i, y_i) =$ the local evidence for x_i (sometimes written simply $\phi_i(x_i)$), and $\psi_{ij}(x_i, x_j) =$ the compatibility between x_i and x_j. In this case we call the MRF "pairwise" because the potential functions are defined pairwise.

The joint probability for a pairwise MRF can be written as

$$p(x_1, \ldots, x_n, y_1, \ldots, y_m) = \frac{1}{Z} \prod_{ij} \psi_{ij}(x_i, x_j) \prod_i \phi_i(x_i),$$

where Z is a normalization constant, also called the *partition function*.

25.2.3 Factor Graphs

In a factor graph (FG) the interactions between variables are represented by the *factor* nodes (the square nodes in figure 25.1c), and the joint probability can be written as a product of all the factors: $p(x) = 1/Z \prod_a f_a(x_a)$. For example, for the FG in figure 25.1c we have

$$p(x_1, \ldots, p_n) = \frac{1}{Z} f_1(x_1) f_2(x_2) f_{13}(x_1, x_3), \ldots$$

It is important to notice that all three graphical models described here (BNs, pairwise MRFs, and FGs) are equivalent in terms of the systems they can model [20]. In particular, the equivalence of pairwise MRFs and FGs can be illustrated by the equivalence between the compatibility functions and the factors:

$$\phi_i(x_i) \text{ in MRF} \Leftrightarrow f_i(x_i) \text{ in FG}$$

$$\psi_{ij}(x_i, x_j) \text{ in MRF} \Leftrightarrow f_{ij}(x_i, x_j) \text{ in FG}.$$

25.3 Belief Propagation (BP)

Belief propagation is a method for approximating the marginal probabilities in a graphical model, in a time linear in the number of variables (nodes). BP is precisely mathematically equivalent for pairwise MRFs, BNs, and FGs. However, it is easier to explain on a pairwise MRF.

Let X_i be the hidden variables. We define a *message* $m_{ij}(x_j)$ from a node i to a node j as an array containing information about what state node j should be in. For example, if x_j can take the values 1, 2, or 3, then the message $m_{ij}(x_j) = (0.7, 0.1, 0.2)$ can be interpreted as follows: node i is telling node j that it should be in state 1 with probability 0.7, in state 2 with probability 0.1, and in state 3 with probability 0.2.

BP is an iterative algorithm. It starts with randomly initialized messages and then it applies the message update rule below until the messages converge:

$$m_{ij}(x_j) \leftarrow \sum_{x_i} \phi_i(x_i)\psi_{ij}(x_i, x_j) \prod_{k \in N(i) \backslash j} m_{ki}(x_i), \qquad (25.1)$$

where $\phi_i(x_i)$ and $\psi_{ij}(x_i, x_j)$ are the potential functions and $N(i)$ is the set of all neighbors of node i. After convergence, the beliefs (the approximate marginals) can be computed as

$$b_i(x_i) = k\phi_i(x_i) \prod_{j \in N(i)} m_{ij}(x_i).$$

When the MRF has no cycles, the computed beliefs are exact! Even when the MRF has cycles, the BP algorithm is still well-defined and empirically it often gives good approximate answers.

25.4 The Connection between Belief Propagation and Free Energy

We can write the Gibbs free energy of a system as $G = E - T \cdot S$, where the entropy is $S = -\sum p(x) \log p(x)$ (x iterates over all possible states of the system).

For a graphical model, if x is an instantiation of all the random variables (or equivalently, the state of the model), we can write the joint probability $p(x)$ as using Boltzmann's law as $p(x) = \frac{1}{Z}e^{-E(x)}$. Z is the partition function. Notice that we consider $T = 1$ and we ignore Boltzmann's constant (which affects only the scale of the units).

Using BP, we approximate the joint $p(x)$ with the belief $b(x)$. To see how close these two distributions are, we can use the Kullback-Leibler distance (cf. chapter 13):

$$D(b(x)||p(x)) = \sum_x b(x) \ln \frac{b(x)}{p(x)} = \sum_x b(x) \ln b(x) + \sum_x b(x)E(x) + \ln Z.$$

Notice that the second equality was obtained by substituting $p(x)$ as defined using Boltzmann's law.

Two important properties of the KL distance are that it is never negative, and it becomes zero if and only if the two probability distributions are exactly the same. In this case, we have

$$G = \underbrace{\sum_x b(x)E(x)}_{E} + \underbrace{\sum_x b(x)\ln b(x)}_{-S} = -\ln Z.$$

In other words, when the beliefs are exact, the approximate Gibbs free energy achieves its minimal value $-\ln Z$, also called the "Helmholtz free energy" [21].

25.5 Generalized Belief Propagation (GBP)

Computing the Gibbs free energy $G = E - TS = \sum_x b(x)E(x) + T \sum_x b(x)\ln b(x)$ involves summing over an exponential number of terms, which is usually computationally intractable. Instead, several approximations of the free energy have been developed:

• *Mean-field* approximations use only one-node beliefs;

• *Bethe* approximations [22] use one-node and two-node beliefs; and

• *Region-based* approximations are based on the following idea: Break up the graph into a set of regions, compute the free energy over each region and then approximate the total free energy by the sum of the free energies over the regions.

Region-based approximations can be computed using generalized belief propagation (GBP), a message-passing algorithm similar to BP. GBP uses messages between regions of nodes instead of messages between nodes. The regions that exchange messages can be visualized as *region graphs* [21].

Different choices of region graphs give different GBP algorithms. There is usually a tradeoff between complexity and accuracy. However, how to optimally choose the the regions is still an open research question. One heuristic I advise is: try to include at least the shortest cycles inside regions.

Similar to BP, GPB is guaranteed the give the optimal solution when the region graph has no cycles, but it also works very well on many graphs that contain cycles. In [21], Yedidia et al. try to understand why GBP works on graphs with cycles, and they offer some guidelines for constructing region graphs for which some theoretical guarantees can be offered.

25.6 An Application of GBP: Estimating the Free Energy of Protein Structures

Kamisetty et al. [15] use GBP to estimate the free energy of all-atom protein structures. They model the structure of a protein as a probability distribution using a pairwise MRF:

• The *observed* variables are the backbone atom positions X_b^i (continuous).

• The *hidden* variables are the side-chain atom positions X_s^i represented using rotamers (discrete).

• Two variables (atoms) share an edge (interact) if they are closer than a threshold distance, set to 8 Å.

• The *potential functions* are defined as

$$\psi(X_s^{ip}, X_s^{jq}) = \exp(-E(x_s^{ip}, x_s^{jq})/k_b T)$$

$$\psi(X_s^{ip}, X_b^j) = \exp(-E(x_s^{ip}, x_b^j)/k_b T),$$

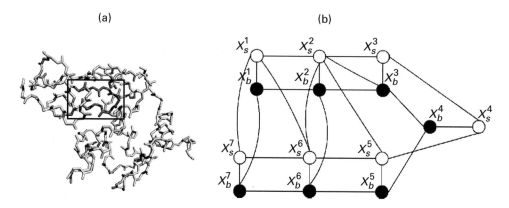

Figure 25.2 (plate 30)

(a) Structure of lysozyme (PDB id: 2lyz) with a few residues highlighted. (b) Part of the random field induced by the highlighted residues: X_s^i's are the hidden variables representing the rotameric state, the visible variables are the backbone atoms in conformations x_b^i [15]. The publisher for this copyrighted material is Mary Ann Liebert, Inc. publishers.

where $E(x_s^{i_p}, x_s^{j_q})$ is the energy of interaction between rotamer state p of residue X_i^s and rotamer state q of residue X_s^j. See figure 25.2 (plate 30).

After building the pairwise MRF, Kamisetty et al. transform the MRF into the equivalent factor graph. Then, they split the factor graph into regions: *big regions* containing two or three variables and *small regions* containing just one variable (see the examples in figure 25.3). The basic idea behind the choice of regions is: place residues that are closely coupled together in the same big region. This idea was borrowed from Aji and McEliece [1], and it provides a balance between the accuracy of the approximation and the complexity of the model.

The region graph is then built by drawing an edge from each big region to all the small regions that contain a strict subset of the big region's nodes. Finally, a GBP algorithm called "the two-way algorithm" [21] is applied on the resulting region graph.

25.6.1 Results and Discussion

Kamisetty et al. use the entropy component of their free energy estimate to distinguish between native protein structures and decoys (structures with similar internal energy to that of the native structure, but otherwise incorrect). They find that when the structures are ranked in decreasing order of their entropy, the native structure is ranked the highest in 87.5% of the datasets that contain decoys with backbones similar to the backbone of the native structure, and 84% of the datasets that include dissimilar backbones.

The authors also compute estimates of the $\Delta\Delta G$ on mutation and their results correlate reasonably well with the experimental values (with correlation coefficients varying from 0.63 to 0.7, and p-values between 1.5×10^{-5} and 0.0063).

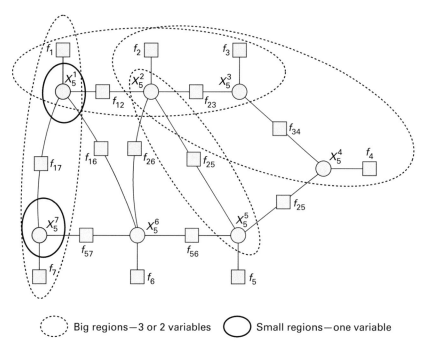

Figure 25.3
The factor graph with regions.

We have learned about a method for estimating side-chain conformational entropy using belief propagation. The K^* algorithm (chapter 12) also models free energies from conformational ensembles. While belief propagation techniques are usually probabilistic or stochastic, the K^* algorithm is deterministic and operates by computing provably good approximations to the partition functions [13, 8, 2].

Up to now, this chapter has considered belief propagation for *ensemble properties* (such as conformational entropy). Belief propagation algorithms are also useful in the search for the *global minimum energy conformation (GMEC)* of protein side-chains, which is an important computational challenge in protein structure prediction and design (see chapters 11–12). We discuss belief propagation algorithms for protein design in the next section.

25.7 Application: Graphical Models for Protein Design

In section 25.6 we introduced a method based on generalized belief propagation to estimate the free energy of a protein structure. Free energy is composed of both an enthalpic term and an entropic term.

Protein design frequently involves searching for a sequence of amino acids that will fold to a target structure or will perform a desired function (chapters 11, 12). Ideally, we would search for the sequence with the lowest free energy, accounting for both enthalpy and entropy since entropy has been shown to be important when scoring mutants of proteins [13, 2, 8].

The entropic component, however, is commonly ignored in protein designs for several reasons. First, it is often assumed that entropy plays a smaller role in the stability and function of a protein [10]. Second, entropy calculations must account for an exponential number of states (conformations) for each sequence. Finally, in most techniques for entropy calculations, like the method introduced in section 25.6 and the K^* algorithm (chapter 12), entropy must be computed at least partially for each sequence, and therefore an exponential number of sequences must be considered.

For the remainder of this chapter, we consider belief propagation as an aid to protein design in which the enthalpy is modeled but the entropic and ensemble properties of section 25.6 and chapter 12 are explicitly ignored. This corresponds to the "classical" inverse folding problem introduced in chapter 11. Recall that this restricted protein design paradigm involves an input model, a protein design algorithm that performs a search over a combinatorially large sequence space, and the output of one or more candidate sequences (chapter 11). Input models are chosen with computational feasibility in mind. A common input model used in protein design assumes a fixed backbone for the protein, a discrete number of conformations possible for each side-chain, and a pairwise additive energy function. This simple model is convenient because it reduces the continuous conformation space to a discrete space with a finite number of conformations. The search over the discrete space, however, remains intractable and the problem of assigning discrete rotamers has been shown to be NP-hard.

Provable algorithms like DEE/A* [12] can solve many instances of the basic problem of finding the global minimum energy conformation. Nevertheless, there are instances of the problem where the state of the art in provable algorithms might not be able to find a solution in a reasonable amount of time.

As we explained at the beginning of this chapter, graphical models are a graph representation of conditional independence between random variables, while Belief Propagation (as presented in section 25.3) is a method to compute the probability distribution of each of the random variables by passing messages between variables. Messages encode the *belief* one neighbor has about another and are propagated during sequential iterations throughout the graph. In protein design, the random variables are residues that can be in a discrete number of states (i.e., rotamers). The neighbors of each random variable in the graphical model are the residues that are close enough to interact.

We will present three variations of the protein design problem for which the use of Belief Propagation as defined in section 25.3, is useful: search for the global minimum energy conformation, enumeration of the m best low-energy sequences, and probabilistic protein design. We build on earlier chapters discussing the protein design problem, energy functions, dead-end elimination (DEE), rotamers, and algorithms for NP-hard problems.

25.7.1 Protein Design Problem

We first define the protein design problem as in chapter 11 and Fromer and Yanover [5], where a fixed backbone and a discrete, finite number of rotamers is assumed. Rotamers of different amino acids are treated uniformly, and we search for a single assignment of rotamers at all of the redesign positions that will have the lowest energy: the global minimum energy conformation (GMEC). We are interested in the amino acid sequence of the GMEC (i.e., the amino acid type of each rotamer). In contrast to chapter 12, under this definition of the protein design problem, the terms "conformation" and "rotamer assignment" are equivalent, since rotamers are always in a single, rigid state.

Let $r = (r_1, ..., r_n)$ denote an assignment of rotamers in n positions. Let $E_i(r_i)$ be the self-energy of rotamer r_i and $E_{ij}(r_i, r_j)$ the pairwise energy between rotamers r_i and r_j. We define the energy for the rotamer assignment r, $E(r)$ as

$$E(r) = \sum_i E_i(r_i) + \sum_{i,j} E_{ij}(r_i, r_j). \tag{25.2}$$

Let $S = (S_1, ..., S_N)$ be an amino acid assignment; let $\tau(r_i)$ be the amino acid type of the rotamer assigned at position i, and $\tau(r)$ the amino acid assignment for rotamer assignment r. We define the minimum energy of a sequence $E(S)$ as

$$E(S) = \min_{\{r \mid \tau(r) = S\}} E(r), \tag{25.3}$$

and the problem of searching for the minimum energy sequence S^* as

$$S^* = \tau(\arg \min_r E(r)). \tag{25.4}$$

Protein Design as an Inference Problem We now recast the protein design problem as an *inference* problem by using Boltzmann-weighted terms for all energy interactions [5]. Let

$$\psi_i(r_i) = e^{\frac{-E_i(r_i)}{T}}, \tag{25.5}$$

$$\psi_{ij}(r_i, r_j) = e^{\frac{-E_{ij}(r_i, r_j)}{T}}, \tag{25.6}$$

where T is the system temperature. We define the probability of each rotamer assignment,

$$\Pr(r) = \Pr(r_1, \ldots, r_n) = \frac{1}{Z} \prod_i \psi_i(r_i) \prod_{i,j} \psi_{ij}(r_i, r_j) = \frac{1}{Z} e^{\frac{-E(r)}{T}}, \tag{25.7}$$

where Z is the normalization factor (i.e., the partition function, which guarantees that the sum of the probabilities over all states equals 1). Finally, analogous to equation (25.4), we define the problem of finding the sequence S^* that contains the rotamer assignment of highest probability:

$$S^* = \tau(\arg \max_r \Pr(r)). \tag{25.8}$$

25.7.2   Graphical Models and Belief Propagation for Protein Design

Long-range interactions in protein design models tend to be weak. The choice of rotamers in two residues that are far away from each other will most likely have little direct influence on each other, but indirect influence is more likely (e.g., a "domino" effect). We take advantage of weak long-range interactions to build a graphical model where an edge exists between a pair of nodes if they have a significant interaction.

In protein design, we must find conformations that have *globally* low energy. Any algorithm that searches for low-energy conformations must account for the effect of local choices on the overall global energy. In belief propagation, a random variable receives a message from each of its neighbors, telling it what they think the probability of each of its rotamers should be. The message encodes both local information (i.e., the intrarotamer energy at each position and the energy interactions with each neighbor), and global information, which is encoded (indirectly) through the relayed effect of messages from neighbors of neighbors. On each iteration, only messages from immediately adjacent neighbors are incorporated into the probabilistic score of each rotamer at a position. But on successive iterations, the *belief* is *propagated* through the entire graph.

Formally [5], a rotamer r_j receives a message from each of its neighbors i containing information about the "belief" that each neighbor has about the probability of residue position j being in state r_j:

$$m_{i \to j}(r_j) = \max_{r_i} \left(e^{\frac{-E_i(r_i) - E_{ij}(r_i, r_j)}{T}} \prod_{k \in N(i) \setminus j} m_{k \to i}(r_i) \right). \tag{25.9}$$

Note that $m_{i \to j}$ is a vector with dimension equal to the number of possible rotamers at j. Initially, all messages are set to 1 (or a normalization of it). Messages are updated on all nodes until convergence or a fixed maximum number of iterations. Within each iteration, nodes can be selected sequentially for updates. If the graphical model is a tree, the belief propagation algorithm will converge. In "loopy" belief propagation (i.e., graphs with loops) there is no guarantee for convergence.

Each rotamer r_i is a member of an exponential number of rotamer assignments. The protein design problem (eq. 25.4) consists of finding the rotamer assignment of lowest energy. In equation (25.9) we take the maximum over all incoming messages to r_j, and the resulting message value is the function of only one of the rotamers in i. This type of belief propagation is called *max-product* belief propagation. Contrast this with the definition in section 25.3, where all the incoming messages are added, which is called *sum-product*. For each rotamer, we will approximate the marginal probability of the lowest-energy (and highest-probability) rotamer assignment that rotamer r_i is part of, as a function of incoming messages and the internal energy of r_i. This is referred to as *max-marginal probability* ($\mathrm{MM}_i(r_i)$):

$$\mathrm{MM}_i(r_i) = e^{\frac{-E_i(r_i)}{t}} \prod_{k \in N(i)} m_{k \to i}(r_i). \tag{25.10}$$

If belief propagation converges, then $\text{MM}_i(r_i)$ approximates the exact max-marginal of r_i, denoted $\text{Pr}_i^\infty(r_i)$, which represents the maximum probability (eq. 25.7) over all rotamer assignments that contain r_i:

$$\text{Pr}_i^\infty(r_i) = \max_{\{r' \mid r_i' = r_i\}} \text{Pr}(r'). \tag{25.11}$$

Finally, we select a rotamer assignment $r^* = (r_1^*, \ldots, r_n^*)$ such that the max-marginal probability of each rotamer in the assignment is maximized over the possible values of r_i:

$$r_i^* = \arg\max_{r_i \in \text{R}_i} \text{MM}_i(r_i), \tag{25.12}$$

where R_i is the set of rotamers allowed in position i. If we can converge to exact max-marginals (i.e., if $\text{MM}_i(r_i) = \text{Pr}_i^\infty(r_i)$, for all r_i), then $S^* = \tau(r^*)$.

25.7.3 Multiple Low-Energy Sequences Through BP

The protein design problem as defined in equation (25.4) searches for a single sequence that contains the GMEC. In practice it might be desirable to have several low-energy sequences for the following reasons. Some parts of energy functions for computational protein design are developed by fitting to experimental data, which might lead to overfitting. Assumptions about proteins in the model, like discrete rotameric states and a fixed backbone, might produce inaccurate results. Provable algorithms that produce combinatorially precise results are usually slower than inexact, sampling methods, but the sequences enumerated by inexact algorithms might not be good. Finally, predicted low-energy structures might be *too* stable, not fold to the target structure, or lack binding affinity to a substrate. By generating a large set of low-energy sequences, we can increase the likelihood that one of them will fold to the target structure and have the desired properties.

The DEE algorithm is efficient in pruning conformations that are provably higher in energy than other conformations. If we wish to avoid pruning of other low-energy sequences, we must choose a threshold value E_w for an energy window so that no conformations within E_w of the lowest-energy conformation are pruned.[1] Provable algorithms generally are not designed to enumerate more than the single lowest-energy sequence (i.e., the GMEC) since they make no distinction between whether two rotamer assignments belong to the same sequence or to different sequences. In DEE/A*, for example, in the worst case, a low-energy sequence with an exponential number of low-energy rotamer assignments might have to be enumerated fully before the next lowest energy sequence can be enumerated.

Fromer and Yanover [5] propose the tBMMF (type-specific Best Max-Marginal First) algorithm, that provably enumerates low-energy sequences in order of increasing energy if the belief

---

[1] Techniques like MinBounds [7] enable more efficient pruning of rotamers with energies above E_w, when E_w is large, but might not be as efficient as DEE with $E_w = 0$. Therefore, problems that might be tractable when searching for a single minimum energy structure, might be slow to solve when multiple low-energy sequences are desired.

propagation algorithm converges to exact max-marginals; unfortunately, exact max-marginals are only guaranteed to converge on tree graphs. Type-dependent dead end elimination (tdDEE) [19] is used to reduce the search space, while guaranteeing that no sequences will be pruned. At least one rotamer of every allowed amino acid at each position is not pruned by tdDEE.

Intuitively, the tBMMF algorithm works as follows. The algorithm first computes the max-marginals as in section 25.7.2 and enumerates the lowest-energy sequence. The max-marginal $MM_i(r_i)$ for each rotamer r_i (eq. 25.10) is an approximation to the probability (eq. 25.7) of the lowest-energy rotamer assignment in which r_i is present. If $S^* = \tau(r^*)$ (eq. 25.12) is the sequence of the rotamer assignment of lowest-energy, the next lowest energy sequence must differ by at least one amino acid from S^*. Let r_i be the rotamer with the lowest max-marginal such that $\tau(r_i) \neq \tau(S_i^*)$. Then we know that the probability of the next lowest energy sequence is that of r_i (the best max-marginal), and that r_i is part of the next lowest energy rotamer assignment.

We cannot, however, know which other rotamers belong to the next lowest-energy rotamer assignment. The tBMMF algorithm therefore partitions the search space into two subspaces: one that contains the amino acid $\tau(r_i)$ as the lone amino acid at position i and another subspace that doesn't contain amino acid $\tau(r_i)$, at position i. BP is then performed on both subspaces; in the subspace that contains $\tau(r_i)$, we should find the second lowest-energy sequence, and we compute the max-marginal of the next lowest-energy sequence in that subspace. In the space constrained to all the amino acids except $\tau(r_i)$, $\tau(r^*)$ should again be the lowest-energy sequence, but a different amino acid should be the next best max-marginal since $\tau(r_i)$ is not present in this subspace. We then choose between the two max-marginals and repeat the process. Using the toy system in figure 25.4, figure 25.5 shows an example of this process. For a detailed description of the algorithm and proof of its correctness under exact max-marginals, refer to Fromer and Yanover [5].

25.7.4 Graphical Models for Probabilistic Protein Design

Phage display is an experimental method where a large number of random peptide sequences (10^9–10^{10}) can be screened for binding to another protein or ligand. Even with libraries this huge, however, we still can't evaluate the exponential number of peptide sequences that arise as we increase the number of design positions and choice of amino acids at each position. With a choice of 20 amino acids per position, even a small peptide with 10 amino acids would have 20^{10} sequences, a thousand times more than what we can handle with current phage display techniques. The distribution of amino acids at each redesign position, however, can be biased toward amino acids that will yield low-energy conformations of the target backbone.

In *probabilistic protein design*, a probability distribution is assigned to each design position, and correspondingly, each amino acid is assigned a probability. Fromer and Yanover [6] show a way to use Belief Propagation to compute probabilities for each amino acid at each position that will yield low-energy sequences.

The study [6] approximates through BP the marginal probabilities of each rotamer at each design position, and then computes the marginal probabilities of each amino acid by summing

(a)

Position #1

	aa	G_1		G_2	
aa	rot.	g_{11}	g_{12}	g_{21}	g_{22}
H_1	h_{11}	**–15**	–11	–6	–3
	h_{12}	–14	–10	**–7**	–2
H_2	h_{21}	–8	–9	0	**–5**
	h_{22}	–12	**–13**	–4	–1

Position #2

(b)

r	$E(r)$	$T(r)$
$(\mathbf{g_{11}, h_{11}})$	**–15**	$(\mathbf{G_1, H_1})$
(g_{11}, h_{12})	–14	(G_1, H_1)
$(\mathbf{g_{12}, h_{22}})$	**–13**	$(\mathbf{G_1, H_2})$
(g_{11}, h_{22})	–12	(G_1, H_2)
(g_{12}, h_{11})	–11	(G_1, H_1)
(g_{12}, h_{12})	–10	(G_1, H_1)
(g_{12}, h_{21})	–9	(G_1, H_2)
(g_{11}, h_{21})	–8	(G_1, H_2)
$(\mathbf{g_{21}, h_{12}})$	**–7**	$(\mathbf{G_2, H_1})$
(g_{21}, h_{11})	–6	(G_2, H_1)
$(\mathbf{g_{22}, h_{21}})$	**–5**	$(\mathbf{G_2, H_2})$
(g_{21}, h_{22})	–4	(G_2, H_2)
(g_{22}, h_{11})	–3	(G_2, H_1)
(g_{22}, h_{12})	–2	(G_2, H_1)
(g_{22}, h_{22})	–1	(G_2, H_2)
(g_{21}, h_{21})	0	(G_2, H_2)

Figure 25.4
A toy design energy matrix. (A) Amino acids G_1 and G_2 are allowed at position 1, while H_1 and H_2 are allowed at position 2. Every amino acid has two possible rotamers. The table contains interaction energies between every rotamer in position 1 and position 2. (B) List of all 16 possible sequences by energy. Reprinted from [5].

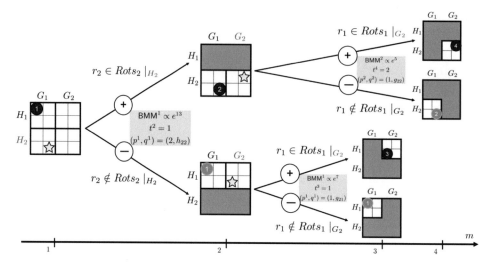

Figure 25.5
The tBMMF algorithm on the example from figure 25.4. Intially BP is run on the full matrix and (g_{11}, h_{11}) is selected as the lowest-energy sequence. Rotamer (g_{12}, h_{22}) is selected as the best max-marginal and used to partition the space in two. On the next iteration, BP is run on both subspaces to find the next lowest energy sequence (which we know contains (g_{12}, h_{22})), and the best max-marginals from both spaces. We select (g_{21}, h_{12}) because it has a higher max-marginal than (g_{22}, h_{21}). The space is partitioned again around (g_{21}, h_{12}) and the third lowest-energy sequence contains (g_{21}, h_{12}). The next best max-marginal is (g_{22}, h_{21}), which we partition to find the fourth best energy sequence. Reprinted from [5].

over the marginals. In the initial state of the belief propagation iteration, each rotamer has the same probability, and every amino acid has a probability proportional to the number of rotamers. As the algorithm progresses through the states and messages are passed, the probability of each rotamer comes closer to the Boltzmann-weighted probability associated with low-energy states.

The method is very similar to what was described in section 25.7.2, but instead of max-product for messages as in equation (25.9), we use sum-product, which was previously introduced in section 25.3:

$$m_{i \to j}(r_j) = \sum_{r_i} \left(e^{\frac{-E_i(r_i) - E_{ij}(r_i, r_j)}{T}} \prod_{k \in N(i) \setminus j} m_{k \to i}(r_i) \right). \tag{25.1}$$

Sum-product is more appropriate for this problem because we are interested in more than just the single lowest-energy sequence. This is in contrast with section 25.7.2 where only the GMEC is of interest, and the max-product probabilities encode the GMEC.

25.7.5 Discussion and Future Directions
All the protein design algorithms described in this section (25.7) use Belief Propagation on models where the backbone is fixed and the side-chains are rigid. Kamisetty et al. [16] extend the free energy calculation of section 25.6 to include a limited number of sampled backbone

conformations, based on the "backrub" motion. However, unlike MinDEE or BD (chapter 12), these algorithms lack the capability to model continuous flexibility in side-chains and backbones. Unlike backrub DEE (`brDEE`, chapter 12), which also uses discrete backbones, these BP techniques cannot guarantee identification of the GMEC or a gap-free enumeration of conformations in order of energy.

None of the algorithms presented in this chapter provides any guarantees on the results when the belief propagation step fails to converge; convergence is guaranteed only on tree graphs, which rarely occur in real protein design problems. Without provable convergence, we have no guarantee that we are generating the lowest-energy structures. If the BP design algorithms' predictions fail to correlate with experimental results, the failure cannot be attributed with certainty to flaws in the input model (e.g., rotamer library, energy function, starting structure), since the inexactness of the BP method might have been responsible.

On the other hand, with a provable gap-free algorithm, any discrepancy between computational predictions and experimental results can be solely attributed to flaws in the model, and not to (for example) undersampling, or inadequate optimization. This makes it easier to feed back experimental results to improve the model. This capability is facilitated by modern statistical parameter estimation or machine learning algorithms. Indeed, one of the strengths of graphical models is that they can be constructed, or "learned" from experimental data [18]. Consider the situation where protein design is coupled with large-scale experimental screening. In this setting, false negatives (predictions that work well experimentally, but were predicted to perform poorly by the algorithm) may be observed. If the design algorithm is provable, then we know we can blame the model for these failures, and, in principle, improve it. This can be a tremendous advantage for a laboratory combining computational and experimental protein design. The tradeoff is that provable algorithms are frequently slower, require more computational resources, or are more complex to implement.

A promising approach is to use Belief Propagation to boost pruning and bounding in DEE/A*. Remarkably, this can be done while maintaining guarantees of completeness and soundness, using a deterministic algorithm. Hong et al. [9] observe the following: DEE methods combined with systematic A* search (DEE/A*) have proven useful, but may not be strong enough as we attempt to solve protein design problems where a large number of similar rotamers is eligible and the network of interactions between residues is dense. In the study [9], an exact solution method, named BroMAP ("branch-and-bound rotamer optimization using maximum a posteriori estimation") is developed for protein design problems.

BroMAP is based on belief propagation, but the algorithm is provable and deterministic. BroMAP reduces the problem size within each node using DEE and pruning by lower bounds from approximate maximum a posteriori (MAP) estimation. The lower bounds are also exploited in branching and subproblem selection for fast discovery of strong upper bounds. BroMAP is able to expand smaller search trees than conventional branch-and-bound methods while performing only a moderate amount of computation in each node, thereby reducing the total running time. Computational results suggest that BroMAP tends to be faster than DEE/A* for large protein design cases. Therefore, BroMAP and other such belief propagation algorithms should be valuable for large protein design problems where DEE/A* struggles, and can also assist DEE/A* in GMEC search. Such algorithms represent the marriage of DEE and belief propagation, and are proving to be a powerful combination.

This chapter discussed only certain aspects of graphical models, Markov random fields, and belief propagation in computational structural biology. More generally, applications of belief propagation and graphical models include learning energy functions, side-chain packing, protein design, and NMR resonance assignment. These problems have been addressed in a number of recent papers [17, 9, 3, 4, 5, 6, 18, 23].

References

[1] S.M. Aji and R.J. McEliece. The generalized distributive law and free energy minimization. *Proceedings of the 39th Allerton Conference on Communication, Control and Computing* (2003):459–467.

[2] C.Y. Chen, I. Georgiev, A.C. Anderson, and B.R. Donald. Computational structure-based redesign of enzyme activity. *Proceedings of the National Academy Science (U S A)* 106;10(2009 Mar 10):3764-3769. Epub 2009 Feb 19.

[3] M. Fromer and J.M. Shifman. Tradeoff between stability and multispecificity in the design of promiscuous proteins. *PLoS Computational Biology* 5;12(2009 Dec):e1000627. Epub 2009 Dec 24.

[4] M. Fromer, C. Yanover, and M. Linial. Design of multispecific protein sequences using probabilistic graphical modeling. *Proteins* 78;3(2010 Feb 15):530–547.

[5] M. Fromer and C. Yanover. Accurate prediction for atomic-level protein design and its application in diversifying the near-optimal sequence space. *Proteins* 75;3(2009 May 15):682–705.

[6] M. Fromer and C. Yanover. A computational framework to empower probabilistic protein design. *Bioinformatics* 24;13(2008 Jul 1):i214-222.

[7] I. Georgiev, R. H. Lilien, and B. R. Donald. Improved pruning algorithms and divide-and-conquer strategies for dead-end elimination, with application to protein design. *Bioinformatics* 22(2006 Jul):e174-183

[8] I. Georgiev, R. H. Lilien, and B. R. Donald. The minimized dead-end elimination criterion and its application to protein redesign in a hybrid scoring and search algorithm for computing partition functions over molecular ensembles. *Journal of Computational Chemistry* 29;10(2008 Jul 30):1527–1542.

[9] E. J. Hong, S. M. Lippow, B. Tidor, T. Lozano-Pérez. Rotamer optimization for protein design through MAP estimation and problem-size reduction. *Journal of Computational Chemistry* 30;12(2009 Sep):1923-1945.

[10] X. Hu, H. Wang, H. Ke, and B. Kuhlman. High-resolution design of a protein loop. *Proceedings of the National Academy of Sciences (U S A)* 104;45(2007 Nov 6):17668-17673.

[11] T. Lazaridis and M. Karplus. Effective energy functions for protein structure prediction. *Current Opinion in Structural Biology* 10(2000):138–145.

[12] A. R. Leach and A. P. Lemon Exploring the conformational space of protein side-chains using dead-end elimination and the A* algorithm. *Proteins: Structure, Function, and Genetics* 33(1998):227–239.

[13] R. H. Lilien, B. W. Stevens, A. C. Anderson, and B. R. Donald. A novel ensemble-based scoring and search algorithm for protein redesign and its application to modify the substrate specificity of the gramicidin synthetase A phenylalanine adenylation enzyme. *Journal of Computational Biology* 12;6(2005 Jul-Aug):740–761.

[14] P. D. Thomas and K. A. Dill. Statistical potentials extracted from protein structures: How accurate are they? *Journal of Molecular Biology* 257 (1996): 457–469.

[15] H. Kamisetty, E. P. Xing, and C. J. Langmead. Free energy estimates of all-atom protein structures using generalized belief propagation. *Journal of Computational Biology* 15;7(2008 Sep):755–766.

[16] H. Kamisetty, C. Bailey-Kellogg, and C. J. Langmead. A graphical model approach for predicting free energies of association for protein-protein interactions under backbone and side-chain flexibility. *Proceedings of 3DSIG 2009 Structural Bioinformatics and Computational Biophysics* (2009):67–68.

[17] Y. Weiss and W. T. Freeman. Correctness of belief propagation in Gaussian graphical models of arbitrary topology. *Neural Computation* 13;10(2001 Oct):2173–2200.

[18] C. Yanover, O. Schueler-Furman, and Y. Weiss. Minimizing and learning energy functions for side-chain prediction. *Journal of Computational Biology* 15;7(2008 Sep):899–911.

[19] C. Yanover, M. Fromer, and J. M. Shifman Dead-end elimination for multistate protein design. *Journal of Computational Chemistry* 28(2007):2122–2129.

[20] J. S. Yedidia, W. T. Freeman, and Y. Weiss. Understanding belief propagation and its generalizations, TR-2001-22. Mitsubishi Electric Research Laboratories, January 2002.

[21] J. S. Yedidia, W. T. Freeman, and Y. Weiss. (2005) Constructing free-energy approximations and generalized belief propagation algorithms. *IEEE Transactions on Information Theory* 51;7(July 2005): 2282–2312.

[22] J. S. Yedidia, W. T. Freeman, and Y. Weiss. Bethe free energy, Kikuchi approximations, and belief propagation algorithms, TR-2001-10. Mitsubishi Electric Research Laboratories, May 2001.

[23] J. Zeng, P. Zhou, and B. R. Donald. A Markov random field framework for protein side-chain resonance assignment. *Proceedings International Conference on Research in Computational Molecular Biology (RECOMB)*, 2010. Lisbon, Portugal.

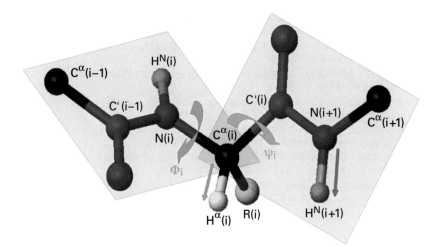

Plate 1

The ϕ and ψ torsion angles of peptide backbones. RDC experiments measure the orientation of internuclear bond vectors, shown as NH^N and $C^\alpha H^\alpha$.

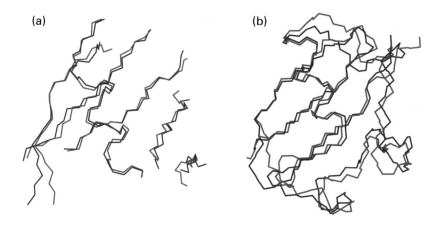

Plate 2

(a) Structure of ubiquitin backbone without loops. The ubiquitin backbone structure (blue) was computed using 37 NH and 39 CH RDCs, 12 hydrogen bonds, and four NOEs; (b) structure of ubiquitin backbone with loops. The ubiquitin backbone structure (blue) was computed by extending the algorithm to handle loop regions along the protein backbone. The structure was computed using 59 NH and 58 CH RDCs, 12 H-bonds, and two unambiguous NOEs. A reference structure (from crystallography) is shown (red).

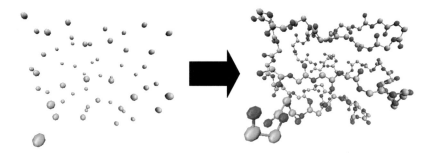

Plate 3
Schematic of going from only C^α atoms to a full backbone model (PDB id: 1DUR). Credit: Kyle Roberts.

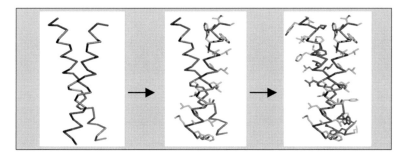

Plate 4
Peptide helix design scheme [6]. Reprinted with permission from AAAS.

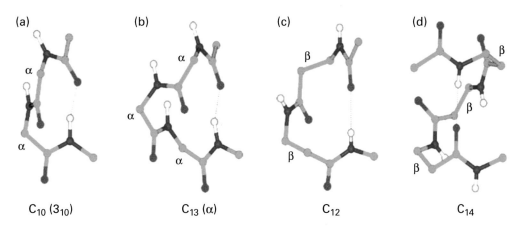

(a)   (b)   (c)   (d)

C_{10} (3_{10}) C_{13} (α) C_{12} C_{14}

Plate 5
Secondary structure of α and β amino acid helices [8]. Reprinted with permission. Copyright 2005 American Chemical Society.

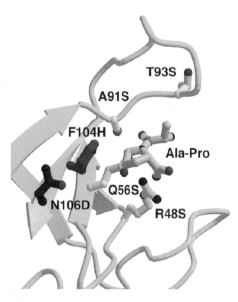

Plate 6
Cyclophilin active site residues mutated in cyproase. Those residues highlighted in yellow were separately mutated to serine in the first round of design. Upon selection of A91S as the best protease, the F104H and N106D mutations were added to reproduce the catalytic triad [2]. Reprinted by permission from Macmillan Publishers Ltd.

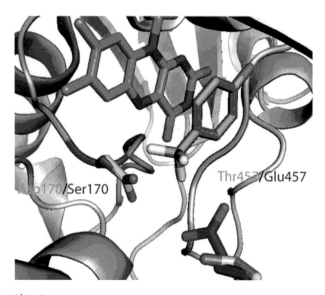

Plate 7
Altering the stereospecificity of vanillyl-alcohol oxidase. Here the flavin cofactor is in orange and a substrate analog in green. Note how the mutations shift the catalytic carboxylate from one side of the substrate to the other [1]. Copyright Wiley-VCH Verlag GmbH & Co. KGaA. Reproduced with permission.

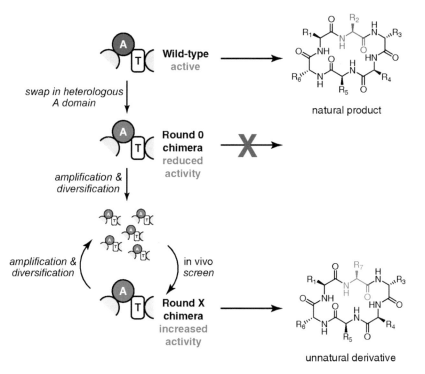

Plate 8
Directed evolution in the rescue of chimeric NRPS activity [3]. Copyright 2007 National Academy of Sciences.

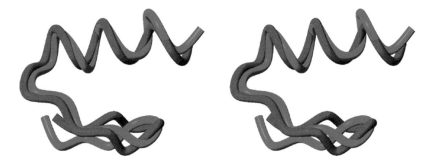

Plate 9
Backbone structure comparison of computed sequence FSD-1 and the target sequence Zif268 [1]. Comparison of the FSD-1 structure (blue) and the design target (red). Stereoview of the best-fit superposition of the restrained energy minimized average NMR structure of FSD-1 and the backbone of Zif268. Residues 3 to 26 are shown. [1]. Reprinted with permission from AAAS.

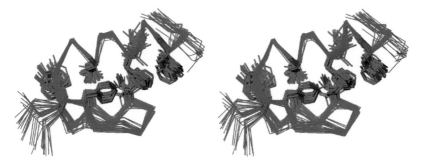

Plate 10
Empirically determined NMR structure ensemble of FSD-1, including side-chains. Stereoview showing the best-fit superposition of the 41 converged simulated annealing structures from X-PLOR. The backbone C^α trace is shown in blue and the side-chain heavy atoms of the hydrophobic residues (Tyr³, Ala⁵, Ile⁷, Phe¹², Leu¹⁸, Phe²¹, Ile²², and Phe²⁵) are shown in magenta. The amino terminus is at the lower left of the figure and the carboxyl terminus is at the upper right of the figure. The structure consists of two antiparallel strands from positions 3 to 6 (back strand) and 9 to 12 (front strand), with a hairpin turn at residues 7 and 8, followed by a helix from positions 15 to 26. The termini, residues 1, 2, 27, and 28 have very few NOE restraints and are disordered [1]. Reprinted with permission from AAAS.

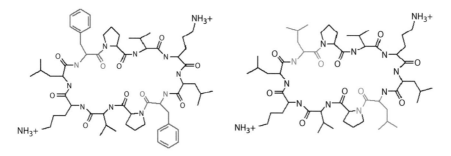

Plate 11
(Top) Gramicidin S synthetase is composed of two NRPS proteins, GrsA (3 domains) and GrsB (13 domains). Gramicidin S is produced in an assemblyline manner where two D-Phe-L-Pro-L-Val-L-Orn-L-Leu peptides are joined and cyclized. (A, Adenylation; T, Thiolation (peptidyl carrier protein); E, Epimerization; C, Condensation; TE, Thioesterase). (Bottom) The GrsA-PheA domain controls incorporation of the first amino acid in the synthesis of the antibiotic gramicidin. (Left) The natural gramicidin construct is shown with the incorporated phenylalanine shown in red. By changing the substrate specificity of the GrsA-PheA domain to accept leucine, it may be possible to create a modified gramicidin (right) where the phenylalanines have been replaced by leucine (blue).

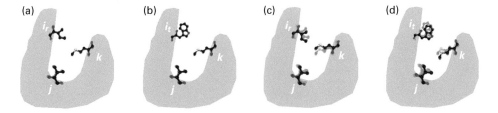

Plate 12

Energy-minimized DEE. Without energy minimization the swapping of rotamer i_r for i_t (panel a to panel b) leaves unchanged the conformations and self and pairwise energies of residues j and k. When energy minimization is allowed, the swapping of rotamer i_r for rotamer i_t (panel c to panel d) may cause the conformations of residues j and k to minimize (i.e., move) to form more energetically favorable interactions (from the faded to the solid conformations in panels c and d).

$$D = D_{max}\mathbf{v}^T\mathbf{Sv}$$

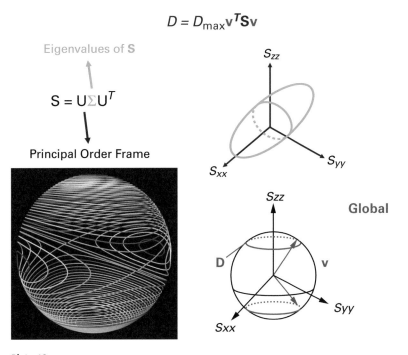

Plate 13

The alignment tensor $\mathbf{S}$ is a symmetric second-rank tensor that may be represented by a real-valued 3×3 matrix that is symmetric and traceless. Hence $\mathbf{S}$ has 5 degrees of freedom and may be decomposed using singular value decomposition (SVD) into a rotation matrix U, called the *principal order frame,* and a diagonal matrix Σ encoding its eigenvalues. The principal axes of U encode the eigenvectors. For a fixed experimental RDC D, the possible orientations of the corresponding internuclear vector $\mathbf{v}$ must lie on one of two *RDC curves* on the two-dimensional sphere S^2. Each curve is the intersection of an ellipsoidal cone with S^2. (Lower left) RDCs curves are shown spaced at 1 Hz intervals. Credit: Vincent Chen and Tony Yan.

Residual Dipolar Coupling

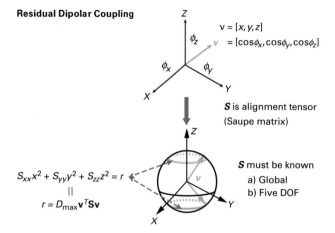

$$v = [x, y, z]$$
$$v = [\cos\phi_x, \cos\phi_y, \cos\phi_z]$$

S is alignment tensor
(Saupe matrix)

S must be known
a) Global
b) Five DOF

$$S_{xx}x^2 + S_{yy}y^2 + S_{zz}z^2 = r$$
$$\|$$
$$r = D_{max}\mathbf{v}^T\mathbf{S}\mathbf{v}$$

Plate 14

It is convenient to express the internuclear vector as a unit vector of the form $\mathbf{v}$, corresponding to its direction cosines. The RDC r can be expressed in Yan-Donald tensor notation [82, 83], as $r = D_{max}\mathbf{v}^T\mathbf{S}\mathbf{v}$, or in a principal order frame that diagonalizes the alignment tensor, namely $r = S_{xx}x^2 + S_{yy}y^2 + S_{zz}z^2$. Here, S_{xx}, S_{yy} and S_{zz} are the three diagonal elements of a diagonalized Saupe matrix $\mathbf{S}$ (the alignment tensor), and x, y and z are, respectively, the x, y, z-components of the unit vector $\mathbf{v}$ in a principal order frame (POF) which diagonalizes $\mathbf{S}$.

RDCs in Two Different Media

$$D = D_{max}\mathbf{v}^T\mathbf{S}\mathbf{v}$$

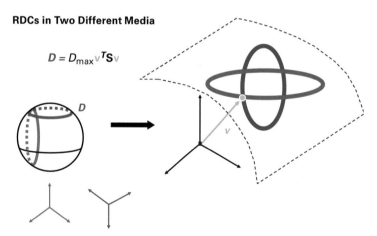

2 Principal Order Frames

Plate 15

A cartoon of the geometry and algebra of RDCs measured in two independent aligning media. If the two principal order frames (POFs) are independent, then the internuclear unit vector $\mathbf{v}$ is constrained to simultaneously lie on the blue RDC curve (from the blue POF) and the red RDC curve (from the red POF). Generically, the blue and red RDC curves will intersect at 0, 2, 4, or 8 points. Here, only one of the two RDC curves is shown for each POF. Suppose that r is the red RDC and that the diagonalized red POF can be represented as (S_{xx}, S_{yy}, S_{zz}). Let $u = 1 - 2(\frac{x}{a})^2$, where x is the x-component of $\mathbf{v}$ and $a^2 = (r - S_{zz})/(S_{xx} - S_{zz})$; see equation (18.5) below and [152, p. 238]. The discrete points corresponding to the RDC curve intersections are calculated exactly by solving a quartic polynomial equation in u, of the form $f_4u^4 + f_3u^3 + f_2u^2 + f_1u + f_0 = 0$ [152], which is also a quartic polynomial equation in x^2.

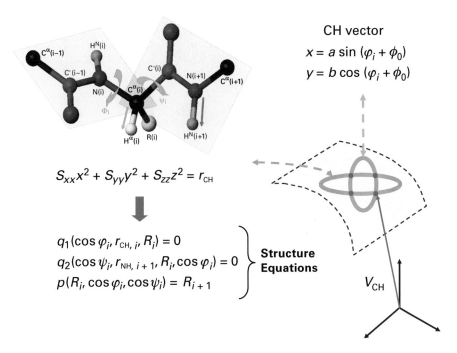

$$S_{xx}x^2 + S_{yy}y^2 + S_{zz}z^2 = r_{CH}$$

CH vector

$$x = a \sin (\varphi_i + \phi_0)$$
$$y = b \cos (\varphi_i + \phi_0)$$

$$q_1(\cos \varphi_i, r_{CH, i}, R_i) = 0$$
$$q_2(\cos \psi_i, r_{NH, i+1}, R_i, \cos \varphi_i) = 0$$
$$p(R_i, \cos \varphi_i, \cos \psi_i) = R_{i+1}$$

Structure Equations

V_{CH}

Plate 16
Given NH and C^α-H^α RDCs measured in one medium, the Wang-Donald Structure Equations yield exact solutions for the (ϕ, ψ) backbone dihedral angles.

(φ, ψ) angles in Conformation Tree

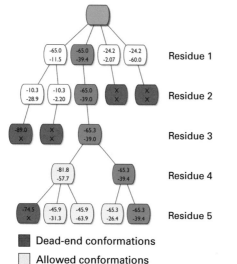

Residue 1
Residue 2
Residue 3
Residue 4
Residue 5

■ Dead-end conformations
□ Allowed conformations
■ Optimal conformation(s)

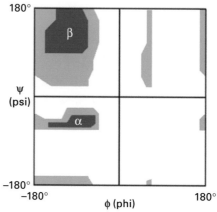

Given RDC error interval, search resulting conformation tree using a depth-first strategy.

Fold via sparse NOEs

Proteins (2006); Prot. Sci. (2007)

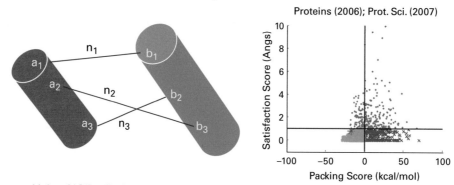

Using NOEs, find a translation* x minimizing:

$$\sum_{i=1}^{\ell} (\|a_i - b_i + x\| - n_i)^2$$

*4-fold discrete orientational symmetry (1 medium)

Plate 17

A *conformation tree* is a data structure used in depth-first search over the exact solutions (roots of polynomials) with backtracking or A* search, to optimally compute the backbone dihedral angles that *globally* best fit the RDC data and an empirical scoring function.

Plate 18

The orientations and conformations of secondary structure elements (SSEs) can be calculated using sparse RDCs. Then the SSEs are packed using sparse NOEs. Packings are scored separately for data fit and molecular mechanics energies [117] to avoid bias. The packing by NOEs also disambiguates the discrete 4-fold orientational degeneracy due to the symmetry of the dipolar operator.

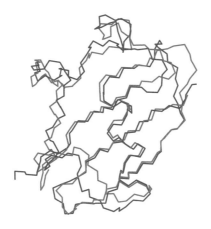

Reference[a]	Program	Technique[b]	Restraints Per Residue[c]	Accuracy[d]
Brown Lab [46]	X-plor	MD/SA	6 RDCs	1.45 Å
Blackledge Lab [62]	SCULPTOR	MD/SA	11 RDCs,	1.00 Å
Bax Lab [32]	MFR	Database	10 RDCs, 5 Chemical shifts	1.21 Å
Baker Lab [125]	RosettaNMR	DataBase/MC	3 RDCs, 5 Chemical shifts	1.65 Å
Baker Lab [125]	RosettaNMR	DataBase/MC	1 RDC	2.75 Å
Donald Lab[e]	RDC-EXACT	Exact Equations	2 RDCs	1.45 Å

Plate 19

Top: Structure of ubiquitin backbone with loops. The ubiquitin backbone structure (blue) was computed by extending RDC-EXACT to handle loop regions along the protein backbone [156]. The structure was computed using 59 NH and 58 CH RDCs (117 out of 137 possible RDCs; 20 are missing), 12 H-bonds, and 2 unambiguous NOEs. The structure has a backbone RMSD of 1.45 Å with the high-resolution X-ray structure (PDB ID: 1UBQ, in magenta) [147]. Bottom: Comparison of RDC-EXACT with previous approaches. (a) References to previously computed ubiquitin backbone structures (including loop regions), (b) algorithmic technique; (c) data requirements; (d) backbone RMSD of structure (for C^{α}, N, and C' backbone atoms) compared to the X-ray structure (1UBQ). The structure computed by RDC-EXACT includes loops and turns, as shown at top. [e]References [152, 151, 156, 155].

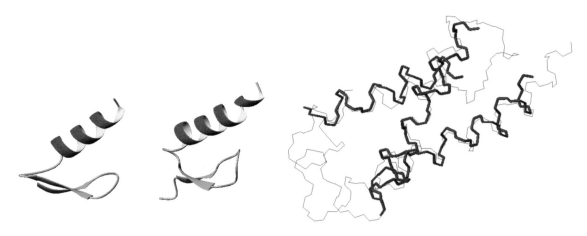

Plate 20
Left: The global fold of polη, computed by RDC-EXACT using 2 RDCs per residue measured in one medium plus 9 NOEs between the helix and β-strands. Center: Comparison of polη to the reference structure, PDB id: 2I5O. The RMSD of the secondary structure elements is 1.28 Å. Right: Global fold of human SRI, computed by RDC-EXACT (thick lines) using two RDCs per residue measured in one medium plus sparse NOEs. The reference structure is shown in thin lines (PDB id: 2A7O [86]).

NMR Structures   NMR vs. X-ray Structure

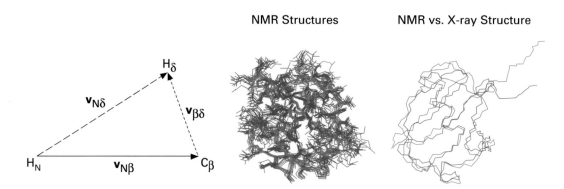

Plate 21
Left: The triangle relationship, defines a decision procedure to filter NOE assignments by fusing information from structure ($\mathbf{v}_{N\beta}$), modeling ($\mathbf{v}_{\beta\delta}$), and experiment ($|\mathbf{v}_{N\delta}|$). An accurate backbone structure is first computed using only 2 RDCs per residue (section 17.1). The vector $\mathbf{v}_{N\beta}$ is computed from this backbone structure. $\mathbf{v}_{\beta\delta}$ is a rotamer ensemble–based intraresidue vector mined from the PDB. The computed length of $\mathbf{v}_{N\delta}$ is compared with the *experimental* NOE distance d_N (measured using NOE cross-peak intensities) to filter ambiguous NOE assignments. The complexity of NOE assignment is $O(n^2 \log n)$, where n is the number of protons in the protein. One cycle of NOE assignment suffices when a well-defined backbone structure is computed using RDC-EXACT. In practice, it took less than one second to assign 1,783 NOE restraints from the NOE peak list picked from both the 3D ^{15}N-edited and ^{13}C-edited NOESY spectra of human ubiquitin. Right and center: The NMR structures computed from the automatically assigned NOEs. The middle panel shows the 12 best NMR structures with no NOE distance violation larger than 0.5 Å. The side-chains are blue; the backbone is magenta. The right panel is the overlay of the NMR average structure (blue) with the 1.8 Å X-ray structure (magenta) [147].

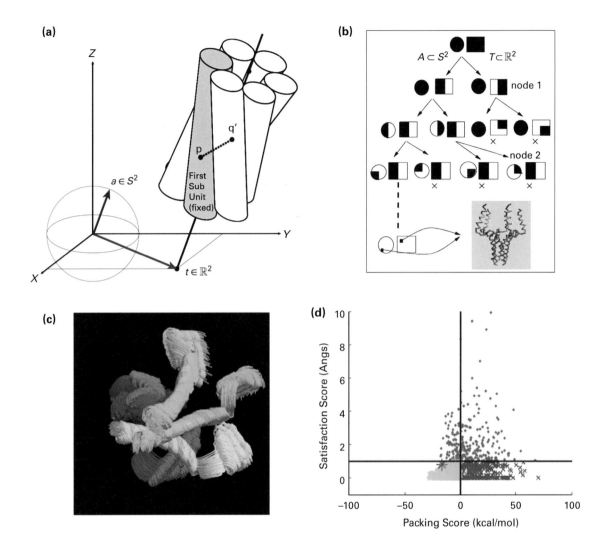

Plate 22

SYMBRANE Algorithm. (a) Given the subunit structure, the relative position $\mathbf{t} \in \mathbb{R}^2$ and orientation $\mathbf{a} \in S^2$ of the symmetry axis uniquely determines structure. Structure determination is a search problem in 4-dimensional *symmetry configuration space (SCS)*, $S^2 \times \mathbb{R}^2$. An NOE is shown between protons $\mathbf{p}$ and $\mathbf{q}'$ [117]. (b) The branch-and-bound algorithm proceeds as a tree search in SCS. The 4D SCS is represented as two 2D regions, a sphere representing the orientation space S^2 and a square representing the translation space $\mathbb{R}^2$. The dark shaded regions at each node of the tree represent the region in SCS being explored ($A \times T \subset S^2 \times \mathbb{R}^2$). Ultimately (bottom left of the tree), the branch-and-bound search returns regions in 4D space representative of structures that possibly satisfy all the restraints. At each node, we test satisfaction of each restraint of form $\mathbf{pq}' \leq d$ by testing intersection between the ball of radius d centered at $\mathbf{p}$ and the convex hull bounding possible positions of $\mathbf{q}'$. If there exists an intersection between the ball and the convex hull for each restraint, further branching is done (node 1); otherwise, the entire node and its subtree are pruned (node 2) [117]. (c) Representative results: complete ensemble of NMR structures of the unphosphorylated human phospholamban pentamer, PDB id: 2HYN. (d) Phospholamban restraint satisfaction score vs. packing score for all structures. The vertical and horizontal lines indicate the cutoffs for WPS structures: 1 Å for the satisfaction score and 0 kcal/mol for the packing score. The green stars and the blue crosses indicate the set of satisfying structures. The magenta points indicate the set of nonsatisfying structures that have been pruned. The green stars indicate the set of WPS structures and the red star indicates the reference structure [117].

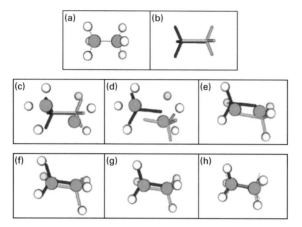

Plate 23

The motion of an ethane molecule as determined by geometric simulation: (a) Initial atomic positions; (b) ghost templates; (c) random atomic displacement; (d) fitting of ghost templates to atoms; (e) refitting of atoms to ghost templates; (f) and (g) further iterations of (d) and (e); (h) until a new conformer is found [3].

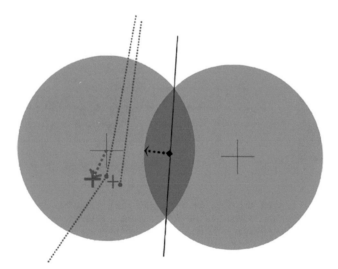

Plate 24

The method to handle steric overlap. When atoms come into contact, the overlap influences the movement of the atoms during the relaxation. The atom on the left is tethered to sites (blue dots) in two ghost templates (shown as blue dotted lines). The center of the atom, shown as a large black cross, would move to the small light blue cross in order to match the midpoint of its two associated ghost template sites. However, the steric overlap (black dashed arrows) causes it to move further away from the touching atom, so that its new position is at the larger blue cross, and the resultant motion is shown by the dashed red arrow [3].

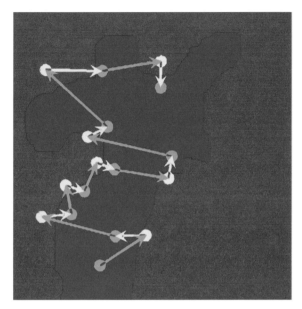

Plate 25
The program flow of the FRODA algorithm. An initial rigidity analysis leads to the creation of the ghost templates. Each iteration of the FRODA loop involves an initial random displacement of the atoms followed by multiple iterations of the fitting loop, in which the geometric and steric constraints are enforced [3].

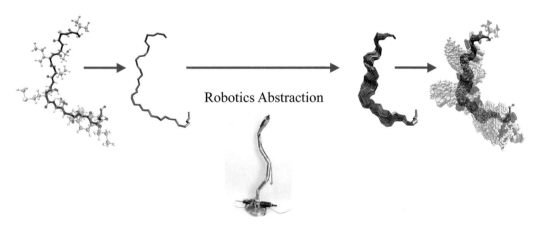

Plate 26
Overview of fragment ensemble method (FEM). (1) Backbone geometric exploration: ensemble of backbones that fit the missing loop; (2) side-chain exploration for a fixed backbone: loop in all-atom detail and without steric clash; (3) all-atom energy refinement: energy-minimized ensemble of loops. Credit: Lydia Kavraki.

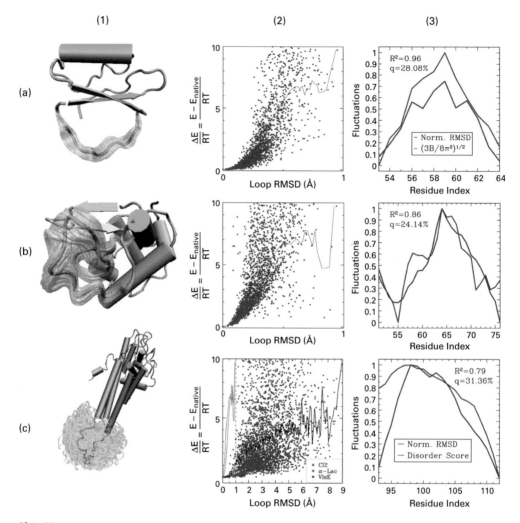

Plate 27

(a1, b1, c1) 5,000 transparent loop conformations versus opaque reference structure (equilibriated native structure for CI2 and α-Lac and lowest energy structure for VlsE). (a2, b2, c2) Energy landscapes associated with generated ensembles are shown by plotting the energetic difference versus the RMSD of each conformation relative to a reference structure. Energy landscapes are shown only for conformations with energy less than 10 RT units away from the reference structure. An average energy profile is computed by distributing conformations in bins every 0.001 Å away from the reference structure and measuring the energy of each bin as an average over its conformations. Average energy profiles obtained for CI2 and α-Lac are very steep compared to the flat average energy profile of VlsE. (a3, b3, c3) Obtained fluctuations versus B factor-derived fluctuations for the CI2 loop, fluctuations in the literature for the α-Lac loop, and disorder scores for the VlsE loop [1].

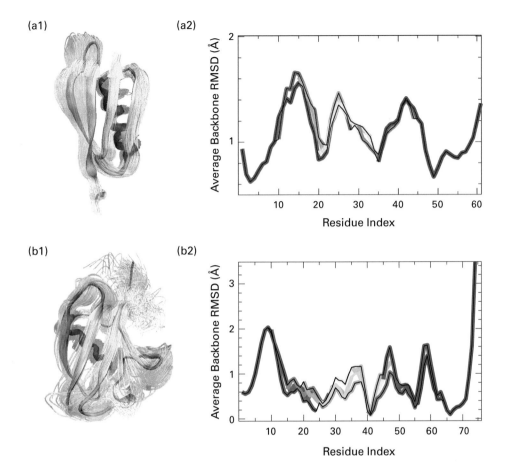

Plate 28

(a1 and b1) Computed native ensembles for protein G and ubiquitin, respectively. (a2 and b2) Average RMSD per residue obtained by combining the local fluctuations of all the different regions. Results for different regions are shown in different colors, from red to blue as a window of 30 residues slides from the N- to the C-terminus of the protein. The black lines mark the highest and lowest RMSD values recorded from all the different windows including each given residue, and provide an estimate for the uncertainty of the procedure. Two consecutive 30-residue windows have an overlap of 25 residues. The results corresponding to the first and last 5 residues of each fragment are discarded since they are biased by the finite size of the window [1].

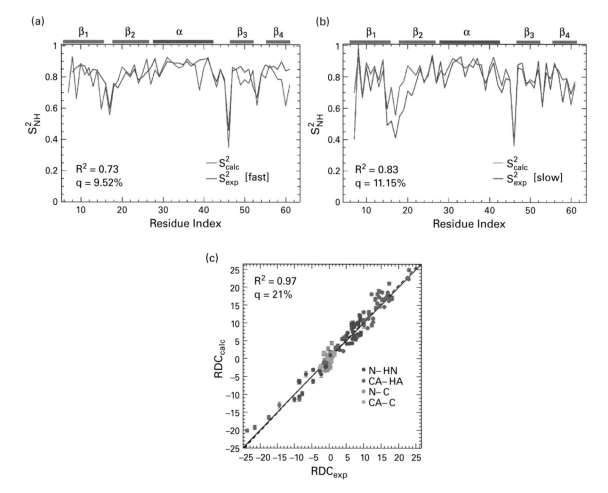

Plate 29
Comparison of NMR relaxation data versus the generated ensemble for protein G. (a) Comparison of S^2 backbone (amide) order parameters calculated over the ensemble (S^2_{calc}) with fast (S^2_{NH}) data obtained from NMR relaxation measurements (S^2_{exp}). (b) Comparison of S^2 backbone (amide) order parameters measured over the ensemble (S^2_{calc}) with slow (S^2_{NH}) data obtained from NMR relaxation measurements (S^2_{exp}). (c) Comparison of residual dipolar coupling (RDC) parameters as obtained in the ensemble (RDC$_{calc}$, on the y-axis), and from NMR relaxation measurements (RDC$_{exp}$, on the x-axis). Results for different bond types are shown in different colors. (a–c) The dashed black line indicates the linear least-squares regression fit on the two sets of data, while the continuous line represents the identity diagonal [1].

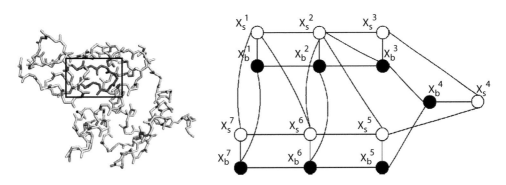

Plate 30

(a) Structure of lysozyme (PDB id: 2lyz) with a few residues highlighted. (b) Part of the random field induced by the highlighted residues: X_s^i's are the hidden variables representing the rotameric state, the visible variables are the backbone atoms in conformations x_b^i [15]. The publisher for this copyrighted material is Mary Ann Liebert, Inc. publishers.

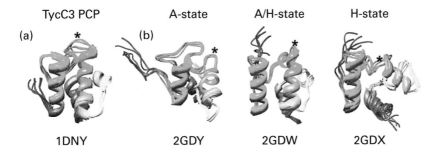

Plate 31

Structural mobility in a PCP: distinct conformations of the third module of tyrocidine synthetase-TycC3-PCP [4]. (a) Canonical structure of TycC3-PCP reported by [5]. (b) Three distinct conformations reported by Koglin et al. [3]: 2GDY for an Apo state (no cofactor attached), 2GDX for a Holo (cofactor attached) state, and 2GDW for an Apo/Holo state. Note that the structure of the A/H state is similar to the previously determined structure [1]. Reprinted with permission. Copyright 2006 American Chemical Society.

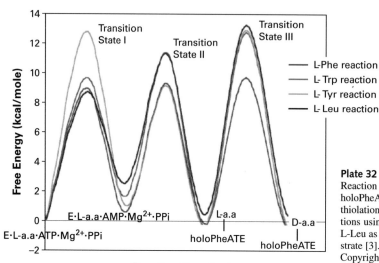

Plate 32
Reaction free energy profiles for holoPheATE-catalyzed adenylation, thiolation, and epimerization reactions using L-Phe, L-Trp, L-Tyr, and L-Leu as the starting amino acid substrate [3]. Reprinted with permission. Copyright 2001 American Chemical Society.

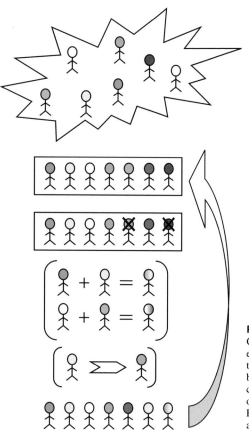

Plate 33
Genetic algorithm with a snapshot of one evolution. In this example seven members comprise the initial population. In the selection phase, two members are found unfit and will be replaced. In the recombination phase (also known as crossover), two member pairs are chosen to reproduce two offspring with combined characteristics from their parents. Finally, in the mutation phase, the pink member is picked and modified to be pinker. Source: See note, p. 271.

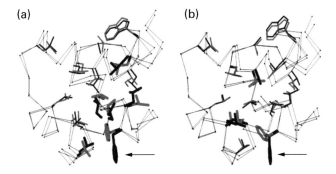

Plate 34
Modeling of cro-rep. Structures are displayed as C^{α} traces with hydrophobic core side-chains. (a) The crystal structures of 434 cro (black) against 434 cro-rep (red). The arrow shows the major difference at Phe44. (b) ROC (black) and SoftROC (red) predictions based on the cro-rep template. While ROC predicts a cro-like orientation of Phe44, SoftROC predicts a core structure more similar to that of cro-rep. Reprinted with permission from Elsevier. [1].

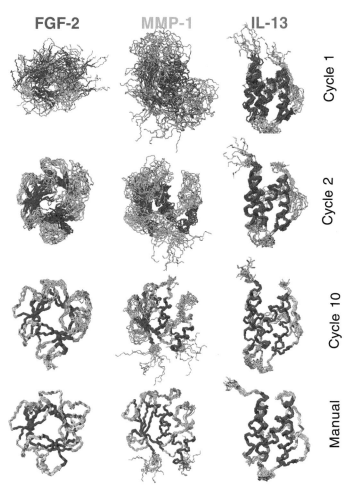

Plate 35
Calculated structures in intermediate and final ensembles. Figure by Janet Huang and Guy Montelione [1].

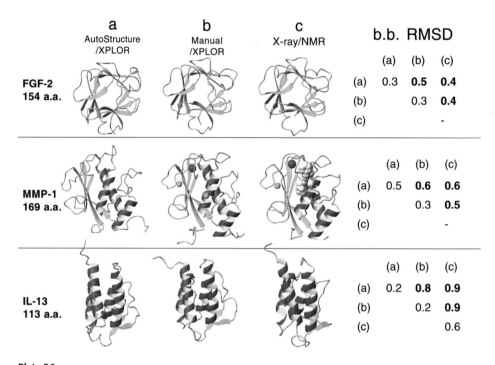

Plate 36
Structures calculated by AutoStructure vs. other methods. The (b) method combines manual assignment analysis with optimization techniques such as SA/MD. Figure by Janet Huang and Guy Montelione.

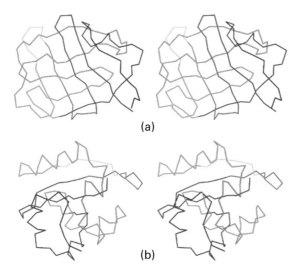

Plate 37
Comparison of an incorrect protein model with its corrected counterparts. The model is stereo C^α traces color-ramped from blue at the N-terminus to red at the C-terminus. (a) Incorrect model of photoactive yellow protein (PDB ID: 1phy) and (b) the corrected model (2phy). In this case, the initial model displayed a β-clam fold, whereas the correct model revealed an α/β protein with a fold similar to that of the SH2 domain. Copyright 2000 IUCr.[2].

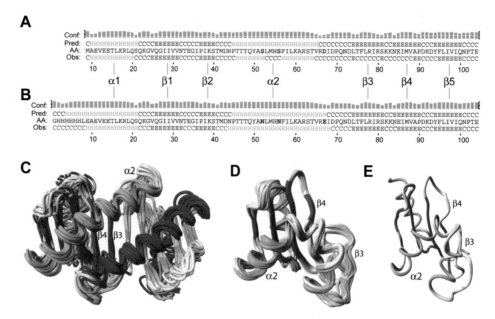

Plate 38

Sequence and structure ensembles of two DLC2A structures. (A and B) The sequences of human (hDLC2A) and mouse (mDLC2A); (C) Ribbon diagram of the structure ensemble of mDLC2A (PDB ID: 1Y4O); (D) ribbon diagram of the structure ensemble of hDLC2A (PDB ID: 1TGQ); (E) the refined average structure of the ensemble calculated using the reconstructed 1TGQ dataset [1].

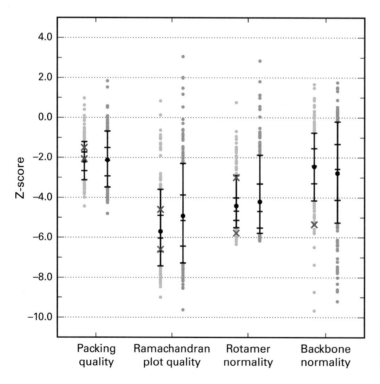

Plate 39

Structure quality Z-scores for a large set of the NMR structures after 2003 [1].

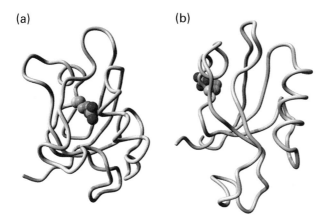

Plate 40
An example of observed structural anomalies [1].

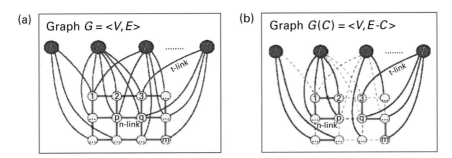

Plate 41
An example of multiway graph cut. Dashed lines in (b) denote the cut edges in the graph [1]. Copyright 1998 IEEE.

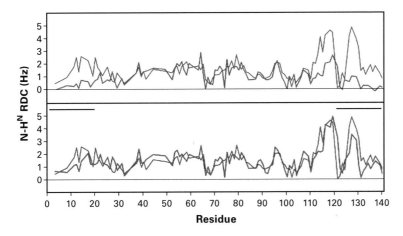

Plate 42
Comparisons between ensemble-averaged RDCs (red) simulated without long-range contacts, and experimental RDCs (blue) for protein αS in (top) Pf1 bacteriophage and (bottom) lyotropic media. Simulated data are scaled to maximize fit in the region 22–112. For illustration purposes, the RDC sign is inverted compared to the usual conventions [1]. Reprinted with permission. Copyright 2005 American Chemical Society.

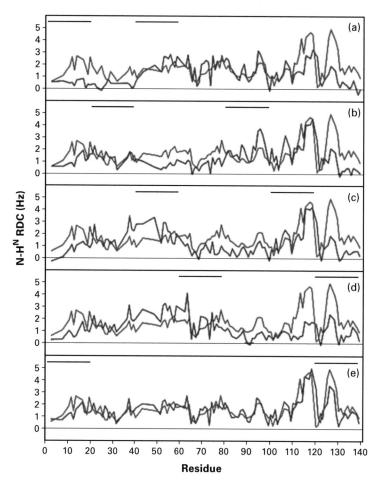

Plate 43

Comparisons between experimental (blue) and simulated RDCs incorporating the long-range contacts (red). Contact regions are indicated by bars. (a) (1–20, 41–60); (b) (21–40, 81–100); (c) (41–60, 101–120); (d) (61–80, 121–140); (e) (1–20, 121–140) [1].

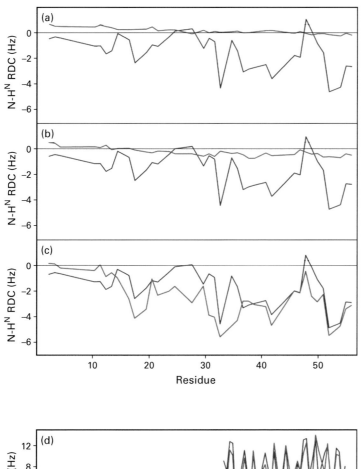

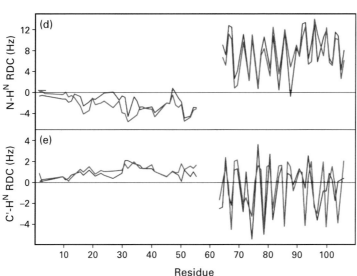

Plate 44
Comparison between simulated (red) and experimental (blue) RDCs for protein PX. PX is numbered from 1 to 109 and contains the 14-residue His tag, such that residues 15–109 correspond to the region 474–568 in the intact C-terminal domain protein. (a) Experimental N-H^N RDCs compared with RDCs simulated in the unfolded domain of PX using random sampling of backbone dihedral angle (ϕ/ψ) space. In this model, (I) no nonbonded interactions were taken into account except for avoiding overlap between the folded and unfolded domains. This model is essentially a random flight chain model of the polymer and reproduced the expected bell-shaped distribution along the chain. (b) Experimental N-H^N RDCs compared with RDCs simulated in the unfolded domain of PX using random sampling of backbone dihedral angle (ϕ/ψ) space. In this model (II) nonbonded interactions between residues in the unfolded domain were taken into account by using a simple steric repulsion model. Overlap between the folded and unfolded domains was also avoided. (c) Experimental N-H^N RDCs compared with RDCs simulated in the unfolded domain of PX using random sampling residue-specific (ϕ/ψ) propensities found in loop regions of a database of folded proteins combined with the simple volume exclusion model (model III). (d) Simulated and experimental N-H^N RDCs from the entire protein by using model III. (e) Simulated and experimental C'-H^N RDCs from the entire protein using model III [2]. Copyright 2005 National Academy of Sciences.

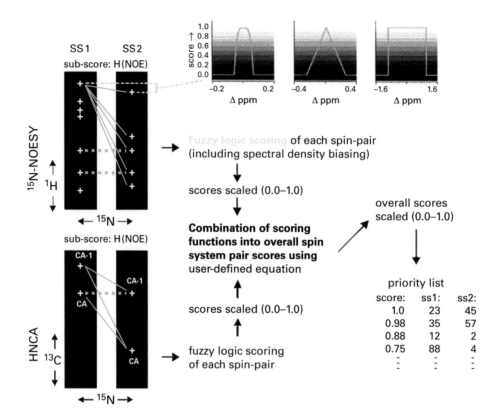

equation	interpretation
Cα	Consider only spin system pairs with compatible Cα (ignore other spins).
H × Cα	Consider only spin system pairs with compatible Cαs and H xpeaks (NOESY) in common.
H × Cα × Cβ	Consider only spin system pairs with compatible Cαs, compatible Cβs, and H xpeaks in common.
Cα × (H + Cβ)	Consider only spin system pairs with compatible Cαs and either compatible Cβs or H xpeaks in common.
H&&Cα	Consider only spin system pairs with compatible Cαs and H xpeaks in common.
H‖Cα	Consider only spin system pairs with compatible Cαs or H xpeaks in common.
H&|Cα	Consider only spin system pairs with compatible Cαs or H xpeaks in common, but especially favor those with both.
Cα² × (H + Cβ + CO)	Consider only spin system pairs with stringently compatible Cαs and at least one of the following: H xpeaks in common, compatable Cβs, or compatible COs.

Plate 45

The spin system pairing scoring. (Top) Diagram of spin-system pair scoring. The spins of both spin systems (represented here by crosspeaks in spectra) are compared using a user-defined scoring function. Spin comparisons are signified by lines connecting relevant points in the NMR spectra. Model subscoring functions are plotted in gold at the top of the figure. The subscores are calculated by adding together all of the relevant comparison scores of each spin of one spin system (ss1) with each spin of another spin system (ss2). After the individual subscore types (Cα and H(NOE) are shown here) are calculated, they are then combined to form an overall spin-system pair score. The manner in which they are combined is determined by a user-defined scoring equation. The overall scores are recorded in the ''priority list.'' (Bottom) Examples of scoring equations and literal interpretation. The left column shows examples of valid scoring equations demonstrating the quasi-operators ''&&,'' ''‖,'' and ''&\|.'' ''xpeak'' = crosspeak. The right column describes the operations of the scoring equations [1]. Reprinted with permission from Elsevier.

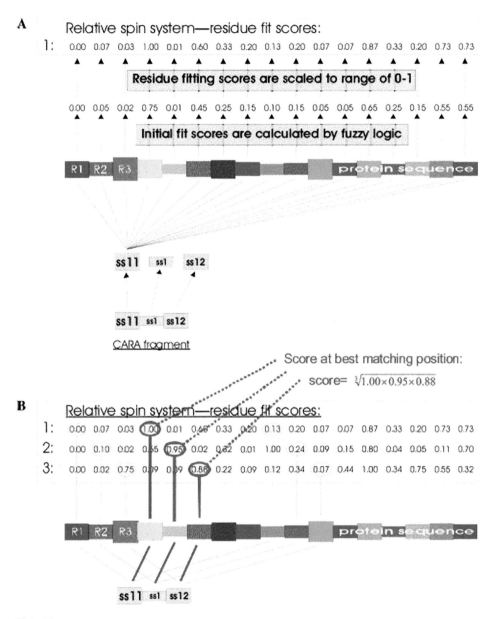

Plate 46

The computation of fitness of a fragment. Matching of spin system fragments to the protein sequence proceeds in two stages, (A) individual spin system-sequence matching, and (B) fragment-sequence matching. (A) Each spin system of each fragment is first fit, using fuzzy logic, to each residue position of the protein sequence. In this diagram, comparison of a single spin system (ss11) to each position of the protein sequence is signified by solid black lines. The fit scores for each spin system are then scaled to a range of 0–1. The final fitting scores for each spin system to the protein sequence are shown here in rows 1, 2, and 3, in the ''Relative spin system residue fit scores'' boxes. (B) Fragment-sequence fitting. Fragment scores are calculated as the geometric average of the single spin system relative fits for consecutive positions within the protein sequence. [1]. Reprinted with permission from Elsevier.

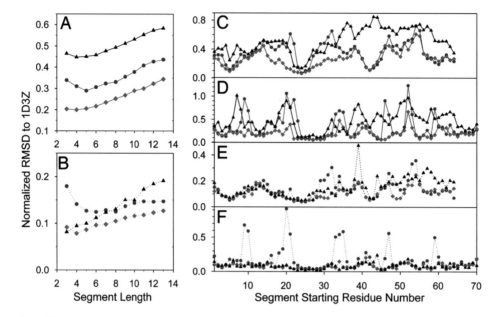

Plate 47

Plots of normalized accuracy of database fragments selected for ubiquitin. For each ubiquitin segment, 200 fragment candidates of the same length were selected using either the standard Rosetta procedure (filled triangles), or an MFR search of the 5,665-protein structural database, assigned by the programs DC (filled circles) or SPARTA (filled diamonds). For all panels, coordinate RMSDS (N, C^α, and C^β) between query segment and selected fragments are normalized with respect to randomly selected fragments. (A and B) Average (A) and lowest (B) normalized RMSD of 200 selected fragments, as a function of fragment size, relative to the X-ray coordinates of the corresponding ubiquitin segment, averaged over all (overlapped) consecutive segments. (C and D) Average normalized RMSD of 200 nine-residue (C) and three-residue (D) fragments relative to the X-ray coordinates, as a function of position in the ubiquitin sequence. (E and F) Lowest normalized RMSD of any of these selected nine-residue (E) or three-residue (F) fragments. Reprinted from [1]. Copyright 2008 National Academy of Sciences.

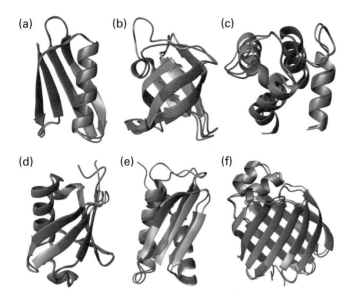

Plate 48

Six structures generated by CS-Rosetta; the ordered secondary structure is very similar but the loops differ significantly between experimental and predicted structures. Backbone ribbon representations of the lowest-energy CS-Rosetta structure (red) superimposed on the experimental X-ray/NMR structures (blue), with superposition optimized for ordered residues. (a) GB3, (b) CspA, (c) calbindin, (d) ubiquitin, (e) DinI, (f) Apo_1afbp. Reprinted from [1].

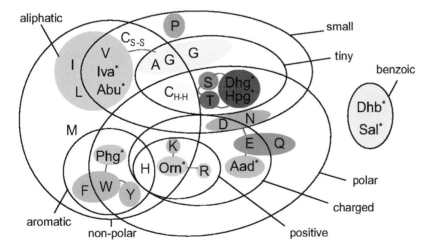

Plate 49

Addressing the side specificity problem by clustering specificities for amino acids with similar physicochemical properties. The colored sets show how similar amino acids have been clustered to composite specificities of A domains. To get larger clusters, several smaller clusters were joined, as indicated by red lines connecting colored sets. An asterisk indicates rare nonproteinogenic amino acids [1]. Reprinted with permission.

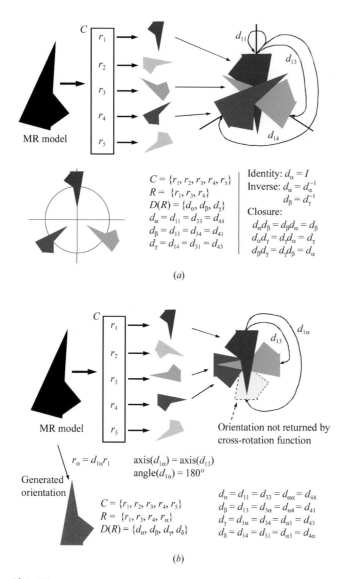

$C = \{r_1, r_2, r_3, r_4, r_5\}$
$R = \{r_1, r_3, r_4\}$
$D(R) = \{d_\alpha, d_\beta, d_\gamma\}$
$d_\alpha = d_{11} = d_{33} = d_{44}$
$d_\beta = d_{13} = d_{34} = d_{41}$
$d_\gamma = d_{14} = d_{31} = d_{43}$

Identity: $d_\alpha = I$
Inverse: $d_\alpha = d_\alpha^{-1}$
$\quad\quad d_\beta = d_\gamma^{-1}$
Closure:
$d_\alpha d_\beta = d_\beta d_\alpha = d_\beta$
$d_\alpha d_\gamma = d_\gamma d_\alpha = d_\gamma$
$d_\beta d_\gamma = d_\gamma d_\beta = d_\alpha$

(a)

$r_\alpha = d_{1\alpha} r_1$

Generated orientation

$\text{axis}(d_{1\alpha}) = \text{axis}(d_{13})$
$\text{angle}(d_{1\alpha}) = 180°$

Orientation not returned by cross-rotation function

$C = \{r_1, r_2, r_3, r_4, r_5\}$
$R = \{r_1, r_3, r_4, r_\alpha\}$
$D(R) = \{d_\alpha, d_\beta, d_\gamma, d_\delta\}$

$d_\alpha = d_{11} = d_{33} = d_{\alpha\alpha} = d_{44}$
$d_\beta = d_{13} = d_{3\alpha} = d_{\alpha4} = d_{41}$
$d_\gamma = d_{1\alpha} = d_{34} = d_{\alpha1} = d_{43}$
$d_\delta = d_{14} = d_{31} = d_{\alpha3} = d_{4\alpha}$

(b)

Plate 50

Two-dimensional examples for threefold (a) and fourfold (b) NCS. (a) A model is shown with the results of a simplified cross-rotation search C containing only five rotations. Orientations corresponding to rotations r_1 (purple), r_3 (green), and r_4 (blue) form an NCS-consistent rotation set R. For clarity, only a few rotation differences (d_{11}, d_{13}, and d_{14}) are shown in the upper right overlapping-orientation figure. The rotation differences, $D(R)$, form a complete rotation difference set and satisfy the group properties of associativity (not shown), identity, inverse, and closure. (b) A fourfold NCS example is shown using similar notation to that in (a). In this example, only three of the four NCS-consistent rotations (r_1, r_3, and r_4) are contained in the cross-rotation peak list C. The missing rotation $r_\alpha = d_{1\alpha} r_1$ is computed using $d_{1\alpha}$ defined by the axis of the three identified NCS-consistent rotations and the missing angle. The now complete NCS-consistent rotation set R has a complete rotation difference set $D(R)$, which satisfies the subgroup properties.

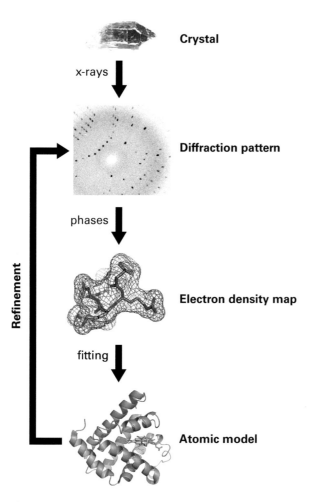

Crystal

x-rays

Diffraction pattern

phases

Electron density map

fitting

Atomic model

Refinement

Plate 51
Schematic diagram for X-ray crystallography in structure determination [6].

26   Ligand Configurational Entropy

To understand protein and ligand flexibility, it is necessary to have a basic grounding in statistical mechanics and the theory of entropy. This chapter presents work on the characterization of the components of ligand conformational entropy [1] and its implications for protein design and drug design. This chapter builds on the understanding of free energy developed in chapter 25.

26.1   Experimental Input

This method in this chapter is not heavily dependent on experimental data. The required inputs consist of structures of a ligand of interest and of the protein it binds to. Standard energy functions are used with mathematical decompositions to characterize changes in entropy on binding based on these structures. The structures used are already solved, so the raw NMR or X-ray data used in structure determination is not considered by the technique in this chapter.

26.2   Entropy

Entropy, in the broad sense, reflects the thermodynamic favorability of disorder in a system. The familiar Gibbs free-energy equation (chapter 25) shows this effect:

$$\Delta G = \Delta H - T \Delta S. \tag{26.1}$$

Thus, a change to a more disordered state (positive ΔS) results in a more negative, more favorable ΔG. One interpretation of disorder is the multiplicity W of a system, that is, the number of states readily available to the system. A toy example of multiplicity is a standard 6-sided die. For such a die, $W = 6$ is the number of states possible, that is, the number of possible rolls. From the multiplicity W of a system, we can determine a measure of the system's entropy:

$$S = -R \ln(W), \tag{26.2}$$

where R is the universal gas constant. If not all states of a system are equally accessible or equally populated, a modified version of equation (26.2) is used:

$$S = -R \sum_j p_j \ln(p_j),$$ (26.3)

where p_j is the probability of state j being occupied, and the summation is over all states.

Thus, the value of ΔS can be determined by comparing multiplicity-derived entropies before and after a reaction.

26.3 Entropy in Ligand Binding

Ligand configurational entropy is an entropy value derived from the total multiplicity of a ligand molecule. Configurational entropy can be considered as the combination of two other types of entropy: *conformational entropy* and *vibrational entropy*.

26.3.1 Conformational Entropy

Conformational entropy has also been covered, in some sense, in our chapters on K^*, belief propagation, and rotamer search/pruning. Conformational entropy reflects the number of conformations that a molecule can assume. In proteins, these conformations are often represented with rotamer libraries, reflecting the rotations that side-chains can undergo. Ligands may be treated similarly, with flexible functional groups or other components. In terms of an energy landscape, conformational entropy reflects the number of wells or minima.

Conformational entropy is lost on binding when proximity to a bound molecule causes interactions (most often, steric clashes) that prevent previously available conformations from being assumed. It has often been assumed that this is the main contributor to entropy change on binding.

26.3.2 Vibrational Entropy

Vibrational entropy reflects the motions available, not to side-chains or functional groups, but to a molecule as a whole or to the individual atoms that compose it. Vibrational entropy can be further subdivided. Rotation/translation accounts for motions of the entire molecule. Torsion, angle bending, and angle stretching account for motions of pairs of individual atoms. These latter three entropies may be recognizable as motions important in infrared (IR) spectroscopy. In terms of an energy landscape, vibrational entropy reflects the width of wells or minima.

Vibrational entropy is lost on binding when proximity to a bound molecule causes interactions (most often steric clashes) that restrict previously possible motions. While rotation and translation are clearly restricted on binding, the entropy loss from atomic motions is less intuitive.

26.4 Entropy and Amprenavir

Gilson and coworkers [1] calculated the entropies for amprenavir, an inhibitory drug molecule that binds to HIV protease. They calculated its overall loss of configurational entropy on binding to HIV protease and the individual components of that entropy as discussed earlier. (For the mathematical decomposition of entropy components, see the appendix of [1].)

The authors determined the overall configurational entropy loss on binding to be 26.4 kcal/mol. Of this, at most 4.1 kcal/mol can be accounted for by conformational entropy loss. While 4.1 kcal/mol is not negligible (the free energy difference between a folded and an unfolded protein being typically approximately that value), it is less than a sixth of the total entropy lost. Therefore, vibrational entropy appears to have an unexpectedly dominant role in binding.

The authors also found that the individual components of vibrational entropy (rotation/translation, torsion, bending, stretching) summed to more than the total vibrational entropy. They conclude that there is crosstalk among these components. This means that certain combinations of vibrational motions are allowed or disallowed together, which is reasonably intuitive.

Table 26.1 is taken from Gilson and coworkers [1]. Here the authors presented the correlations among components of vibrational entropy, as determined by a quasiharmonic analysis (essentially a molecular dynamics approach). Rotation/translation, torsion, and angle bending are all fairly strongly correlated to each other. Bond stretching, which contributes vastly to total entropy, is almost unaffected by binding, and is less strongly correlated to the other components.

26.5 Implications for Design

Entropy loss on binding is of interest to drug designers because a negative ΔS value in Eq. (26.1) renders the reaction, in this case inhibition of a target protein, less favorable. Understanding the components of the inevitable loss in entropy may allow the design of drugs that minimize that loss.

Table 26.1
Decomposition of amprenavir's change in configurational entropy upon binding, calculated via submatrices of the covariance matrices of the dominant free and bound energy well

	R/T	Tors	Angle	Stretch
Free				
R/T	−7.07	0	0	0
Tors		54.5	3.6	0.1
Angle			78.1	1.0
Stretch				109.52
Bound				
R/T	7.02	1.5	0.8	0.08
Tors		61.6	4.1	0.13
Angle			80.1	1.2
Stretch				109.55
Difference				
R/T	12.3	1.5	0.8	0.08
Tors		7.1	0.5	0.03
Angle			2.0	0.2
Stretch				0.04

Entropy values multiplied by $-T$ to yield free energy contributions in kcal/mol. R/T-rotational and translation part, including $RT \ln(8\pi^2/C^\circ)$ for the free ligand; Tors-torsional; Ang-angle bend; Stretch-bond stretch. Off-diagonal terms are entropy changes caused by changes in correlations between classes of motions, for example, $S_{\text{tors,stretch}}^{\text{corr}} = S_{\text{tors,stretch}} - S_{\text{tors}} - S_{\text{stretch}}$. Lower triangle terms are not listed because the matrix is symmetric [1]. Copyright 2006 National Academy of Sciences.

The analysis presented here only measures the entropy change of the ligand amprenavir. A true total entropy change on binding would also have to take into account entropy loss in the protein it binds to. However, in considering only the ligand, this study indicates that the entropy loss of the ligand alone can have a substantial effect on binding.

The most substantial contributor to entropy loss appears to be rotation/translation. Unfortunately, it would be very difficult to change the ligand structure to address this. However, both conformational entropy and torsional vibration can be addressed. Fewer or less mobile side groups would reduce the loss of conformational entropy on binding. In addition, any bond with multiple-bond character would reduce the overall torsional entropy of a ligand and thus reduce the potential for entropy loss.

It may also be necessary to reconsider the weighting in energy functions to reflect the dominant role of vibrational entropy more accurately.

Reference

[1] Chia-en A. Chang, Wei Chen, and Michael K. Gilson. Ligand configurational entropy and protein binding. *PNAS* 104;5(2006):1534–1539.

In this chapter, we describe carrier protein structure and recognition in polyketide and nonribosomal peptide biosynthesis [1, 2, 3]. This chapter builds on the earlier lectures on protein flexibility, and motivates, by a specific empirical example, the concepts of conformational change (switch) in proteins, and recognition at protein-protein interfaces. The purpose is to ground the algorithmic discussion by reference to specific systems, and to learn more about NRPS.

27.1   Carrier Proteins

Carrier proteins in polyketide synthase (PKS) and nonribosomal peptide synthetase (NRPS) systems play an important role in synthesis of polyketides and nonribosomal peptides, respectively (figure 27.1). Carrier proteins are relatively small in size (80–100 amino acids; 8–10 kDa) compared to other catalytic domains in a synthase system. The key role of carrier proteins is to serve as an attachment site for monomers (acyl, aminoacyl, or aryl compounds), and intermediates during the sequential synthetic reactions, shuttling intermediates between catalytic modules during the synthesis. Both monomers and intermediates are attached to a carrier protein at a thiol group of the phosphopantetheine cofactor that is covalently attached to a carrier protein. Since there are three kinds of possible building blocks for these two enzyme systems, carrier proteins are categorized into groups accordingly: *acyl carrier proteins (ACPs)*, *peptidyl carrier proteins (PCPs)*, and *aryl carrier proteins (ArCPs)*. Basic reaction steps are shared among different systems. These sequential reaction steps include activation of a monomer, and condensation between an activated monomer and a growing chain.

The PKS and NRPS systems are responsible for syntheses of a vast number of secondary metabolites. Such variability comes from varieties of building blocks, and additional modification steps in each elongation round. Even though final compounds, monomers, and catalytic members of the systems can differ to a great extent, there are some similarities among carrier proteins of different systems. Specifically, many carrier proteins have a similar general fold motif, and hence are characterized into a superfamily. Figure 27.2a (plate 31) shows this prototypical four-helix bundle structure of carrier proteins.

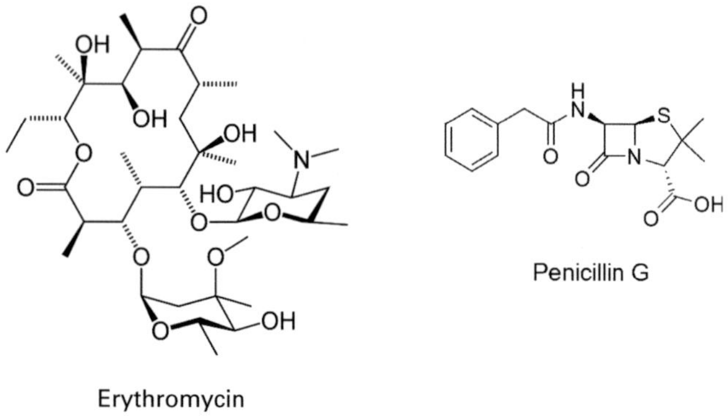

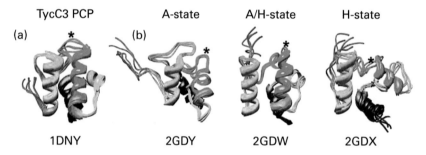

Erythromycin

Penicillin G

Figure 27.1
Two examples of a polyketide compound (erythromycin) and a nonribosomal peptide (penicillin).

TycC3 PCP	A-state	A/H-state	H-state
(a) *	(b) *	*	*
1DNY	2GDY	2GDW	2GDX

Figure 27.2 (plate 31)
Structural mobility in a PCP: distinct conformations of the third module of tyrocidine synthetase-TycC3-PCP [4]. (a) Canonical structure of TycC3-PCP reported by [5]. (b) Three distinct conformations reported by Koglin et al. [3]: 2GDY for an Apo state (no cofactor attached), 2GDX for a Holo (cofactor attached) state, and 2GDW for an Apo/Holo state. Note that the structure of the A/H state is similar to the previously determined structure [1]. Reprinted with permission. Copyright 2006 American Chemical Society.

The structure portrayed in figure 27.2a (plate 31) is not the only conformation of this carrier protein, TycC3. NMR studies have shown that some carrier proteins in PKS or NRPS systems have multiple distinct conformations, and dynamic changes among conformations modulate protein interactions during the synthesis. Proper and specific protein-protein interactions (PPIs) between carrier proteins and other catalytic domains are essential for sequential reactions to occur. However, the carrier proteins are small compared to other catalytic proteins that they interact with, and movements of the phosphopentatheine arm are not in a sufficient range for interactions. This is where conformational mobility of carrier proteins comes into play, modulating interactions during the well-timed sequential synthetic reactions.

Combinatorial mutagenesis studies on some PKS/NRPS systems have revealed that the interacting surface on some carrier proteins is localized. Only a few key residues are responsible for interacting with a catalytic domain, and the key residues are distinct and specific for interactions with different catalytic domains in these systems. These results give an insight into possibilities to create a carrier protein with a noncognate interaction, and hence a possible novel final compound. However, we still need to know more about whether these principles hold true in other systems, and to have better understanding of protein-protein recognition and interactions between carrier proteins and other catalytic domains for this engineering purpose.

Dötsch and coworkers [3] showed how NMR studies revealed three distinct conformations, encoding a conformational switch, for a carrier protein. These switches modulate PPIs in NRPS. The study [3] demonstrates the power of NMR to elucidate the connection between dynamics and biological protein function.

References

[1] J. R. Lai, A. Koglin, and C. T. Walsh. Carrier protein structure and recognition in polyketide and nonribosomal peptide biosynthesis. *Biochemistry* 45;50(2006 Dec 19):14869–14879.

[2] J. Lai, M. A. Fischbach, D. Liu, and C. T. Walsh. A protein interaction surface in nonribosomal peptide synthesis mapped by combinatorial mutagenesis and selection. *PNAS* 103;11(2006):5314–5319.

[3] A. Koglin, M. R. Mofid, F. Lohr, B. Schafer, V. V. Rogov, M. Blum, et al. Conformational switches modulate protein interactions in peptide antibiotic synthetases. *Science* 312(2006):273–276.

[4] J. R. Lai, M. A. Fischbach, D. R. Liu, and C. T. Walsh, Localized protein interaction surfaces on the EntB carrier protein revealed by combinatorial mutagenesis and selection. *Journal of American Chemical Society* 128(2006):11002–11003.

[5] T. Weber, R. Baumgartner, C. Renner, M. A. Marahiel, and T. A. Holak. Solution structure of PCP, a prototype for the peptidyl carrier domains of modular peptide synthetases. *Structure* 8(2000):407–418.

In this chapter, we study the kinetics of the initiation module PheATE (GrsA) of gramicidin S synthetase mainly based on studies by Walsh and co-workers [3, 7, 8]. These enzymes are used repeatedly as examples in this book, and this chapter describes their kinetics, which is provided to give a more complete understanding of their function. It also provides a primer on some of the experimental methods used to validate computational enzyme designs.

28.1   Background

Peptidic-type products, including ribosomally synthesized peptides and nonribosomally synthesized peptides, constitute a large family of natural products that are widely used for developing therapeutic drugs with novel activities. The basic structure of nonribosomally synthesized peptides is constituted by joining several small molecule building blocks. Not only the 20 standard amino acids, but also a vast number of different molecules can be incorporated into the building blocks, which provides a huge amount of diversity. The synthesis of nonribosomal peptides is carried out by the nonribosomal peptide synthetases (NRPSs). These enzymes are modularly organized multidomain complexes, from which different domains catalyze the incorporation of each building block into the elongated peptide product. The antibiotic gramicidin S is synthesized by the gramicidin S synthetase, which contains two multimodular subunits in which the GrsA subunit recognizes and initiates the incorporation of the first amino acid Phe, while the GrsB subunit elongates the rest of 4 amino acids to form the peptide chain (see figure 12.1). The initial module of the gramicidin S synthetase (GrsA) contains three domains: the N-terminal adenylation (A) domain followed by a small thiolation (T) domain and a C-terminal epimerase (E) domain. The A domain recognizes and activates L-phenylalanine (L-Phe) by adenylation to form a covalently bound L-phenylalanyl-adenosine-5′-monophosphate diester (L-Phe-AMP). The T domain uses an HS-phosphopantetheine arm (Ppant) to carry the L-Phe-AMP to the E domain, where the L-Phe-S-4′-Ppant-acyl enzyme complex is epimerized to D-Phe-S-4′-Ppant-acyl enzyme, which can be recognized by the GrsB subunit. The ATE domain module provides a well-established model for the amino acid incorporation by the nonribosomal peptide synthetases. Nevertheless, how they specifically select Phe as the substrate remains unclear. Computational structure-based protein redesign provides an incisive tool to dissect this nonribosomal code (chapter 12).

In this lecture, we use the following terms to distinguish the apo from the holo form of the multifunctional enzyme: *apoPheATE* (without Ppant link to T domain), and *holoPheATE* (with Ppant link to T domain).

28.2   Binding of the Amino Acid Substrate to the A Domain of GrsA

The catalysis of adenylation by the A domain requires binding of magnesium, ATP, and amino acid substrate into the binding pocket. Magnesium coordinates the binding of the ATP, which may induce a conformational change to facilitate association of the amino acid substrate. To determine the equilibrium binding constants, a fluorescence titration method was used to monitor the change of the fluorescence signal, which is proportional to the concentration of the substrate-enzyme complex. Percentage of fluorescence difference (ΔF) was measured as a function of substrate concentration $[S]$. The dissociation constant (K_d) can be calculated from the slope by plotting ΔF against $\Delta F/[S]$ or by directly fitting into the equation:

$$\Delta F = \frac{\Delta F_{max}[S]}{[S] + K_d}.$$

The fluorescence signal decreases as more substrate binds to the enzyme and the graph shows a hyperbolic curve. The results show similar K_d values with L-Phe and D-Phe, which suggest that the binding pocket can accommodate both stereoisomers equally well. The results also indicate that the binding pocket is large enough to accommodate most other amino acids, and is not very sensitive to the size and charge of the side-chains.

28.3   Aminoacyl-AMP Formation Catalyzed by the A Domain

Both steady-state and pre–steady-state kinetics were used by Luo and Walsh [8] to monitor the rate of adenylation. Michaelis-Menten constants K_m, k_{cat} and k_{cat}/K_m were obtained by the steady-state methods while the pre–steady-state gave microscopic rate constant. The reaction follows a reversible kinetics model (reprinted from [3]):

$$\text{PheA·a.a·ATP·Mg}^{2+} \underset{k_{-1}}{\overset{k_1}{\rightleftharpoons}} \text{PheA·a.a-AMP·Mg}^{2+} + \text{PPi.}$$

28.3.1   The Steady-State Assays

ATP-PPi Exchange Assay   The reactions are initiated by addition of [^{32}P]-pyrophosphate. Radioactive phosphate is then incorporated into ATP by the reverse reaction. The formation of radioactive ATP is separated and counted for radioactivity and converted into initial velocity. Because of the reverse measurement of ATP, both the forward and reverse rates of reaction are included.

Continuous PPi Release Assay   During the reaction process, inorganic pyrophosphate (PPi) is released into solution. The PPi release can be detected by employing a coupled enzyme, inorganic pyrophosphatase (PPi-ase), which converts PPi into Pi. The Pi product, together with 2-amino-6-mercapto-7-methylpurine ribonucleoside (MesG) are then catalyzed by purine nucleoside phosphorylase (PNP) to form 2-amino-6-mercapto-7-methylpurine, which has an absorbance at 360 nm. The initial velocity is calculated by monitoring the absorbance at 360 nm during the reaction.

Both methods show similar catalytic efficiency toward L-Phe and D-Phe but at least 10-fold less efficiency toward L-Trp, L-Tyr, and L-Leu. The results also show that apoPheATE and holoPheATE enzyme have similar catalytic efficiency of adenylation. The ATP-PPi exchange assay measures both forward and reverse reaction, which doesn't represent the true rate of the reaction we want to measure. However, it is still the traditional way to compare the adenylation activity. The continuous assay is limited by the slower off-rates when the a.a.-AMP intermediate occupies the active site of the enzyme during multiple turnover.

28.3.2   The Pre-Steady-State Assay

Single turnover methods were used to determine the rate of transient kinetics. In order to approach the single turnover condition, excess enzyme is used to saturate the substrate so that each substrate can undergo only one turnover during catalysis. Because of the fast rate, a rapid-quench flow apparatus is used to rapidly quench the reaction with millisecond timescale. Radiolabeled amino acid substrate is used and converted to radioactive aminoacyl-AMP (a.a.-AMP) product. A cellulose thin-layer chromatography (TLC) plate is used to separate the product from the substrate and the amount of radioactivity was counted and converted to real concentration of each species. The microscopic kinetic constants (k_1 and k_{-1}) are thus obtained.

The observed rate reflects the rate of the catalytic step and is rate limiting. The rate constants for a.a.-AMP formation are reduced 0.7-fold for D-Phe-AMP, 2.5-fold for L-Trp-AMP, 910-fold for L-Tyr-AMP, and 20-fold for L-Leu-AMP compared to that of L-Phe-AMP.

28.4   Loading of the Amino Acid Substrate to the T Domain

After formation of the aminoacyl-AMP, the intermediate can be recognized and loaded to the T domain by attaching to the cofactor 4′-phosphopantetheine (Ppant). Because the amino acid substrate now is covalently bound to the T domain, radioactivity can be detected by trichloroacetic acid (TCA) precipitation to separate substrate, a.a.-AMP and a.a.-PheATE complex. The results showed that 88% of the L-Phe was attached to the T domain during thiolation. Among the 17 amino acids tested, L-Phe, D-Phe, L-Trp, L-Tyr and L-Leu showed significant loading on the T domain. The schematic diagram shows the reversible reaction model for loading of the aminoacyl-AMP to the T domain with the release of AMP (Reprinted from [3]):

$$\text{holoPheATE} \cdot \text{C}^{14}\text{-L-a.a-AMP} \cdot \text{Mg}^{2+} \underset{k_{-2}}{\overset{k_2}{\rightleftharpoons}} \overset{\text{C}^{14}\text{-L-a.a}}{\underset{|}{\text{holoPheATE}}} + \text{AMP}.$$

28.5 Epimerization of the L-Form Substrate-Enzyme Complex to D-Form by the E Domain

The final step is to convert the T domain–attached L-form amino acid to a D-form amino acid. This is done by the epimerase domain (E domain). The 4′-phosphopantetheine cofactor carries the L-form amino acid from the A domain to the E domain for its epimerization. After the reaction, D-form amino acid is attached to the 4′-phosphopantetheine cofactor on the T domain. The reaction assay was done in the same manner as the single turnover assay for adenylation but the holoPheATE was used to ensure the whole process was complete. The reaction was rapidly quenched at varying times. Protein with both L-form and D-form amino acid attached to it was TCA precipitated to be separated from substrate and a.a-AMP intermediate. Chiral TLC was used to distinguish L- and D-form amino acid from the protein pellet while TLC was used to separate substrate from a.a.-AMP in supernatant. The schematic diagram shows the model of the reversible reaction for the whole reaction pathway (Reprinted from [3]):

$$\text{holoPheATE} \cdot \text{L-a.a} \cdot \text{ATP} \cdot \text{Mg}^{2+} \underset{k_{-1}}{\overset{k_1}{\rightleftharpoons}}$$

$$\text{holoPheATE} \cdot \text{L-a.a-AMP} \cdot \text{Mg}^{2+} \cdot \text{PPi} \underset{k_{-2}}{\overset{k_2}{\rightleftharpoons}}$$

$$\overset{\text{L-a.a}}{\underset{|}{\text{holoPheATE}}} \underset{k_{-3}}{\overset{k_3}{\rightleftharpoons}} \overset{\text{D-a.a}}{\underset{|}{\text{holoPheATE}}}$$

28.6 Free Energy Profiles for HoloPheATE Catalysis

The microscopic rate constants and the thermodynamic equilibrium constants can be used to calculate the energy for the three catalysis steps. Free energy is calculated using the following equation:

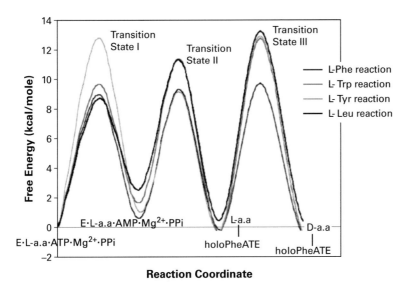

Figure 28.1 (plate 32)
Reaction free energy profiles for holoPheATE-catalyzed adenylation, thiolation, and epimerization reactions using L-Phe, L-Trp, L-Tyr, and L-Leu as the starting amino acid substrate [3]. Reprinted with permission. Copyright 2001 American Chemical Society.

$$\Delta G = -RT[\ln(k_{obs}) - \ln(k_B/Th)]$$

$$\Delta G = -RT \ln K_{eq}.$$

Here, h is Planck's constant. The equilibrium constants (K_d) are used to calculate the ground-state free energy and the rate constants are used to calculate the transition-state free energy. The constructed energy profiles show an energy barrier selective for the noncognate amino acids in the adenylation and the epimerization steps. When compared to the native substrate L-Phe, the first two steps show a net selectivity of 1/1200 for L-Tyr, 1/608 for L-Trp, and 1/245 for L-Leu. The E domain contributes another major selectivity with 1/300 for L-Tyr-S-enzyme, 1/545 for L-Trp-S-enzyme, and 1/245 for L-Leu-S-enzyme (figure 28.1, plate 32).

References

[1] J. Grunewald and M. A. Marahiel. Chemoenzymatic and template-directed synthesis of bioactive macrocyclic peptides, *Microbiol Mol Biol Rev* 70(2006):121–146.

[2] T. Stachelhaus and M. A. Marahiel. Modular structure of peptide synthetases revealed by dissection of the multifunctional enzyme GrsA, *Journal of Biological Chemistry* 270(1995):6163–6169.

[3] L. Luo, M. D. Burkart, T. Stachelhaus, and C. T. Walsh. Substrate recognition and selection by the initiation module PheATE of gramicidin S synthetase. *Journal of the American Chemical Society* 123(2001):11208–11218.

[4] R. H. Lilien, B. W. Stevens, A. C. Anderson, and B. R. Donald. A novel ensemble-based scoring and search algorithm for protein redesign and its application to modify the substrate specificity of the gramicidin synthetase a phenylalanine adenylation enzyme, *Journal of Computational Biology* 12(2005):740–761.

[5] E. Conti, T. Stachelhaus, M. A. Marahiel, and P. Brick. Structural basis for the activation of phenylalanine in the non-ribosomal biosynthesis of gramicidin S. *EMBO Journal* 16(1997):4174–4183.

[6] T. Stachelhaus, H. D. Mootz, and M. A. Marahiel. The specificity-conferring code of adenylation domains in nonribosomal peptide synthetases, *Chemistry & Biology* 6(1999):493–505.

[7] J. R. Lai, A. Koglin, and C. T. Walsh. Carrier protein structure and recognition in polyketide and nonribosomal peptide biosynthesis. *Biochemistry* 45(2006):14869–14879.

[8] L. Luo and C. T. Walsh. Kinetic analysis of three activated phenylalanyl intermediates generated by the initiation module PheATE of gramicidin S synthetase. *Biochemistry* 40(2001):5329–5337.

We return now to consider an application of graph algorithms to NMR and drug design. In this chapter, we describe protein-ligand "NOE matching," a high-throughput method for binding pose evaluation that does not require protein NMR resonance assignments [1]. This chapter provides another example of the powerful Hungarian algorithm in computational structural biology.

29.1   Background

Determining binding pose is important in biochemistry, especially in drug design. However, for larger proteins, assigning resonances in proteins becomes much more difficult as spectra become much more cluttered, requiring a multitude of additional NMR experiments. An algorithm presented in Claus and coworkers [1] is called *NOE matching*, and addresses this problem, whose objective is to compute a protein-ligand binding pose without requiring protein NMR resonance assignments.

The data used in NOE matching is from three-dimensional (3D) ^{13}C-edited, ^{13}C/^{15}N-filtered HSQC-NOESY spectra, where only the ^{1}H NMR assignments of the bound ligand are essential. A 3D ^{13}C-edited, ^{13}C/^{15}N-filtered HSQC-NOESY, referred to as an *3D X-filtered* NOESY, is the 3D combination of the HSQC with a NOESY experiment, which exploits the nuclear Overhauser effect (NOE). Simply, in an NOE, the resultant cross peak intensity measures the interatomic distance; heteronuclear single quantum correlation (HSQC) experiments record when two specific different types of atoms are covalently bound. The results of a 3D X-filtered NOESY experiment, as demonstrated in figure 29.1, can be represented as a 3D plot, where the x-y plane contains HSQC of a protein CH pair (with ^{13}C and ^{1}H frequencies of the protein being x- and y- axes, respectively), and the z-axis showing the ligand ^{1}H frequency. The volume of peaks in the NOESY spectrum measure the distance between ligand ^{1}H and protein ^{1}H. Therefore, the 3D graph describes which protein CH pairs are near to which ligand H, and measures explicitly distances of ligand H's to unassigned protein CHs.

Protein structure can often be determined through X-ray crystallography or be approximated reasonably well by homology modeling. In the work [1], the structures of the test cases come from crystallography, and the protein coordinates are from the Protein Data Bank (PDB). For our

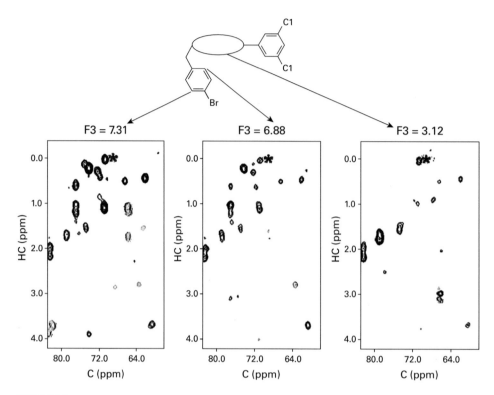

Figure 29.1
An example of an X-filtered NOESY spectrum, with the protein CH HSQC as the x-y plane, and the z-axis (F3) referring to slices of the spectrum at the resonant frequencies of different ligand ^{1}H frequencies. Reprinted with permission. Copyright 2006 American Chemical Society. [1].

purposes, the most important feature of the X-filtered NOESY experiment is that by exploiting stable isotopic labeling, the pulse sequence ensures that all observed NOEs are intermolecular.

29.2 Methods

Claus and coworkers [1] determine the binding pose without protein resonance assignments. The recorded NOESY data is discretized according to signal strength, and compared with a back-calculated HSQC-NOESY spectrum based on a large, calculated ensemble of possible ligand poses. The computer-generated ligand pose whose synthetic NOESY spectrum is the best "match" with the experimental data, is assumed to be the correct binding pose. This NOE matching algorithm includes four steps:

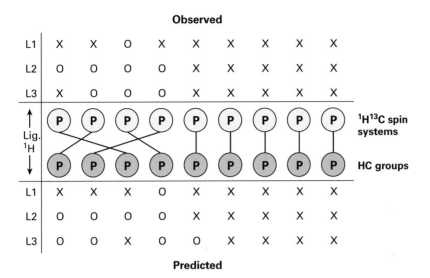

Figure 29.2
Equally partitioned bipartite graph representing a hypothetical instance of the 3D X-filtered NOESY bipartite graph-weighted matching problem, with $N = 9$ [1]. Reprinted with permission. Copyright 2006 American Chemical Society.

Step 1. Trial Pose Generation　Many trial poses are generated either using DOCK, or simulated annealing using the protein and the available trial poses of a similar ligand (with that ligand substituted for the compound of interest).

Step 2. Preparation of Experimental X-filtered NOESY Peak List　The peaks of the experimental X-filtered NOESY are manually binned according to peak intensity from 1 to 4 (from least intense to most intense). Heuristic rules or higher-dimensional NMR experiments are used to determine peak shifts of the ^{13}C's.

Step 3. Prediction of Simulated X-filtered NOESY Spectra　In a fast simulation, using the assigned ligand ^1H frequencies, distances from each ligand ^1H to every protein CH pair within 5 Å are classified as having intensity 1 to 4, referring to distances of 5 Å, 4 Å, 3 Å, and 2.5 Å, respectively. They are placed in the x-y plane using the mean chemical shifts of protein CH moieties available from the BioMagResBank. The algorithm is designed to "overpredict" peaks, such that there are more simulated peaks than experimental peaks.

Step 4. Spectrum Matching and Pose Scoring　The predicted peaks are matched to the experimental peaks and a correlation score is generated. This is done using a variation of the bipartite graph-matching algorithm, as shown in figure 29.2. Each protein CH peak (considered a node in this graph) is separated into experimental and simulated vertices. Edge weights were determined by scoring the differences in peak intensities for each ligand ^1H; this bipartite graph-matching problem is solved in $O(n^3)$ time using the Hungarian algorithm, and poses are scored as the sum of the edge weights resulting from the optimization. A number of "unknown peak" nodes are generated as a place to store the worst-case matches of the simulated peaks.

The rules for determining edge weights are available in Claus and coworkers [1], but a global picture of these can easily be seen: There is no weight penalty for perfect matching of the intensity and the x, y coordinates (within a margin of error), but the penalties are added as follows (in increasing order):

1. Experimental peak and predicted peak of a different intensity. In a "loose" setting, this penalty obtains only if the difference is greater than 1; in a "tight" setting, all differences effect a small penalty.

2. Simulated peak with no experimental peak, small penalty, since the algorithm aims to generate more simulated peaks than experimental.

3. Experimental peak with no predicted peak, larger penalty.

This makes intuitive sense, since the algorithm essentially bins experimental peak intensities into four groups (allowing for small differences in intensity); likewise, the algorithm generates more simulated peaks than the experimental data, and therefore simulated peaks with no experimental peaks are not penalized as much as missing predicted peaks.

29.3   Results and Discussion

The results of the NOE matching algorithm are promising, as the authors observed a strong linear relationship between the cost of the pose and the RMSD to the experimentally determined binding poses, as shown in figure 29.3. More specifically, the average RMSD to the mean coordinates of ligand for the test case of mFABP/1 (muscle fatty acid–binding protein) is 0.42 Å, and that for the test case of LFA-1/2 (leukocyte function–associated antigen) is 0.19 Å.

Possible extensions to the algorithm include using this method as a fast filtering tool to eliminate poses that can definitely not be the binding pose, and using a higher-order CH frequency approximation based on quantum mechanical methods for a more accurate search on a smaller space. In addition to these, the study [1] suggests using more NMR experiments, which may be added to the algorithm, to more accurately assign resonances based on similarity to the generated data.

In conclusion, 'NOE matching' provides a fast filtering method for determining the binding poses of a ligand and protein, using minimal NMR data and with no need of protein resonance assignments. Although the paper [1] did not show any data for more difficult protein-ligand complexes, the results of the algorithm are accurate in the test cases reported.

Caveats In other applications, the Hungarian algorithm can be used reliably only when there is almost no noise in the edge weights. It would be interesting to evaluate the effect of noise on

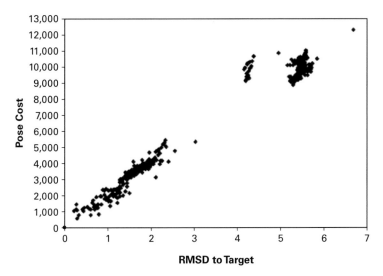

Figure 29.3
Cost$_{pose}$ versus the RMSD (Å) to the target pose for mFABP/1 with synthetic (ideal) experimental and predicted 3D X-filtered NOESY data. This graph shows a fairly linear relationship. Reprinted with permission. Copyright 2006 American Chemical Society. [1].

the performance of this application of the Hungarian algorithm. Can you design a more accurate matching algorithm?

Reference

[1] K. L. Constantine, M. E. Davis, W. J. Metzler, L. Mueller, and B. L. Claus. Protein-ligand NOE matching: A high-throughput method for binding pose evaluation that does not require protein NMR resonance assignments. *Journal of the American Chemical Society* 128;22(2006 Jun 7):7252–7263.

30 Side-Chain and Backbone Flexibility in Protein Core Design

Having discussed protein design and protein flexibility in several chapters, we now consider another way that researchers have combined these two algorithmic and modeling themes. In this chapter, we discuss soft repacking of cores (SoftROC) [1], an optimization technique that combines genetic algorithms (GA) and Monte Carlo (MC) procedures with Metropolis sampling and simulated annealing. This technique is applied to protein core structure prediction with flexibility for both the backbone dihedral angles (DAs) and the core side-chain rotamers.

This chapter builds on, and complements, our earlier discussion of provable, DEE-based algorithms for protein design with backbone flexibility (section 12.6 of chapter 12).

30.1 Protein Modeling with Fixed or Flexible Backbone

Many protein modeling algorithms are based on the assumption that the backbone structure is given and fixed. A few of these methods have achieved success by experimental confirmations [2, 3, 4]. Despite the reduced complexity and the nontrivial task of constructing or selecting an appropriate force field for that backbone, these methods have certain limitations. Most important, stable hydrophobic core structures with significant backbone relaxation have been found in crystallographic studies [5, 6, 7]. These structures are disallowed under the fixed backbone assumption. Further, these algorithms are usually unable to contribute in de novo protein design because apparently no precise starting backbone is available.

An earlier work [8] employing backbone flexibility targeted coiled-coil core motifs. The method uses the symmetric geometry of such structures to alleviate the increase in complexity introduced from additional backbone parameters, and impressive success was achieved in both prediction and design. Another study [9] approximated backbone flexibility by systematically generating multiple backbone variants and then feeding them into a fixed-backbone algorithm. This method successfully designed a series of novel variants that have similar properties to those of the wild-type protein. The limitation of both methods is that the generalizability still remains unclear. Thus, the authors of SoftROC provide a possibly more general solution by introducing simultaneous optimization on backbone and core side-chains, and solving it with a combination of fast non-deterministic algorithms. The algorithms are heuristic, and are not provable, but they provide an interesting study in design with flexibility. For provable algorithms for protein design with side-chain and backbone flexibility, see section 12.6 (chapter 12).

30.2 SoftROC

The SoftROC algorithm is based on torsional modeling, which modifies dihedral angles (DAs) instead of Cartesian coordinates. It comprises three steps: (1) generating initial backbone models; (2) finding a locally optimal model using genetic algorithm; and (3) refining the model with a specialized Monte Carlo sampling.

30.2.1 First Step: Initializing Backbone Population

First, the backbone structure of each member in the initial population is generated by randomly choosing ϕ and ψ angles within $3°$, and ω values within $1°$, of those from the template protein. Then, to avoid error propagation throughout the entire amino acid chain, an initial Monte Carlo (MC) optimization on the folding energy defined by the Amber force field [10] is applied.

The core side-chain identities and DAs are generated by randomly choosing a rotamer from a rotamer library [11], and then choosing each DA to be within $30°$ of that from the rotamer. For noncore side-chains within the core sequence, these values are fixed to be values from the template protein. For all the other side-chains, only the fixed DAs of C^{α} and C^{β} from the template are included.

30.2.2 Second Step: Optimization with Genetic Algorithm

The genetic algorithm (GA) is a series of evolutions (iterations) on a population with predefined stopping criteria. In each evolution, every member in the population is evaluated by a fitness function. Then, some pairs of members (usually the fitter ones) reproduce offspring via recombination, that is, exchanging values between each pair. The offspring are again evaluated, and the total population is subject to selection, where only the fitter members survive. Finally, the resulting population is subject to random mutations to maintain diversity. An illustration of GA is shown in figure 30.1 (plate 33).

In a typical run of SoftROC, a population of 200 initial torsional models is fed into the GA algorithm, where the population evolves during 300 evolutions. In each evolution, SoftROC performs the following variant of GA procedures:

Fitness Evaluation The fitness function p is defined as the normalized Boltzmann probability, whose temperature parameter decays linearly from 10,000 K to 100 K (also known as *simulated annealing*) to avoid early convergence to local optima.

Stochastic Selection Each model is allowed to reproduce offspring proportional to the fitness function. It is stochastic in that every model has a chance to reproduce offspring, and none of the parent models, whether fit or not, is kept in the next evolution.

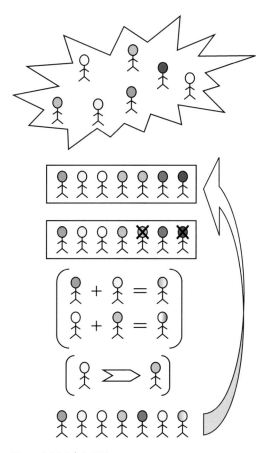

Figure 30.1 (plate 33)
Genetic algorithm with a snapshot of one evolution. In this example seven members comprise the initial population. In the selection phase, two members are found unfit and will be replaced. In the recombination phase (also known as crossover), two member pairs are chosen to reproduce two offspring with combined characteristics from their parents. Finally, in the mutation phase, the pink member is picked and modified to be pinker.

Uniform Recombination   Each model pairs with 200 randomly selected other models. Only one of the two recombined offspring is kept, and its values come from either parent randomly and uniformly. From the property of normalized probability, the number of offspring sums up to 200, which means the population size remains invariant over evolutions.

Mutation   Each new model is subject to random changes of three types: (1) A 3% chance to change rotamer classes and/or side-chain identities within a 10° range; (2) a 10% chance to change side-chain DAs within a 5° range; and (3) a 0.5% chance to change backbone DAs within a 3° range. A special restriction after 280 iterations disallows rotamer class change and allows only a 1° range of backbone DA changes.

The model in the final population with the lowest energy serves as the input to the ensuing MC refinement.

30.2.3   Third Step: Refining the Model with Monte Carlo Sampling

Monte Carlo with Metropolis sampling is an iterative stochastic optimization approach. In the tth iteration, a sample x' is obtained by a random movement from x^t according to a proposed probability distribution $Q(x'; x^t)$. This movement is accepted as x^{t+1} with probability u, where u is defined as

$$u = \frac{P(x')Q(x^t|x')}{P(x^t)Q(x'|x^t)}.$$

If x' is not accepted, x^{t+1} is set to x^t.

In SoftROC, $P(x)$ is the Boltzmann probability, whose temperature parameter decays linearly from $200\,K$ to $0\,K$. $Q(x'; x^t)$ is a mixture of unchanging and uniform changing within a range. For each side-chain DA, the changing probability is 10% and changes are restricted within 3° range. For each backbone DA, the changing probability is 2%, within a 1° range. Rotamer class changes are disallowed throughout the MC process.

30.2.4   Final Model

SoftROC performs 20 runs of the aforementioned steps. After a first set of ten runs, the rotamer library is trimmed by omitting those classes not included in any of the ten models. The same trimming is performed after an additional five runs. The last five runs are then performed with the trimmed library. The model with the lowest energy among the 20 runs is selected as the final model.

30.3   Issues on Energy Calculations

In the SoftROC paper, the authors first apply the conventional AMBER force-field [10] potential with OPLS (Optimized Potential for Liquid Simulations) [14] to calculate the energy terms used in Boltzmann probabilities and the objective functions in both GA and MC optimizations. However, they found that the backbone deviation becomes very large when allowing backbone flexibility. Therefore, they try to solve this problem by augmenting/replacing the AMBER/OPLS potential in backbone energy calculation with another estimation called PLG (Preservation of Local Geometry). In SoftROC, the PLG potential is included with a weight of 2.0 relative to the other energy terms calculated via AMBER/OPLS potentials.

PLG Definition   Basically, PLG approximates the sum of local root mean square deviations (RMSDs) from template substructures. A sliding window of size 5 is maintained along the target residues. For each atom in each window, the distances between it and the other four atoms are evaluated and compared with those in the corresponding window along the template residues. The differences in distances are then summed up for all five atoms to yield one value, and the square root of this value is taken and summed over all sliding windows to yield the PLG estimation.

In principle, there is no reason not to employ the conventional AMBER/OPLS potentials; using PLG instead, seems like forcing the predicted backbone local structures close to the template. The backbone deviation problem might be due to the recombination and/or selection strategies used in SoftROC. In the uniform recombination phase in SoftROC, the backbone DAs for each residue are chosen randomly from one of the two different models. Such choices yield an interweave that is likely to possess higher potential than any parent. Also, in the stochastic selection phase, all parents are eliminated despite the possibility that they could be of lower energy than some offspring. From these observations, applying other GA variants, such as one- or two-point recombination and selection on both parents and offspring, might be viable to reduce backbone deviations without introducing PLG.

30.4 Results: Comparison to ROC Variants

30.4.1 ROC Settings

ROC (repacking of cores) [3] is a prototype of SoftROC and is written by the same authors. ROC assumes a fixed backbone and includes only the GA step of SoftROC. The authors compare the performance of SoftROC with three settings of ROC: ROC with an existing library [11], ROC with the library expansion including all rotamers with $\pm 15°$ DA differences, and ROC with a library expansion including all rotamers with $\pm 50°$ DA differences at $5°$ increments. The last setting is considered to allow full flexibility of side-chains, and it is also used in the GA and MC algorithms in SoftROC.

Note that the running time of each run of ROC/SoftROC varies from 1 minute to 6 hours (table 30.1). The expanded rotamer library might contribute the most to the increase of running time.

30.4.2 Experiments on 434 cro

The wild-type crystal structure of 434 cro [12] is used as the template for all methods. The SoftROC paper reports the plots of calculated energies against the melting temperatures T_m, along with the correlations between them. Low correlations are seen in the results of ROC with rigid rotamer library ($R = 0.04$) and SoftROC with AMBER/OPLS backbone potentials ($R \approx 0$).

Table 30.1
Averaged running time for SoftROC and three ROC variants

Method	ROC, Rigid lib.	ROC, $\pm 15°$ lib.	ROC, $\pm 50°$ lib.	SoftROC, $\pm 50°$ lib.
Hour(s) for 1 run	$\sim 1/60$	$\sim 1/60$	~ 1	~ 6

(a)   (b)

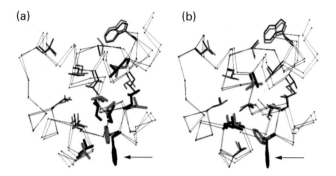

Figure 30.2 (plate 34)
Modeling of cro-rep. Structures are displayed as C$^{\alpha}$ traces with hydrophobic core side-chains. (a) The crystal structures of 434 cro (black) against 434 cro-rep (red). The arrow shows the major difference at Phe44. (b) ROC (black) and SoftROC (red) predictions based on the cro-rep template. While ROC predicts a cro-like orientation of Phe44, SoftROC predicts a core structure more similar to that of cro-rep. [1] Reprinted with permission from Elsevier.

The correlations for ROC with partial and full rotamer flexibility are much higher ($R = 0.54$ and 0.88, respectively). SoftROC with PLG backbone energies yields moderate correlations ($R = 0.75$). From the plots, it is hard to tell whether any method stands out because the variations from different runs are very high, except those of ROC with a rigid library.

The structures for many 434 cro variants have not yet been determined, and therefore it is difficult to compare predictions. Thus, the SoftROC paper compares the computational models of cro-rep, a known variant in which the native core of 434 cro is replaced with that of 434 repressor. According to figure 30.2 (plate 34), while ROC with full rotamer flexibility failed to change Phe44 from cro-like to cro-rep-like, SoftROC predicted the change as well as several other cro-rep side-chains.

30.4.3 Experiments on T4 Lysozyme
The wild-type crystal structure 2LZM [13] and eight of its variants (I-VIII, in decreasing stability order) were used as the template for all methods. The plots of predicted and experimental stabilities still have high variances, which provide little information for performance comparisons. Therefore, the SoftROC paper proposes another metric called *prediction penalties,* which is the sum of the number of atoms misplaced by each DA incorrectly predicted by at least 40°. The authors chose this metric because it should normalize the effect of incorrect side-chain predictions on residues of different sizes. The plot of the penalties against the actual RMSD between wild-type and the eight variants is shown in figure 30.3, which indicates that ROC with full rotamer flexibility performs better than SoftROC only for some of the stabler variants. A further examination in the SoftROC paper shows that only SoftROC predicts the orientations of Leu129 and Trp153 in variant II, where the latter residue is bigger than the corresponding wild-type residue, and therefore a major change in the structure is expected.

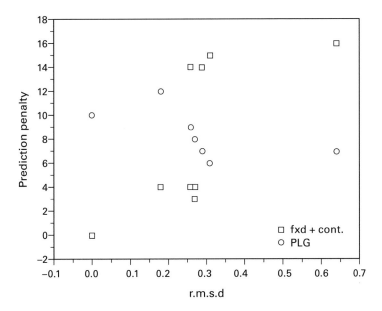

Figure 30.3
Side-chain prediction penalties for T4 lysozyme and variants. A penalty of zero implies a very good prediction. Open circles denote penalties for SoftROC predictions, and open squares denote penalties for ROC predictions [1].

Note

Figure 30.1 (plate 33) source unknows. All attempts have been made to find the rightsholder. If you have information, please contact the author or publisher.

References

[1] J. R. Desjarlais and T. M. Handel. Side-chain and backbone flexibility in protein core design. *Journal of Molecular Biology* 290(1999): 305–318.

[2] B. I. Dahiyat and S. L. Mayo. Protein design automation. *Protein Science* 5(1996): 895–903.

[3] J. R. Desjarlais and T. M. Handel. De novo design of the hydrophobic cores of proteins. *Protein Science* 4(1995): 2006–2018.

[4] J. H. Hurley, W. A. Basse, and B. W. Matthews. Design and structural analysis of alternative hydrophobic core packing arrangements in bacteriophage T4 lysozyme. *Journal of Molecular Biology* 224(1992): 1143–1159.

[5] E. P. Baldwin, O. Hajiseyedjavadi, W. A. Baase, and B. W. Matthews. The role of backbone flexibility in the accommodation of variants that repack the core of T4 lysozyme. *Science* 262(1993): 1715–1718.

[6] A. E. Eriksson, W. A. Baase, X. J. Zhang, D. W. Heinz, M. Blaber, E. P. Baldwin, and B. W. Matthews. Response of a protein structure to cavity-creating mutations and its relation to the hydrophobic effect. *Science* 255(1992): 178–183.

[7] W. A. Lim, A. Hodel, R. T. Sauer, and F. M. Richards. The crystal structure of a mutant protein with altered but improved hydrophobic core packing. *Proceedings of the National Academy of Sciences (USA)* 91(1994): 423–427.

[8] P. B. Harbury, B. Tidor, and P. S. Kim. Repacking protein cores with backbone freedom: Structure prediction for coiled coils. *Proceedings of the National Academy of Sciences (USA)* 92(1995): 8408–8412.

[9] A. Su and S. L. Mayo. Coupling backbone flexibility and amino acid sequence selection in protein design. *Protein Science* 6(1997): 1701–1707.

[10] S. J. Weiner, P. A. Kollman, D. A. Case, U. C. Singh, C. Ghio, G. Alagona, et al. A new force-field for molecular mechanical simulation of nucleic acids and proteins. *Journal of the American Chemical Society* 106(1984): 765–784.

[11] P. Tuffery, C. Etchebest, S. Hazout, and R. Lavery. A new approach to the rapid determination of protein side-chain conformations. *Journal of Biomolecular Structure & Dynamics* 8(1991): 1267–1289.

[12] A. Mondragon, C. Wolberger, and S. C. Harrison. Structure of phage 434 Cro protein at 2.35A resolution. *Journal of Molecular Biology* 205(1991): 179–188.

[13] L. H. Weaver and B. W. Matthews. Structure of bacteriophage T4 lysozyme refined at 1.7A resolution. *Journal of Molecular Biology* 193(1987): 189–199.

[14] W. L. Jorgensen and J. Tirado-Rives. The OPLS potential functions for proteins. Energy minimizations for crystals of cyclic peptides and crambin. *Journal of the American Chemical Society* 110(1988): 1657–1666.

31 Distance Geometry

We now discuss, in several chapters, the theme of distance geometry, which relates to graph realization and NOE assignment. This topic touches on graph theory, algorithms, and topology, and is a classical and important subfield of computational structural biology.

31.1 The Molecule Problem

Consider a graph $G = (V, E)$ where V is the set of atoms in a molecule. Attach an edge $e_{ij} \in E$ with weight d_{ij} between two vertices i and j if the distance between them is d_{ij}. This situation arises when we know the distance, either from empirical measurements such as NOEs, or from modeling (e.g., crystal structures of covalent bonds). The molecule problem is now the following. Given G, can one compute the three-dimensional structure of the molecule? Formally, one wishes to construct a *satisfying realization* of the graph. A graph realization $\mathbf{P}$ is an assignment of coordinates in $\mathbb{R}^d$ to each vertex.

This problem could be formulated as a global optimization problem in the following way. Given a graph G with weight d_{ij} over each edge e_{ij}, we are asked to assign a coordinate $\mathbf{p}_i \in \mathbf{P}$ for each vertex i, minimizing the following objective function:

$$F(\mathbf{P}) = \sum_{e_{ij}} (|\mathbf{p}_i - \mathbf{p}_j|^2 - d_{ij}^2)^2,$$

where $\mathbf{p}_i \in \mathbf{P}$ is the coordinate realization of vertex i, d_{ij} is the distance between vertices i and j, and $|\cdot|$ is the Euclidean norm. This objective function calculates the sum of squared deviations of the realized distances from the known distances. It will have a global minimum of zero when all the distance constraints are satisfied. The reason why the molecule problem is hard to solve in this form is that there are an exponential number of local minima. One approach to circumvent this problem is to break up the monolithic optimization problem into smaller problems and solve them independently. While this divide and conquer approach cannot always be applied to arbitrary optimization problems, Hendrickson [1, 2] shows it is possible for the molecule problem.

31.2   Divide and Conquer

The question is then, how can a divide and conquer approach be applied to the molecule problem? The answer lies in the observation that realized subgraphs can be combined to form larger realizations. Imagine that it were possible to split our graph G into smaller subgraphs that are more easily realizable (this is sometimes possible, as we will see later). Each realization can then be treated as a rigid body and combined with other rigid bodies to form the final solution. We can find this solution by keeping one body fixed and searching the space of rotations and translations of the other bodies relative to the first. This search can be done with another global optimization step, but its success depends on the crucial assumption that each subgraph is uniquely realizable. That is, only one satisfying assignment of coordinates to the graph can exist. We then concern ourselves with partitioning our original graph G in such a way that the subgraphs are uniquely realizable.

For this approach to work, we must make assumptions about the distances given. First, the distances must be known exactly. This is not feasible for real NOESY spectra, which reveals this work as somewhat of a theoretical exercise. However, these algorithms could be applied if exact distances could be inferred (or even measured directly) from molecules in the future. Moreover, it may be that solving a problem without noise or error in the data can provide insight into the real problem with experimental uncertainty. The second assumption is that the source of the distance constraints (which can also be viewed as a realization of G) must comprise a set of algebraically independent coordinates. That is, we need a *generic* realization. This means the orignal atom positions cannot satisfy any polynomial equation. An example of a configuration excluded by this requirement is when four or more atoms lie on the same plane. It may seem like a strong requirement of the data, but in practice, almost all realizations are generic.

31.3   Conditions for Unique Realizibility

What properties does a graph need to be uniquely realizable? Hendrickson shows there are three conditions a graph must satisfy to have a unique realization in $\mathbb{R}^d$:

1. $(d+1)$ vertex-connectivity;

2. Redundant rigidity; and

3. The *stress matrix* has nullity $(d+1)$.

Hendrickson also shows the first two conditions are necessary conditions and the third condition is sufficient.

Graph connectivity is a well-studied property of graphs and is not reviewed here. Redundant rigidity is a modification of the concept of graph rigidity (see chapter 21, section 21.2), which can be thought of as follows. Imagine if one were to physically pick up a graph with one's hands and try to bend it. A rigid graph would not allow any motions (or flexings) but a flexible graph would. Figure 31.1 illustrates this concept.

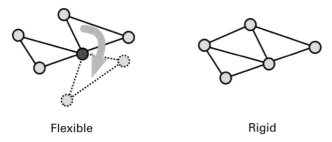

Flexible   Rigid

Figure 31.1
Flexible vs. Rigid graphs in $\mathbb{R}^2$. On the left, the graph is flexible because the right portion is free to rotate about the dark vertex. On the right, no such dark vertex could exist. Credit: Jeff Martin and Bruce Donald.

Redundant rigidity is then a property that describes a graph that remains rigid even after any single edge is removed.

To characterize the *stress* of a graph, we turn to mechanical engineering and static systems. If one were to view a graph as a bridge, the edges could become steel beams and the vertices could become joints. It is often useful to calculate the equilibrium of such a system. To do this, we assign weights to each edge. Then we are interested in finding values for the weights such that, for any vertex, the weighted sum of all the edge vectors starting from this vertex is zero. This weight that we assigned is called a *stress* of the graph. The stress itself is a scalar quantity, but the direction of the edge it describes can give it a vector interpretation. It is the sum of these stress vectors, then, that we wish to be zero and not the stresses directly. Let ω_{ij} be the stress of an edge e_{ij}. The equilibrium condition is then

$$\sum_{j=1}^{n} \omega_{ij}(\mathbf{p}_i - \mathbf{p}_j) = 0$$

for all i where, again, $\mathbf{p}_i$ is the position of vertex i in $\mathbb{R}^d$. We can then construct the stress matrix

$$\Omega = \begin{cases} -\omega_{ij} & \text{if } i \neq j \\ \sum_k \omega_{ik} & \text{if } i = j \end{cases}$$

and examine its nullity to determine unique realizibility.

31.4   Graph Partitioning

Once we know conditions for uniquely realizable graphs, we then turn to partitioning a graph such that the subgraphs satisfy those conditions. Again, vertex connectivity is a well-studied problem and efficient algorithms exist for partitioning graphs in that sense. One challenge Hendrickson addresses is how to partition graphs so that their subgraphs are redundantly rigid. His algorithm is based on the following intuition. An edge that is not redundant in the whole graph cannot be redundant in any subgraph. Therefore, one can remove such a nonredundant edge from the graph and examine its flexibility. Regions of the graph that are not flexible can then be targeted for subgraph partitioning.

A related problem is that, in the preceding case, we were interested in finding the maximal subgraphs that were still redundantly rigid; however, sometimes these maximal graphs are still too large for feasible processing. In this case, we want to further break up the graph and, in practice, Hendrickson found the optimal subgraph size to be 15 vertices. If it is any smaller, the divide and conquer approach would spend more computational time on the merge step; any larger, and solving the realization problem on the subgraph would become more difficult. Graphs can be broken up by finding a vertex separator. Then two graph pieces will have a number of vertices in common, which will simplify the recombination of their realizations later in the merge step.

Once graphs are found that pass both of the first two conditions (section 31.3), the algorithm checks for the third condition. Note that the algorithm has no recourse for subgraphs that do not pass the third condition. Hendrickson states that in $\mathbb{R}^3$, no such graph is known that satisfies the first two conditions and not the third. Were such a graph to be found, Hendrickson instructed the algorithm to return it so he could study its properties.

31.5 Realizing Subgraphs

Finally, to realize these subgraphs, we formulate the problem as a global optimization (similar to the one described in section 31.1). Up until this point, we have focused on reducing the complexity of the realization problem while still preserving the guarantee that our answer is correct. This is because we must rely on a global optimization to perform the actual realization step. The distance geometry problem has been proven to be strongly NP-hard [3], so the cost of performing the optimization is high even for much smaller problems. Therefore, Hendrickson makes every effort to minimize the complexity of the optimization problem before relying on the optimizer to compute the answer.

Hendrickson also uses another intuition to simplify realization. Imagine a set of four vertices have already been realized in $\mathbb{R}^3$. Then, to completely specify the position of a fifth vertex, one merely needs distances to the four already-realized vertices. One can then "grow" the *chunk* of realized vertices by attaching other vertices given an adequate number of distance restraints. In this context, "chunk" refers to a connected group of realized vertices. This part of the realization process can be accomplished combinatorially for a relatively low cost compared to optimization. Thus, by realizing chunks or "easy bits" of the graph first, the final problem that must be ultimately optimized is greatly simplified. Finally, the optimization problem concerns itself with the task of fixing an arbitrary chunk and trying to find the positions of other chunks that satisfy the distance restraints completely. Once the subgraphs are realized, we can merge them in the same way by using the optimizer. Once all realizations are successfully merged, we can then know the full structure of the "protein" whose geometry the graph encodes.

31.6 Conclusion

Hendrickson states that his approach requires exact interatomic distances to correctly solve the protein structure. He also states that the algorithm is intolerant to noisy or miscalculated restraints. While this removes some of the immediate applicability for solving structures, it does lay theoretical groundwork for future work in the area.

In some cases, there may simply not be enough distance restraints to completely specify the full structure of a protein. This could be a catastrophic input for some algorithms that assume complete data. However, Hendrickson's divide and conquer approach allows his algorithm to recurse on subgraphs and solve them independently. In some cases, this subgraph might correspond to a meaningful region of the protein such as a binding site or a particular functional domain. Conversely, the absence of redundant rigidity might give evidence that the protein is flexible or that more experiments are required to define the native structure.

References

[1] B. Hendrickson. The molecule problem: Exploiting structure in global optimization. *SIAM Journal of Computing* 5;4(November 1995):835–857.

[2] B. Hendrickson. Conditions for unique graph realizations. *SIAM Journal of Computing* 21;1(February 1992):65–84.

[3] J. B. Saxe. Embeddability of weighted graphs in k-space is strongly NP-hard. Tech. Report, Computer Science Department, Carnegie-Mellon University, Pittsburgh, PA, 1979.

32   Distance Geometry: NP-Hard, NP-Hard to Approximate

NMR spectroscopy exploiting the nuclear Overhauser effect (NOE) gives us a way to measure distances between atoms. This is very handy: Given all the distances between every pair of atoms in a protein, it should, in theory, be possible to reconstruct the protein's structure. However, NOEs are both *noisy* and *sparse*. As a result, it is computationally difficult to produce a structural model of a protein, from NOE data alone. In fact, as Saxe [1] proves, it is NP-complete to do so (if you solve this problem, you solve every other NP problem), and NP-complete to even approximate it.

What does this mean? It means that there should be no provably fast way to solve the problem we know as *distance geometry*, which Saxe [1] refers to as "*k*-embeddability." The application to protein structure determination is obvious. And given that this problem is proven to be NP-complete to solve and even to approximate, algorithms attempting to solve for distance geometry are provably unable to provide both reasonably fast, and provably good results. As a result, researchers in the field have developed either provably good but slow algorithms, or reasonably fast but not provably accurate algorithms.

Saxe's results [1] do not mean that heuristic, stochastic approaches such as simulated annealing/molecular dynamics (SA/MD) cannot run quickly or compute accurate structures. But it does say that these NOE-based approaches cannot be *proven* to do so. On the other hand, we have seen that RDC-based algorithms enjoy guarantees of efficiency, soundness, and completeness that are forbidden to NP-hard problems such as distance geometry (chapters 15–18). This is because RDCs are global, while NOEs are local.

32.1   Introduction

32.1.1   Review: Reductions

A Nondeterministic Polynomial-time (NP) problem is a problem that, given a solution together with sufficient support, can be *verified* in polynomial time. Problems that are NP-hard are believed to be harder than all the other NP problems.

The basic idea behind a reduction is the ability to show that, given a problem of one form (usually that of an NP-hard problem), we can efficiently convert it into a problem of the form we want to prove is NP-hard, and the solution to the new form can be transferred back. If we can do this in polynomial time, and the transformed problem is in P, then we've just developed a solution for a nonpolynomial time problem whose time complexity is merely the sum of two polynomials, which is still a polynomial! The idea is that "if you can solve this problem in polynomial time, you can solve an NP-hard problem in polynomial time as well!"

The implications of any such proof are obvious: Unless a polynomial-time solution for an NP-complete problem exists (and none have been found yet), any problem that can be proved to be NP-complete can thus be considered intractable, in the sense that unless P = NP, no polynomial-time solution exists. There is much evidence that P ≠ NP.

32.1.2   NP-Hard Problems
In his paper [1], J. B. Saxe reduces two different problems, named *Partition* and *3SAT*, to the problem of distance geometry (which he refers to as "k-embeddability").

To describe the problems, as Saxe defines them:

Partition: Given a set of integers, can it be divided into two subsets such that their sums are equal?

3SAT: Given an expression in first-order logic, is there a set of Boolean variable assignments that makes the expression evaluate to true?

k-embeddability: Given a graph $G = (V, E, w)$, where w assigns each edge in E a weight, is there a function f that assigns integer positions to every vertex v in V into k-dimensional space such that for every edge $e = (v_1, v_2)$, $\|f(v_1) - f(v_2)\| = w(e)$?

32.2   Reduction from Partition to 1-Embeddability

Given any set of integers $\{a_1, a_2, a_3, \ldots a_n\}$, a cycle of n vertices such that each edge (v_i, v_j) has weight a_i can be constructed in linear time. If this cycle is embeddable in 1 dimension, all edges where $v_i > v_j$ can be grouped into one set and all edges where $v_j < v_i$ can be grouped into the other set, such that the two sets have an equal sum. This is because, when traversing the vertices of a cycle, the first vertex visited is returned to, by definition of a cycle. The implication of this is that any distance traveled in one direction must be traveled back, to return to the starting vertex's assigned location.

32.3   Reduction from 3SAT to {1,2} 1-Embeddability

Using edge lengths of only 1 and 2, Saxe shows that it is possible to construct a graph for which the output (if there is any function for which the graph becomes embeddable in one dimension) corresponds to the logical truth value of whether an expression can be satisfied. See figures 32.1 and 32.2.

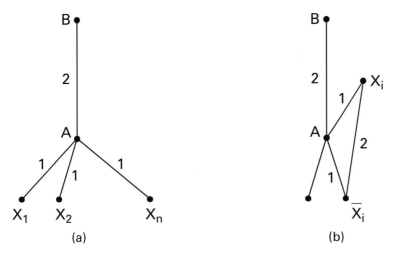

Figure 32.1
Representing variables and clauses in 1-embeddable graphs. (a) Implementation of variables; (b) implementation of a negative literal. Variables can only be assigned values of $\{-1, 1\}$, which corresponds with the true/false values of 3SAT variables [1].

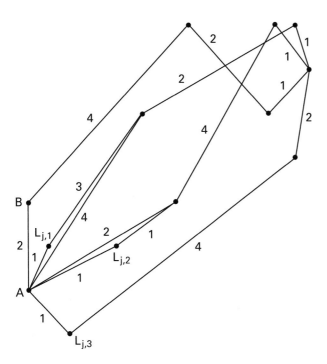

Figure 32.2
The gadget to implement a disjunctive 3SAT clause. On inspection one can confirm that at least one of the three variables in the graph must be assigned to $+1$ for there to exist a function that can embed the graph into one dimension [1].

32.4   Reduction from 3SAT to Integer 1-Embeddability

Up to this point, Saxe only shows that to embed graphs of edge weights {1,2} is NP-hard. With the use of these two gadgets, however, embedding any graph of lengths {1,2,3,4} can be shown to be NP-hard, and, inductively, any integer-weighted graph can be broken down until it consists entirely of edge lengths {1,2} (see figure 32.3). Thus, he shows that the 1-embeddability of any integer-weighted graph is NP-hard.

32.5   Adding Dimensions

By introducing two more gadgets, Saxe shows that it is possible to convert a graph that is embeddable in 1 dimension into a graph that must be embedded in two dimensions; in his words, he "adds a dimension" to the graph (see figure 32.4). To move the graph into additional dimensions, other gadgets could be used as well.

32.6   Approximation

32.6.1   Definition of ε-Approximate k-Embeddability

A graph $G = \langle V, E, w \rangle$ is said to be ε-approximately k-embeddable if there exists a function f such that for all vertices $v_i, v_j \in V$, $\| f(v_i) - f(v_j) \| \le w(e_i)(1 + \varepsilon)$.

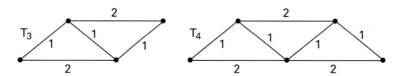

Figure 32.3
By replacing all edges of length 3 and all edges of length 4 with the gadgets shown above, all graphs can be reduced to {1,2} edge-weighted graphs [1].

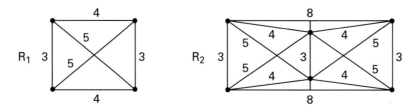

Figure 32.4
By replacing all edges of length 1 and all edges of length 2 with the gadgets R1 and R2, respectively, any 1-dimensional graph can be converted into a graph that is representable in only 2 or 3 dimensions. It is left to the reader to contrive gadgets for additional dimensions if so desired [1].

Saxe shows the reduction of 3SAT to real number edge-weighted embeddability. He observes that the gadget used to reduce {1,2} 1-embeddability relies only on cycles of length no greater than 16. From this, it follows that the the total distance traveled from any point in that cycle can be no longer than $\frac{16}{2} = 8$, and therefore for any real number $\varepsilon < \frac{1}{8}$, the resulting embeddability problem is equivalent to exact integral k-embeddability. As a result of this, he claims that given any ε-approximate k-embeddability problem, there exists a rational number k such that for all $\varepsilon < k$, the problem reduces intrinsically to the exact k-embeddability of that graph, and is thus NP-hard as well.

Reference

[1] J. B., Saxe. Embeddability of weighted graphs in k-space is strongly NP-hard. *Proceedings of the 17th Allerton Conference on Communications, Control, and Computing,* (1979): 480–489.

33 A Topology-Constrained Network Algorithm for NOESY Data Interpretation

We now move from distance geometry to the related, practical problem of NOE assignment, which is also informed by graph topology. In this chapter, we discuss a nuclear Overhauser effect (NOE) assignment approach applied in the software AutoStucture [1], which formulates the problem into a topology-constrained distance network, and then iteratively refines the distance network that interprets the NOESY data and restrains the structure calculations.

33.1 Algorithms

The flow chart of AutoStructure's NOESY data interpretation approach for structure calculation in [1] is shown in figure 33.1. The procedure is divided into two parts: *initial fold analysis* and *iterative structure analysis*. The initial fold analysis stage obtains an optimal distance network HG_{NOE} and calculates the initial global fold. The iterative structure analysis stage iteratively refines the NOE-constrained distance network and intermediate structures.

Table 33.1 lists the graph theory formulation used in the algorithm, which is illustrated in figure 33.2.

Step 1: Preparation of Input Data   The input data for AutoStructure includes the (1) protein amino acid sequence, (2) resonance assignments, (3) NOESY peak list from third-party peak-picking software, (4) scalar coupling data (optional), (5) amide ^{1}H exchange data (optional), and (6) other constraint data, such as RDCs.

Step 2: Construction of Initial Ambiguous Network G_{ANOE}^{0}   The initial ambiguous NOE distance network is formed by matching frequencies of NOESY cross peaks with chemical shifts in the resonance assignment list within error match tolerances (table 33.1).

Step 3: Input Data Validation and Heuristic Distance Network Initialization   To validate the input NOESY data, AutoStructure first constructs a local distance network G_{local}, which basically contains the geometric proximity information based on bond-connections solely indicated from protein sequence and resonance list. Then M_{score}, a measurement of input data quality, is defined as

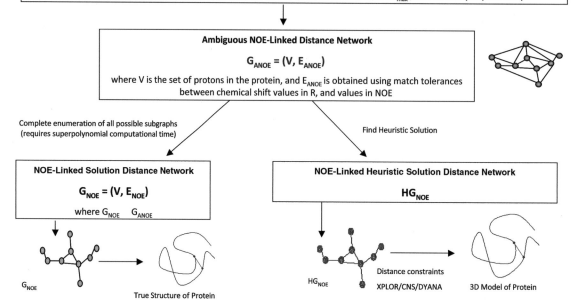

Figure 33.1
AutoStructure's formulation of the NOESY interpretation problem using graph theory. Figure by Janet Huang and Guy Montelione.

Table 33.1
Summary of key parameters [1]

Symbol	Definition	Default Value
d_{NOE}	The maximum distance observed in NOESY spectrum	5.0 Å
Δ^i_{err}	Error match tolerances for the ith dimension	H: 0.05 ppm $^{15}N/^{13}C$: 0.5 ppm
Δ^i_{allow}	Allowable match tolerances for the ith dimension	H: 0.04 ppm $^{15}N/^{13}C$: 0.4 ppm
Δ^i_{good}	Good match tolerances for the ith dimension	H: 0.03 ppm $^{15}N/^{13}C$: 0.3 ppm
Δ_{sym}	Threshold for symmetric peaks	H: 0.03 ppm $^{15}N/^{13}C$: 0.3 ppm
$dvio_{min}$	Threshold of the minimum distance violation	0.1 Å
ms_{min}	Threshold of the minimum model-support score	0.4
ms_{high}	Threshold of the high model-support score	0.7

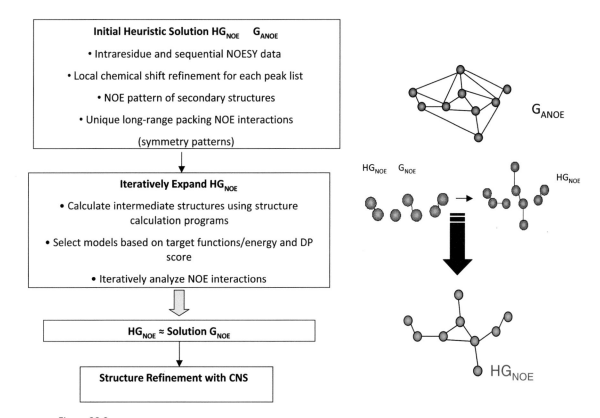

Figure 33.2
Autostructure is a topology-constrained bottom-up approach for finding the heuristic solution, HG_{NOE}. Figure by Janet Huang and Guy Montelione.

$$M_{score} = \frac{\text{Number of peaks in } G_{local} \text{ but not in } G^0_{ANOE}}{\text{Total number of peaks in } G_{local}}.$$

If M_{score} is larger than certain threshold, AutoStructure will require users to refine the input data. The heuristic distance network HG_{NOE} is obtained from G^0_{ANOE} by using narrower error tolerances and adding obvious pairs of NOE interactions, such as $H^\alpha H^N(i, i+1)$, $H^\beta H^N(i, i+1)$ and $H^N H^N(i, i+1)$.

Step 4: Refine Resonance Assignment List Based on HG_{NOE} and Prune G^0_{ANOE}   In this step, resonances are refined based on the mean value of all linked NOE cross peaks. Then the refined resonance list is used to improve the NOE distance network.

Step 5: Generation of Initial HG_{NOE}   This step is divided into two parts. In the first part, the NOE patterns of secondary structure elements are added into the distance network HG_{NOE}. The secondary structures are identified based on the following information: characteristic NOE patterns, chemical shift index, and scalar coupling data. In the second part, the NOE distance network is refined by symmetry checking and identification of unique connections. *Potential contact support scores*, which measure the confidence of an assigned proton pair, are used to filter out weakly supported but "apparently unique" proton pairs.

Step 6: Construct Initial Structure and Refine Self-Consistent HG_{NOE}   The following constraint information is fed into other software such as XPLOR/CNS or DYANA to calculate the initial structure: distance constraints calibrated from peak intensity, dihedral angle constraints (obtained by using a conformational grid search approach from local NOEs and scalar coupling data), H-bond distance constraints (obtained from helix and beta-sheet NOE contact patterns and slow amide hydrogen exchange data). Then the heuristic distance network HG_{NOE} is refined based on the initial structure.

Step 7: Iterative Fold Analysis and Refinement of HG_{NOE} Using Intermediate Structures   First, the NOE constraints are refined to be consistent with the topology constraints from helical-packing and beta-sheet packing geometries. Then a *model-support (ms) score* for each peak is computed based on the appearing frequency of short distances in G_{min} graphs, which are computed from all intermediate structures. Only those NOE assignments with high ms scores are considered to be well-supported from the intermediate structures and added into the heuristic NOE distance network. In less well-defined or loosely packed regions, loose criteria such as a larger upper-bound for distance limit and loose symmetric checking are used (cf. section 18.1.3). At the same

Table 33.2

Summary of experimental NMR data sets used for validation of AutoStructure [1]

Protein	FGF-2	MMP-1	IL-13
Fold class	β	α/β	α
Size (residues)	154	169	113
Assigned chemical shifts (set R)			
All (% completeness)	94.7	90.5	93.7
Side-chain atoms (% completeness)	91.5	85.4	90.3
Aromatic atoms (% completeness)	93.8	83.6	92.7
Completeness of stereospecific isopropyl methyl proton assignments	16/84 (19.1%)	6/80 (7.5%)	56/88 (63.6%)
Completeness of stereospecific β methylene proton assignments	156/444 (35.1%)	86/408 (21.1%)	80/284 (28.2%)
NOESY spectra (set NOE)			
Peaks picked from 3D(^{15}N) spectrum	1353	2409	1994
Assignable	1255	2003	1509
Peaks picked from 3D(^{13}C) spectrum	3702	3043	4143
Assignable	3588	2970	3943
Input validation—M score			
3D(^{15}N) spectrum (%)	12	4	7
3D(^{13}C) spectrum (%)	15	24	6
Overall (%)	14	21	6

time, suspect long-range orphan contacts, that is long-range contacts whose neighboring residues have no long-range contacts, are removed from HG_{NOE}.

Step 8: Assessment of Final Structure   The following measurements are used to assess final structures: *recall score*, *precision score*, *F-measure score*, and *discriminating power (DP)* score. *Recall score* is the fraction of NOE cross peaks consistent with the resulting structures. *Precision score* is the fraction of back-calculated interproton interactions that are observed in peak list. *F-measure score* is the overall fit between the resulting structures and experimental data. *Discriminating power (DP)* score, which calculates the difference in F-measure scores between the

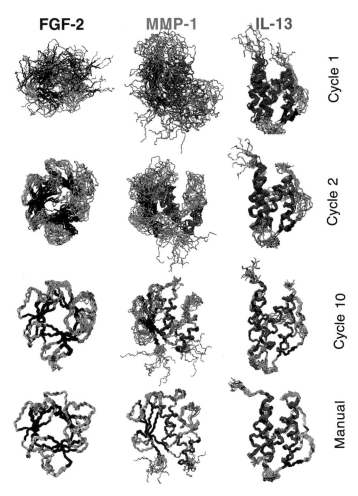

Figure 33.3 (plate 35)
Calculated structures in intermediate and final ensembles. Figure by Janet Huang and Guy Montelione.

query structure and "random coil" structures (a freely rotating chain model), is used to estimate the correctness of the overall fold.

The approach in step 8 has been extended into a protocol called *RPF* score (Recall, Precision, and F-measure) for NMR structure validation [2]. An improved version of AutoStructure (AS), called ASDP, uses RPF to define a DP score for intermediates in the AS trajectory, which permits selecting more accurate intermediate models to resolve ambiguous assignments. This results in more accurate interpretation of the NOESY data, and better convergence.

As described in step 8, RPF scores [2] assess how well the query 3D structures fit the experimental NOESY peak list and resonance assignment data. RPF scores compute the goodness-of-fit of the query structures to these experimental data, and can be used as a guide for further structure refinements. RPF also calculates DP scores (step 8). DP score has been used by CS-Rosetta for model selection (see chapter 45 and [3]).

The second-generation version of AutoStructure, ASDP, uses DP scores from RPF to select models generated in AutoStructure's iterative cycles. Only the top models with highest DP scores and lowest target function (CYANA) or conformational energy (XPLOR) values are selected for iterative AutoStructure NOE analysis. Using a DP filter for model selection improves the accuracy

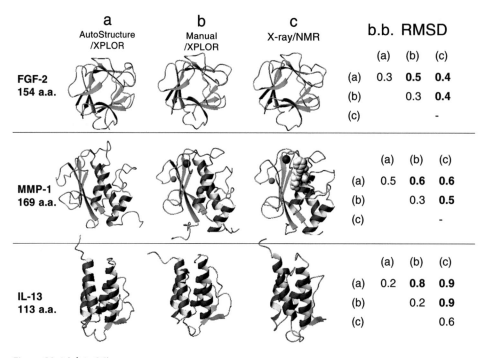

Figure 33.4 (plate 36)
Structures calculated by AutoStructure vs. other methods. The (b) method combines manual assignment analysis with optimization techniques such as SA/MD. Figure by Janet Huang and Guy Montelione.

of the selected intermediate structures for iterative NOE analysis and therefore improves the accuracy of NOESY crosspeak assignments and final models.

33.2 Results

AutoStructure has been evaluated using three different human protein NMR test data sets: FGF-2, IL-13, and MMP-1, ranging in size from 113 to 169 amino acid residues. The experimental NMR data used in the tests are given in table 33.2. Figure 33.3 (plate 35) illustrates the structures

Table 33.3
Comparisons of computed structures with other methods [1]

Quality	FGF-2	MMP-1	IL13
A. AutoStructure models			
NOE crosspeaks assigned			
^{15}N-NOESY	1147	1613	1214
^{13}C-NOESY	3160	2522	3254
Constraints			
NOE distance constraints	2058	2088	2705
Hydrogen bond distance constraints	76	86	76
Dihedral angle constraints	177	221	165
RPF scores	94.2/92.1/93.2/85.4	88.6/89.2/88.9/79.5	87.3/96.9/91.9/79.8
RMSD			
Sec. Strut. Region (b.b./heavy)	0.3/0.9	0.5/1.0	0.2/0.7
Ordered region (b.b./heavy)	0.7/1.3	1.3/1.8	0.6/1.1
Ramachandran plot summary from PROCHECK	75.0/23.9/1.1/0.0	81.7/18.0/0.3/0.0	90.4/9.3/0.1/0.1
B. Manual analysis models			
Constraints			
NOE distance constraints	2105	2078	1979
Hydrogen bond distance constraints	50	84	50
Dihedral angle constraints	330	426	302
RPF scores	94.5/92.6/93.5/86.2	88.9/89.6/89.2/80.7	82.5/97.1/89.2/72.3
RMSD			
Sec. Strut. region (b.b./heavy)	0.3/0.7	0.3/0.6	0.2/0.7
Ordered region (b.b./heavy)	0.4/0.8	0.4/0.8	0.4/0.8
Ramachandran plot summary from PROCHECK	77.5/21.5/1.0/0.0	90.1/9.8/0.31/0.0	91.1/8.0/0.9/0.0
C. Comparisons between AutoStructure and manual analysis			
% Distance constraints of AutoStructure that are violated by >1.0 Å in manual analysis models	3.3	3.7	5.8
% Distance constraints of manual analysis that are violated by >1.0 Å in AutoStructure analysis models	4.2	5.1	3.9
Mean Coordinate Differences			
Sec. Strut. Region (b.b./heavy)	0.5/0.6	0.6/0.8	0.8/0.9
Ordered region (b.b./heavy)	0.8/1.0	1.2/1.5	1.4/1.6

calculated in intermediate and final ensembles. Figure 33.4 (plate 36) and table 33.3 compare structures computed by AutoStructure vs. those from other approaches. The AutoStructure program has been used by several NMR groups, and more than two dozen protein structures have been determined using AutoStructure. ASDP was assessed in the first CASD-NMR workshop [4], where it performed well.

References

[1] Y. J. Huang, R. Tejero, R. Powers, and G. T. Montelione. A topology-constrained distance network algorithm for protein structure determination from NOESY data. *Proteins* 62;3 (2006 Mar 15):587–603.

[2] Y. J. Huang, R. Powers, and G. T. Montelione. Protein NMR recall, precision, and F-measure scores (RPF scores): Structure quality assessment measures based on information retrieval statistics. *Journal of the American Chemical Society* 127, 1665–74 (2005).

[3] Raman, S. Huang, Y.J., Mao, B., Rossi, P., Aramini, J.M., Liu, G., Montelione G.T. , and Baker, D. Accurate Automated Protein NMR Structure Determination Using Unassigned NOESY Data. *Journal of the American Chemical Society* 132, 202–207 (2010).

[4] A. Rosato, A. Bagaria, D. Baker, B. Bardiaux, A. Cavalli, J. F. Doreleijers, et al. CASD-NMR: A rolling experiment for the critical assessment of automated structure determination of proteins from NMR data. *Nature Methods* 6(2009): 625–626.

34 MARS: An Algorithm for Backbone Resonance Assignment

In this chapter, we describe MARS, an algorithm for backbone resonance assignment of proteins, and its extension for backbone assignment of proteins with known structure using residual dipolar couplings (RDCs) [1, 2].

34.1 MARS—Backbone Assignment of Proteins

34.1.1 Backbone Resonance Assignment

Backbone resonance assignment is a prerequisite for structure determination of proteins by NMR. In assignment process, the chemical shifts $^{1}H_i^N$, $^{15}N_i$, $^{13}C_i^\alpha$, $^{13}C_i^\beta$ of residue i and $^{13}C_{i-1}^\alpha$, $^{13}C_{i-1}^\beta$ of residue $i-1$ are generally assembled into arrays called *pseudoresidues (PRs)*, and each PR is associated with a single $^{1}H^N$, ^{15}N root.

Many previous backbone resonance assignment approaches can be grouped into two categories: The first group comprises optimization algorithms that try to minimize a global pseudoenergy function, or in other words to maximize a global "goodness of fit"; the second is based on best-first search strategies. Each approach has disadvantages: The optimization algorithms can be trapped into local minima, and the best-first search strategies are prone to propagation of errors made in the initial phases of the assignment process. In contrast, MARS heuristically and stochastically attempts to simultaneously optimize the local and global quality of assignment to minimize propagation of initial assignment errors and to extract reliable assignments.

34.1.2 Method

One common implementation of resonance assignment of $^{13}C/^{15}N$-labeled proteins comprises five steps: (1) identify and filter peaks, and reference resonances across different spectra; (2) group resonances into PRs; (3) identify the amino acid type of PRs; (4) find and link sequential PRs into segments; (5) map PR segment onto the primary sequence. MARS does not perform the first two steps. Jung and Zweckstetter [1] recommend using some other analysis software, such as FELIX, AURELIA, XEASY, SPARKY, and NMRView, for peak picking and referencing of multiple NMR spectra. In order to avoid an unreasonable high number of PRs and to improve reliability, they [1] suggest partially discarding ambiguous peaks and reinserting suspicious peaks in later iterations when inspecting PRs.

The input data of MARS are as follows: (1) the primary sequence of protein; (2) secondary structure predicted from the software PSIPRED; (3) the ASCII text file of defining assignment parameters; (4) observed intra- and interresidual chemical shifts grouped into PRs. In addition, sequential information from HNCO/HN(CA)CO and H^N-H^N NOESY spectra and the amino acid type for a PR can also be included in MARS assignment.

The assignment procedure, as shown in figure 34.1, consists of four steps:

1. *Establishing sequential connectivity*   Initially, each PR is assumed to be connected to every other PR and only connectivitities not in agreement with experimental intra- and interresidual chemical shifts are removed. In this step, all matching shifts are equally accepted, that is, there is no preference for the "best match" to avoid a bias from insignificant chemical shift differences. Moreover, PRs are not classified according to the number of chemical shifts they contain or the

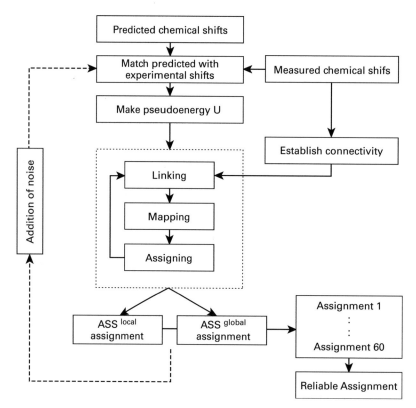

Figure 34.1
Overview of the MARS assignment procedure [1].

intensity of their corresponding NMR resonances. Therefore, PRs strongly affected by chemical exchange or by the presence of a paramagnetic ion can be fully utilized.

2. *Calculating pseudoenergy*   Pseudoenergy (i.e., a scoring function) is calculated by comparing the experimental C^α and C^β chemical shifts with values that were obtained for each residue from a statistical analysis of chemical shifts deposited in the BMRB [3]. In MARS, this process is further improved by using chemical shift distributions that are corrected for neighbor residue effects [4] and the type of secondary structure of each PR predicted from the secondary structure prediction program PSIPRED [5]. MARS calculates for all experimentally observed PRs the deviation of their experimental chemical shifts from predicted values according to

$$D(i, j) = \sum_{k=1}^{N_{CS}} \left(\frac{\delta(i)_k^{exp} - \delta(j)_k}{\sigma_k} \right)^2 , \tag{34.1}$$

where $\delta(i)_k^{exp}$ is the measured chemical shift to type k of PR i, $\delta(j)_k$ is the predicted chemical shift of type k of residue j, and σ_k^2 is the variance of the statistical chemical shift distribution that is used for calculating $\delta(j)_k$. By ranking all residues j according to their chemical shift deviation with respect to PR i, MARS calculates a pseudoenergy $U(i, j)$ for each PR i. Note that this scoring function does not rely directly on chemical shift deviation. This is claimed to make MARS robust against unusual chemical shifts, because the exact fit of calculated to experimental chemical shifts is not as important as the overall quality of the chemical shift fit.

3. *Exhaustive search for sequential connectivity and mapping*   The PRs are initially assigned randomly to the protein sequence. In order to refine the assignment, MARS randomly selects a starting PR, and searches in the forward direction of the primary sequence for all PR segments of length five that can be assembled based on the available connectivity information. Then, MARS exhaustively maps all these segments on the protein sequence, and selects a best-fit segment SEG_{for} by calculating a scoring according to

$$U_i^m(j) = \sum_{k=i}^{i+n} U(k, j_i). \tag{34.2}$$

To validate SEG_{for}, MARS repeats the same procedure but from the last position of SEG_{for} and in the backward direction of the primary sequence. If the calculated segment $SEG_{back} = SEG_{for}$, then the assignment is updated according to this resulting segment. The assignment is further refined by exhaustive search for best-fit segments of length 4, 3, and 2.

4. *Identifying reliable assignments*   The algorithm results in a final optimization assignment mainly derived by the local fit of fragments. In addition, pseudoenergy values $U(i, j)$, which qualitatively describe the mapping of a single residue j to PR i, have been changed during the process. Thus, a global assignment can be extracted from the refined $U(i, j)$. This global assignment is compared with the local assignment and only consistent assignments are retained.

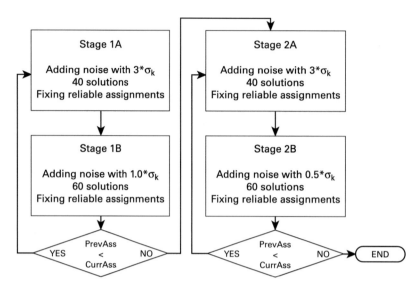

Figure 34.2
Empirically optimized scheme for avoiding errors due to inaccuracies in predicted chemical shifts in MARS [1].

A major factor influencing the assignment is the quality of chemical shifts predicted from the primary sequence. To overcome this problem, MARS repeats the complete assignment many times and adds some noise into assignment procedure to make the resulting assignment more robust (figure 34.2).

34.1.3   Results and Discussion

MARS was tested using either real data or chemical shifts deposited in the BMRB, on 14 proteins (table 34.1) ranging in size from 71-residue Z domain of Staphylococcal protein A to 723-residue malate synthase G (MSG). For small proteins, using C^{α}/C^{β} connectivity information, the assignment accuracy of MARS is nearly 100%. In case that only C^{α} chemical shift information is available, the assignment accuracy of MARS remains at a high level and all reliable assignments reported by MARS are correct. The test results also indicate that MARS could compute reliable assignments for big proteins like the 723-residue MSG and for proteins with incomplete chemical shift data. MARS is claimed to be useful for proteins with very high degeneracy such as partially or fully unfolded proteins and proteins with missing or erroneous chemical shift data.

34.2   Backbone Assignment with Known Structure Using RDCs

In this section, an extension of MARS is discussed, in which resonance assignment was enhanced when experimental RDCs are matched to values back-calculated from a known three-dimensional (3D) structure.

Table 34.1
Proteins and data quality used for testing MARS [1]

Protein	BMRB code	# of residues	# of PRO/GLY	$C_i^\alpha / C_{i-1}^\alpha$ (%)[a]	$C_i^\beta / C_{i-1}^\beta$ (%)[a]	C_i' / C_{i-1}' (%)[a]	$H_i^\alpha / H_{i-1}^\alpha$ (%)[a]
Malate synthase G	5471	723	31/51	95/95	94/94	94/95	–
Maltose-binding protein	4354	370	21/29	96/96	95/96	–	–
Rous sarcoma virus capsid	4384	262	23/20	92/92	89/91	92/93	–
Human carbonic anhydrase I	4022	260	17/16	100/100	100/100	95/96	–
N-terminal domain of enzyme I (EIN)	4106	259	4/15	96/97	96/97	–	–
E-cadherin domains II and III	4457	227	14/12	78/63	78/63	–	–
Human prion protein	4402	210	15/43	98/97	98/97	–	–
Superoxide dismutase	4341	192	8/14	64/64	62/63	48/61	–
Calmodulin/M13 complex	547	148	2/11	99/99	–	99/99	–
Profilin	4082	139	4/16	99/99	100/98	–	–
E. coli EmrE	4136	110	5/12	86/84	57/60	73/77	–
Human ubiquitin	–	76	3/6	100/100	100/100	–	–
Z domain	–	71	3/0	90/96	51/82	–	89/100
Tir1 10	–	110	12/15	100/100	100/100	100/100	–

[a] Percentage of available chemical shifts of a given type.

34.2.1   Method

When dipolar couplings are to be used in assignment, a PDB file of a known 3D structure or homology model of the protein as well as the resolution of that structure are supplied as input to MARS. To include RDCs in the process of mapping PR segments onto the protein sequence, MARS calculates for all experimentally observed PRs the deviation of their experimental RDCs and chemical shifts from predicted values according to

$$D(i, j) = w \sum_{k=1}^{N_{CS}} \left(\frac{\delta(i)_k^{exp} - \delta(j)_k}{\sigma_k^{CS}} \right)^2 + \sum_{l=1}^{N_{RDC}} \left(\frac{RDC(i)_l^{exp} - RDC(j)_l}{\sigma_l^{RDC}} \right)^2 , \tag{34.3}$$

where w is a weighting factor measuring the different reliability of calculated chemical shifts and RDCs. Empirical optimization resulted in $w = 3.3$, thereby downscaling the contribution of RDCs. In addition, the RDC normalization constant σ_l^{RDC} is adjusted according to the resolution R_{struc} as $\sigma_l^{RDC} = c_{RDC} R_{struc} D_a^{HN}$, or σ_l^{RDC} can also be set to a fixed value of 0.21 for R_{struc} ranging from 1.4 to 2.4 Å without strongly affecting the assignment result.

When the internuclear vectors are distributed reasonably uniformly in orientation (chapter 6), the magnitude and rhombicity of a molecular alignment tensor $\mathbf{A}$ can be obtained without assignment from a histogram of experimental RDCs [6].

In MARS, four different methods are available to extract the orientation of the alignment tensor: (1) shape and charge/shape prediction of molecular alignment tensors; (2) SVD (after an initial assignment step using only chemical shifts); (3) exhaustive back-calculation; (4) a grid search to optimize the fit of experimental chemical shifts and RDCs to values predicted from the 3D structure.

In order to improve the accuracy of assignment, two complete MARS assignment runs are performed for RDC-enhanced assignment. The first run is to back-calculate the dipolar couplings from the 3D structure using the approximate alignment tensor, and the second run is to refine the alignment tensor. The test results suggest that with this two-step procedure the final assignment score obtained by MARS is almost independent from the method that was chosen to get a first estimate of the alignment tensor.

Furthermore, some additional noise is added to the assignment procedure of RDCs and chemical shifts to improve the reliability of the assignment results. This procedure is different from adding noise to chemical shifts; however, the amount of noise added to back-calculated RDCs is kept fixed at a large amount of five times σ_l^{RDC} to avoid wrong assignments.

34.2.2   Results and Discussion

The RDC-enhanced assignment was applied to three proteins: 76-residue ubiquitin, 259-residue N-terminal domain of enzyme I of the phosphoenolpyruvate (EIN), and 370-residue two-domain maltose-binding protein (MBP). In order to evaluate how much information is required for successful assignment, tests were performed using different data, such as assignment without sequential connectivity information using only dipolar coupling/chemical shift matching, assignment with

Table 34.2
RDC-enhanced assignment of ubiquitin and MBP for varying amounts of data [1]

(a) RDC-enhanced assignment of ubiquitin

RDCs	Chemical Shifts for Linking	Chemical Shifts for Matching	Assignment Score (%) 1UBQ Total correct	Reliable	Wrong reliable	1AAR Total correct	Reliable	Wrong reliable
Without sequential connectivity information								
—	—	$C'_{i-1}, C^{\alpha}_{i-1}, C^{\beta}_{i-1}$	36.1	19.5	5.6	36.1	19.5	5.6
$^1D_{NH}$	—	$C'_{i-1}, C^{\alpha}_{i-1}, C^{\beta}_{i-1}$	47.2	16.7	1.4	38.9	9.7	1.4
$^1D_{NH}, ^1D_{CaC'}$	—	$C'_{i-1}, C^{\alpha}_{i-1}, C^{\beta}_{i-1}$	76.4	44.4	0.0	68.1	33.4	2.8
$^1D_{NH}, ^1D_{CaC'}, ^1D_{NC'}$	—	$C'_{i-1}, C^{\alpha}_{i-1}, C^{\beta}_{i-1}$	91.7	55.6	0.0	83.3	50.0	0.0
With sequential connectivity information								
—	C^{α}	$C'_{i-1}, C^{\alpha}_{i-1}, C^{\alpha}_{i}$	80.6	25.0	0.0	80.6	25.0	0.0
$^1D_{NH}$	C^{α}	$C'_{i-1}, C^{\alpha}_{i-1}, C^{\alpha}_{i}$	93.1	51.4	0.0	97.2	37.5	0.0
$^1D_{NH}, ^1D_{CaC'}$	C^{α}	$C'_{i-1}, C^{\alpha}_{i-1}, C^{\alpha}_{i}$	100.0	90.3	0.0	100.0	73.6	0.0
$^1D_{NH}, ^1D_{CaC'}, ^1D_{NC'}$	C^{α}	$C'_{i-1}, C^{\alpha}_{i-1}, C^{\alpha}_{i}$	100.0	100.0	0.0	100.0	100.0	0.0

(b) RDC-enhanced assignment of 370-residue MBP

RDCs	Chemical Shifts for Linking	Chemical Shifts for Mapping	Missing Chemical Shifts (%)	Assignment Score (%) Total correct	Reliable	Wrong reliable
—	C^{α}, C^{β}	$C'_{i-1}, C^{\alpha}_{i-1}, C^{\alpha}_{i}, C^{\beta}_{i-1}, C^{\beta}_{i}$	4	95.8	87.2	0.0
$^1D_{NH}$	C^{α}, C^{β}	$C'_{i-1}, C^{\alpha}_{i-1}, C^{\alpha}_{i}, C^{\beta}_{i-1}, C^{\beta}_{i}$	4	98.5	94.6	0.0
$^1D_{NH}, ^1D_{CaC'}$	C^{α}, C^{β}	$C'_{i-1}, C^{\alpha}_{i-1}, C^{\alpha}_{i}, C^{\beta}_{i-1}, C^{\beta}_{i}$	4	98.5	93.4	0.0
$^1D_{NH}, ^1D_{CaC'}, ^1D_{NC'}$	C^{α}, C^{β}	$C'_{i-1}, C^{\alpha}_{i-1}, C^{\alpha}_{i}, C^{\beta}_{i-1}, C^{\beta}_{i}$	4	99.1	94.6	0.0
—	C^{α}, C^{β}	$C'_{i-1}, C^{\alpha}_{i-1}, C^{\alpha}_{i}, C^{\beta}_{i-1}, C^{\beta}_{i}$	20	82.7	44.2	0.7
$^1D_{NH}$	C^{α}, C^{β}	$C'_{i-1}, C^{\alpha}_{i-1}, C^{\alpha}_{i}, C^{\beta}_{i-1}, C^{\beta}_{i}$	20	88.8	51.4	0.0
$^1D_{NH}, ^1D_{CaC'}$	C^{α}, C^{β}	$C'_{i-1}, C^{\alpha}_{i-1}, C^{\alpha}_{i}, C^{\beta}_{i-1}, C^{\beta}_{i}$	20	95.0	58.0	0.4
$^1D_{NH}, ^1D_{CaC'}, ^1D_{NC'}$	C^{α}, C^{β}	$C'_{i-1}, C^{\alpha}_{i-1}, C^{\alpha}_{i}, C^{\beta}_{i-1}, C^{\beta}_{i}$	20	94.3	62.6	0.0

$^1D_{CaC'}$ denotes a C^{α}-C' RDC.

only C^α sequential connectivity information, and assignment using C^α/C^β chemical shifts. The effect of including one, two, or three types of RDCs was also tested. As shown in table 34.2, in case of a small protein, such as ubiquitin, MARS performed assignment of more than 90% of backbone resonances without the need for sequential connectivity information if 3 RDCs per residue are available; for bigger proteins such as MBP, the combination of sequential connectivity information with RDC-matching enables more residues to be assigned reliably and backbone assignment to be more robust against missing data.

In conclusion, the RDC-enhanced assignment method with known structure is applicable to both small and big proteins. In particular, the RDC-enhanced assignment is useful for large proteins where chemical shift data are often missing or for a substantial portion of residues the chemical shift degeneracy is too high to allow unambiguous assignment, or chemical shift and sequential connectivity only provide a few reliable assignments.

References

[1] Y. S. Jung, and M. Zweckstetter. MARS — robust automatic backbone assignment of proteins. *Journal of Biomolecular NMR*. 30;1 (2004 Sep): 11–23.

[2] Z. S. Jung, and M. Zweckstetter. Backbone assignment of proteins with known structure using residual dipolar couplings. *Journal of Biomolecular NMR*. 30;1 (2004 Sep): 25–35.

[3] J. F. Doreleijers, S. Mading, D. Maziuk, K. Sojourner, L. Yin, J. Zhu, et al. BioMagResBank database with sets of experimental NMR constraints corresponding to the structures of over 1400 biomolecules deposited in the Protein Data Bank. *Journal of Biomolecular NMR*, 26 (2003 October): 139–46.

[4] Y. J. Wang, and O. Jardetzky. Investigation of the neighboring residue effects on protein chemical shifts. *Journal of the American Chemical Society*. 124;47 (2002 Nov 27): 14075–14084.

[5] L. J. McGuffin, K. Bryson, and D. T. Jones. The PSIPRED protein structure prediction server. *Bioinformatics* 16 (2000): 404–405.

[6] G. M. Clore, A. M. Gronenborn, and A. Bax. A robust method for determining the magnitude of the fully asymmetric alignment tensor of oriented macromolecules in the absence of structural information. *Journal of Magnetic Resonance* 133 (1998): 216–221.

35   Errors in Structure Determination by NMR Spectroscopy

This chapter describes the different types and sources of errors in the determination of structures by NMR. We address concerns in terms of experimental data collection as well as touch on algorithmic problems, highlighting the possibility of errors in the deposited structures to the Protein Data Bank (PDB) [1, 2, 3]. Improved algorithms, and the use of RDC data, have the potential to compute structures without such errors.

35.1   Errors in Published Protein Folds

There are many examples of proteins for which incorrect structures were published, and some of these have been revised or replaced by a new structure in the PDB. One example, as shown in figure 35.1 (plate 37), is phosphoactive yellow protein (PDB id: 2PHY, replacing 1PHY), which was determined by X-ray crystallography. Phosphoactive yellow protein was incorrectly solved due to problems in electron density interpretation, and subsequently a major change in the fold was determined [2].

Currently, X-ray crystallography and nuclear magnetic resonance (NMR) spectroscopy are two main techniques for high-resolution biomolecular structure determination, and all resulting models from these experiments are derived from their underlying experimental data. In this process, different problems may lead to errors. Purifying the wrong protein, very noisy data (and thus, for example, misassigning spectral signals in NMR), and problems in the structure-determination algorithm and error-checking themselves are all possible sources of errors [1, 2], and the results can be as disastrous as an incorrect fold.

35.2   Case Study: Dynein Light Chain

Certain problems arose in building a homology model for the protein dynein light chain (DLC2A). Using a BLAST search, one finds two NMR homologous structures, 1Y4O from mouse and 1TGQ from human, in the PDB with more than 95% sequence identity. Surprisingly, a first visual inspection of both structures reveals striking differences. As illustrated in figure 35.2 (plate 38), compared to the human version of DLC2A as a monomer, the mouse version appeared in the PDB

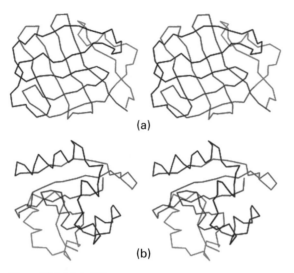

(a)

(b)

Figure 35.1 (plate 37)

Comparison of an incorrect protein model with its corrected counterparts. The model is stereo C$^\alpha$ traces color-ramped from blue at the N-terminus to red at the C-terminus. (a) Incorrect model of photoactive yellow protein (PDB ID: 1phy) and (b) the corrected model (2phy). In this case, the initial model displayed a β-clam fold, whereas the correct model revealed an α/β protein with a fold similar to that of the SH2 domain [2]. Copyright 2000 IUCr.

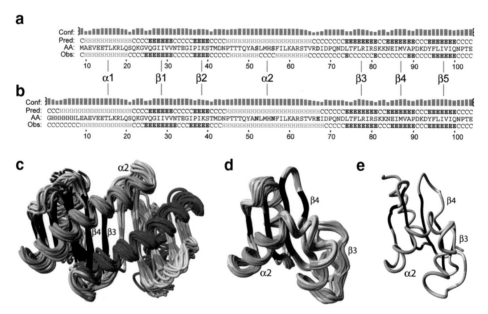

Figure 35.2 (plate 38)

Sequence and structure ensembles of two DLC2A structures. (a and b) The sequences of human (hDLC2A) and mouse (mDLC2A); (c) Ribbon diagram of the structure ensemble of mDLC2A (PDB ID: 1Y4O); (d) ribbon diagram of the structure ensemble of hDLC2A (PDB ID: 1TGQ); (e) the refined average structure of the ensemble calculated using the reconstructed 1TGQ dataset [1].

as a dimer. Given the fact that two sequences sharing over 30% sequence identity in 100 aligned residues are very likely to have the same fold [3], it is extremely unlikely that both mouse and human DLC2A structures are correct.

This large difference between the two ensembles probably originates from the oligomeric state of the two structures (see sections 17.5 and 18.1.4). The structures of these two proteins were derived from NMR spectroscopy; in particular, the presence of the tertiary structures were assessed using a proton-nitrogen correlation (^{15}N-HSQC) spectrum. It is not straightforward to determine the oligomeric state of a protein from its ^{15}N-HSQC NMR spectra alone, and typically assessments must be made from estimates of the protein's relaxation rates. Therefore, if the oligomeric state of a protein is not known or is incorrectly known, the NMR spectra of a dimeric protein could easily be interpreted as originating from a monomer. In this case, the assignment of human DLC2A (PDB ID: 1TGQ) was based on the assumption of monomeric structure, and thus it incorrectly interpretes the intermolecular contacts as *intramolecular* contacts. To further test this hypothesis, if we switch the 72 original intermolecular NOEs of the mouse DLC2A to be 36 intramolecular restraints (since those for the 1TGQ ensemble were not available), the subsequently computed structure has a similar fold to the monomer originally derived for the human ortholog [1]. In addition, it has been shown experimentally that dynein light chain exists as the dimer.

35.3 Identifying the Problems: Problems in Identifiers

Table 35.1 lists various typical "checks" used in determining the appropriateness of a structure [1]. The 1TGQ structure (monomer, human DLC2A) was subjected to a series of refinements to improve the structure, while keeping it in its monomer form. These conditions include setting rules for dihedral restraints, as well as looking at the rotamer normality and backbone conformation.[1] Observe that the only clear correlation between the incorrect structure and the various criteria is the rotamer normality; however, the backbone conformation does not change at all between the unrefined and refined 1TGQ structures.

Moreover, figure 35.3 (plate 39) plots the structure quality Z-scores for a large set of NMR structures deposited after 2003. As can be seen, 1TGQ is scored low on a distribution of these NMR ensembles [1]. As another example, figure 35.4 (plate 40) shows the PDZ domain of PTP-Bas (hydrolase), where panel (a) indicates an arginine side-chain in hydrophobic core, and panel (b) shows a homologous domain to the PTP-Bas, with a strangely solvent-exposed arginine side-chain. In this case, RDC data could likely be used to distinguish between these backbones, one of which may be incorrect.

---

[1] The ROTCHK command will compare the χ_1 rotamer of all residues with the distribution of observed rotamers for the same residue type in a similar local backbone conformation in the database. Also, the backbone conformation check will check whether there are fragments of 5 C$^\alpha$ coordinates that are not represented by other structures in the WHAT IF database.

Table 35.1

Average quality indicators of the 1Y4O and 1TGQ structure ensembles before and after refinement in an explicit solvent model [1]

Criteria	Characteristic	1Y4O (Original)	1Y4O (Refined)	1TGQ (Original)	1TGQ (Refined)
Agreement with experimental data	RMS violation 1Y4O distance restrains (Å)	0.0129	0.0097	0.607	0.0284
	Violations > 0.5 Å 1Y4O distance restraints	0	0	63	0
	RMS violation 1TGQ$_{sim}$ restraints (Å)	12.8	12.6	0.521	0.0231
	Violations > 0.5 Å 1TGQ$_{sim}$ restraints	32	32	4	0
	RMS violation 1Y4O dihedral restraints (°)	0.497	0.336	25.0	1.59
	Violations > 5° 1Y4O dihedral restraints	0	0	34	4
PROCHECK validation results[a]	Most favored regions	91.2	90.5	67.7	85.8
	Additionally allowed regions	8.4	9.0	27.3	12.8
	Generously allowed regions	0.2	0.2	4.7	0.5
	Disallowed regions	0.2	0.3	0.2	0.9
WHAT IF structure Z-scores[b]	Packing quality	−0.4	0.1	−2.1	−1.5
	Ramachandran plot appearance	−3.6	−3.3	−6.6	−4.6
	χ_1/χ_2 rotamer normality	−0.3	−0.7	−5.8	−3.0
	Backbone conformation	−0.8	−1.1	−5.4	−5.4

[a]Percentage of residues present in the four different regions of the Ramachandran plot.
[b]A Z-score [31,32] is defined as the deviation from the average value for this indicator observed in a database of high-resolution crystal structures, expressed in units of the standard deviation of this database-derived average. Typically, Z-scores below a value of −3 are considered poor, those below −4 are considered bad.

DOI: 10.1371/journal.pcbi.0020009.t001

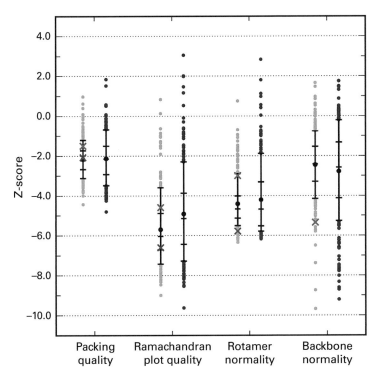

Figure 35.3 (plate 39)
Structure quality Z-scores for a large set of the NMR structures after 2003 [1].

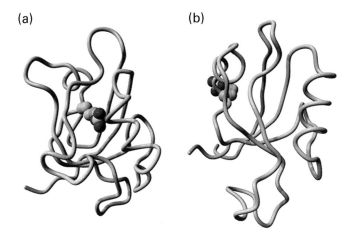

Figure 35.4 (plate 40)
An example of observed structural anomalies [1].

References

[1] S. B. Nabuurs, C. A. Spronk, G. W. Vuister, and G. Vriend. Traditional biomolecular structure determination by NMR spectroscopy allows for major errors. *PLoS Computational Biology* 2;2 (2006 Feb):e9.

[2] G. J. Kleywegt. Validation of protein crystal structures. *Acta Crystallographica D Biological Crystallography* 56;3 (2000 Mar): 249–265.

[3] B. Rost. Twilight zone of protein sequence alignments. *Protein Engineering* 12;2 (1999 Feb): 85–94.

36 SemiDefinite Programming and Distance Geometry with Orientation Constraints

This chapter introduces Semidefinite programming and its applications in two areas: the side-chain positioning problem and the sensor network localization problem. We present the problem of distance geometry with orientation constraints: we first introduce the graph embedding problem with angle information, and then revisit the problem of protein backbone determination using the global orientational constraints from RDC data.

36.1 SemiDefinite Programming and Two Applications

36.1.1 Overview of SemiDefinite Programming

Semidefinite programming (SDP) has been one of the most exciting developments in mathematical programming since the 1990s [4]. Semidefinite programming is a special case of convex programming, in which we must optimize a linear function of symmetric matrices subject to linear equality constraints, and where the matrix must be positive semidefinite.[1] For any additive error $\varepsilon > 0$, semidefinite programs can be solved in polynomial time, in terms of both the size of input and $\log(1/\varepsilon)$, using the *ellipsoid algorithm* or *interior-point method*.

36.1.2 Application in the Side-Chain Positioning Problem

In the side-chain positioning problem (See chapters 11–12), we are given a backbone of length p, and a set of side-chains $\{i_r\}$, also called *rotamers*, for each residue position i, and our objective is to find a sequence of side-chains that minimizes the following energy function:

$$\mathcal{E} = E_0 + \sum_i E(i_r) + \sum_{i<j} E(i_r, j_s),$$

where E_0 denotes the self-energy of the backbone, $E(i_r)$ denotes the energy resulting from the interaction between the backbone and the rotamer i_r at position i, and $E(i_r, j_s)$ denotes the energy resulting from the interaction between chosen rotamers i_r and j_s.

---

[1] Suppose that X is an $n \times n$ matrix. Then X is a *positive semidefinite (PSD)* matrix if $v^T X v \geq 0$ for any $v \in \mathbb{R}^n$.

The side-chain positioning problem can also be formulated in graph-theoretic terms. Let G denote a p-partite graph with node set $V = V_1 \bigcup \cdots \bigcup V_p$, where V_i represents the set of rotamers $\{i_r\}$ at position i. Each node $u \in V_i$ is associated with a weight $E_{uu} = E(i_r)$. Each edge between nodes $u \in V_i$ (corresponding to i_r) and $v \in V_j$ (corresponding to j_s) ($i \neq j$) is assigned a weight $E_{uv} = E(i_r, j_s)$. Now, the side-chain positioning problem is converted into the problem of finding a node per V_i to minimize the weight sum of the induced subgraph.

The side-chain positioning problem has been proven to be NP-hard [5] and even hard to approximate [3]. It can be directly formulated into the following integer quadratic programming problem:

$$\text{minimize} \sum_{(u,v) \in G} E_{uv} x_u x_v \tag{36.1}$$

$$\text{subject to} \sum_{u \in V_i} x_u = 1 \text{ for } i = 1, \ldots, p.$$

$$x_u \in \{0, 1\}.$$

We next show how to relax this integer quadratic programming problem to a semidefinite programming formulation [3]. First, we add a new singleton vertex set, $V_0 = \{u_0\}$, into graph G. The formulation in (36.1) can be converted into the following, equivalent program:

$$\text{minimize} \sum_{(u,v) \in G} E_{uv} x_u x_v \tag{36.2}$$

$$\text{subject to} \sum_{u,v \in V_i} x_u x_v = 1 \text{ for } i = 0, 1, \ldots, p.$$

$$\sum_{u \in V_i} x_{u_0} x_u = 1 \text{ for } i = 0, 1, \ldots, p$$

$$x_u x_u = x_{u_0} x_u$$

$$x_u \in \{0, 1\} \text{ for all } u.$$

We relax the requirement $x_u \in \{0, 1\}$ to $0 \leq x_u^T x_v \leq 1$ for all u and v, and then obtain the following semidefinite programming formulation:

$$\text{minimize} \sum_{(u,v) \in G} E_{uv} x_u x_v \tag{36.3}$$

$$\text{subject to } x_{uu} = x_{u_0 u} \text{ and } x_{uv} \geq 0$$

$$\sum_{u \in V_i} x_{u_0 u} = \sum_{u,v \in V_i} x_{u,v} = 1 \text{ for } i = 0, 1, \ldots, p.$$

$$X \text{ is PSD.}$$

We can obtain a solution of the preceding SDP system within any desired level of accuracy ε in polynomial time by using the ellipsoid algorithm or interior-point method. After that, we use a randomized rounding scheme to round each vector $0 \le x_u \le 1$ to 0 or 1. The basic idea is to sample each position from a specified probability distribution in which long vectors have a higher probability of being selected. For more details of the rounding scheme, see section 3 in Biswas et al. [2].

36.1.3 Application in the Sensor Network Localization Problem

It is interesting that SDP can also be used to solve the other problems, such as those in sensor networks. We give one example now.

In a sensor network in $\mathbb{R}^2$, we are given n sensors and m anchors. An *anchor* is a node whose position a_k, $k = 1, \ldots, m$, in $\mathbb{R}^2$ is known. We also know N_x, a set of Euclidean distances d_{ij} between two sensors x_i and x_j, and N_a, a set of Euclidean distances d_{jk} between a sensor x_j and an anchor a_k. The *sensor network localization problem* is to find all x_j, $j = 1, 2, \ldots, n$, that satisfy all the distance constraints

$$\|x_j - x_i\|^2 = d_{ji}^2, \forall (j, i) \in N_x$$

$$\|x_j - a_k\|^2 = d_{jk}^2, \forall (j, k) \in N_a.$$

Let $X = (x_1, x_2, \ldots, x_n)$. Then the *sensor network localization problem* can be formulated into the following matrix form:

find $X \in \mathbb{R}^{2 \times n}, Y \in \mathbb{R}^{n \times n}$ (36.4)

subject to $(e_i - e_j)^T Y (e_i - e_j) = d_{ji}^2, \forall (j, i) \in N_x$

$$\begin{pmatrix} a_k \\ -e_j \end{pmatrix}^T \begin{pmatrix} I_2 & X \\ X^T & Y \end{pmatrix} \begin{pmatrix} a_k \\ -e_j \end{pmatrix} = d_{jk}^2, \forall (j, k) \in N_a.$$

$$Y = X^T X.$$

This form can be relaxed to a semidefinite program by changing constraint $Y = X^T X$ to $Y \succeq X^T X$. Here, the symbol $\succeq$ means positive, semidefinite:

find $Z \in \mathbb{R}^{(n+2) \times (n+2)}$ (36.5)

subject to $(1, 0, 0)^T Z (1, 0, 0) = 1$

$$(0, 1, 0)^T Z (0, 1, 0) = 1$$

$$(1, 1, 0)^T Z (1, 1, 0) = 2$$

$$(0, e_i - e_j)^T Z (0, e_i - e_j) = d_{ji}^2, \forall (j, i) \in N_x$$

$$(a_k, -e_j)^T Z (a_k, -e_j) = d_{ji}^2, \forall (j, i) \in N_a$$

$$Z = \begin{pmatrix} I_2 & X \\ X^T & Y \end{pmatrix} \succeq 0.$$

The ε-solution of this SDP system—that is, the solution within ε error distance from the optimal one—can be computed in time $O(\sqrt{n+k}(n^3 + n^2k + k^3)\log\frac{1}{\varepsilon})$ using the interior-point algorithm, where $k = 3 + |N_x| + |N_a|$.

36.2   Distance Geometry with Orientation Constraints

We now discuss the graph embedding problem with local angle information, and then compare with the protein structure determination problem based on global RDC orientational constraints.

36.2.1   Graph Embedding with Angle Information

In l_s^k space, the distance between two points $(x_1, \ldots, x_k)$ and $(y_1, \ldots, y_k)$ in k-dimensional space is measured according to the l_s norm, that is,

$$\|(x_1, \ldots, x_k) - (y_1, \ldots, y_k)\| = \sqrt[s]{(x_1 - y_1)^s + \cdots + (x_k - y_k)^s}.$$

In a graph embedding problem, we are given a graph $G = (P, E)$, and lengths $D(p, q)$ for all edges $\{p, q\} \in E$. Our objective is to embed G into a l_s^k space by a map $f \colon P \to l_s^k$ such that the following graph distortion is minimized:

$$\max_{(p,q)\in E} \|f(p) - f(q)\|_s / D(p, q).$$

Badoiu et al. [1] considered the embedding problem assuming multiplicative errors on edges and additive errors on angles. The naïve setup of the problem is nonconvex and difficult to solve. The following shows how to relax the problem to a convex problem at the cost of some error ε [1].

Consider a distance geometry problem with orientational restraints, where the angle of an edge (p, q) with the x-axis is given up to angular error γ. As shown in figure 36.1, the distance and angle information for the edge (p, q) specifies a nonconvex feasible region for q, given a fixed point p. We can add an edge (a, b) and relax the feasible region to be convex. Next, a linear program can be used to solve the problem in polynomial time. The maximum expansion error is, at most, $(1+\varepsilon)/\cos\gamma$ after computing the maximum distance in the cutoff region.

Exercise 36.1   Compare the approximation algorithm above to the lower bounds in chapter 32 and [6], which prove there is an obstruction to obtaining an approximation for graph embedding without orientational constraints. How does the addition of this orientation information enable a provably good, polynomial-time approximation algorithm? In protein structure determination, how could such information be measured experimentally?

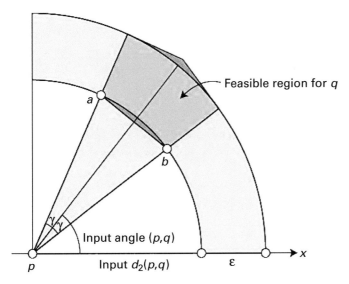

Figure 36.1
The relaxation in the embedding problem. Feasible region of a point q with respect to p given the ℓ_2 distance within a multiplicative error ε and given the angle to the x axis within an additive angular error γ [1].

36.2.2 Protein Structure Determination from RDCs

We close with an open problem. Let us review RDC-based protein structure determination (chapters 15–18). Given a bond vector v, its corresponding RDC r can be measured by

$$r = D_{\max} v^T S v,$$

where $D_{\max}$ is a physical constant, and S is the alignment tensor.

Let (x, y, z) be the coordinates of bond vector v. We have

$$r = S_{xx} x^2 + S_{yy} y^2 + S_{zz} z^2,$$

where S_{xx}, S_{yy}, and S_{zz} are the three diagonal elements of S in the principal order frame.

Suppose we are given RDCs for a bond vector v in two independent media, then we have the following relations:

$$r = S_{xx} x^2 + S_{yy} y^2 + S_{zz} z^2,$$
$$r' = S'_{xx} x'^2 + S'_{yy} y'^2 + S'_{zz} z'^2,$$
$$\begin{pmatrix} x' \\ y' \\ z' \end{pmatrix} = R_{12} \begin{pmatrix} x \\ y \\ z \end{pmatrix},$$

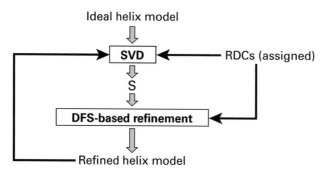

Figure 36.2
The computation of alignment tensors.

where the third equation stands for the transformation relationship (a rigid rotation) between the two alignment tensors. This system is solvable, and by symmetry, there are up to 8 possible solutions (see chapters 15–18).

Suppose we are given N-H RDCs in two media for a peptide chain. We can compute all possible N-H bond vectors according to the preceding equations. If we can fix the position of the first peptide plane, there are two possible (ϕ, ψ) angles for each pair of bond vectors in two adjacent peptide planes, based on trigonometry and inverse kinematics computation. Thus, we have a discrete and finite set of all possible conformations for this peptide chain. However, two issues still remain to be addressed: First, how do we fix the first peptide plane, and second how do we compute the alignment tensors S_1 and S_2. In [7], an ideal helix model is initially used to represent a helix in the protein structure, as shown in figure 36.2. Then an initial alignment tensor is calculated using SVD to fit $r_i \approx D_{\max} v_i^T S v_i$, that is, minimize $\sum |r_i - D_{\max} v_i^T S v_i|$. Based on the computed alignment tensor, a DFS-based method is used to search for the structure, based on which we can recompute the alignment tensor. The orientation of the first peptide plane can be obtained based on the computed alignment tensor and two known bond vectors, N-C$^\alpha$ and N-H in the plane.

A divide-and-conquer strategy is used to determine the overall protein structure. We first divide the whole protein into several fragments (according to secondary structure elements). The conformation of each fragment can be computed based on the preceding protocol. Finally, we assemble all fragments by pure translations based on sparse NOE distance constraints.

Time Complexity   Let n denote the number of residue in the protein. The dividing step takes $O(n)$ time. The structure determination of each fragment takes $O(\frac{n}{m} \exp(m))$, where m is the maximum length of fragments. The assembling step runs in $O(n)$ time. Since $m = O(1)$ [8], the total running time is $O(n)$. This algorithm is therefore optimal in terms of the geometric and combinatorial complexity n. See chapters 15 through 18 and Wang et al. [8].

Open Problem RDCs are a geometric measurement similar to the angular restraints with error γ in figure 36.1 and section 36.2.1. Use SDP and the concept of distance geometry with angle restraints to model RDC-based structure determination. The angular restraints come from RDCs, and the distance restraints from NOEs. What is the best algorithm you can derive, in terms of complexity? Implement and test your algorithm, using both synthetic and real NMR data.

References

[1] M. Badoiu, and E. D. Demaine, M. T. Hajiaghayi, and P. Indyk. Low-dimensional embedding with extra information. *Proceedings of the Twentieth Annual Symposium on Computational Geometry,* 2004.

[2] P. Biswas and T. Lian and T. Wang, and Y. Ye. Semidefinite Programming based algorithms for sensor network localization. *ACM Transactions on Sensor Networks (TOSN),* (2006).

[3] Bernard Chazelle, Carl Kingsford, and Mona Singh. A semidefinite programming approach to side-chain positioning with new rounding strategies. *INFORMS Journal on Computing* 16;4(2004): 380–392.

[4] R.M. Freund, Introduction to Semidefinite Programming. Sloan School, MIT, DSPACE online publication (1999) accessed: 2009, url: http://dspace.mit.edu/bitstream/handle/1721.1/35259/15-094Spring-2002/NR/rdonlyres/Sloan-School-of-Management/15-094Systems-Optimization–Models-and-ComputationSpring2002/A849A5EB-FBF7-4631-8B52-CBBC667E74EB/0/sdpintro.pdf

[5] N. A. Pierce and E. Winfree. Protein design is NP-hard. *Protein Engineering,* 15;10(2003):779–782, 2002.

[6] J. B. Saxe. Embeddability of weighted graphs in k-space is strongly NP-hard. *Proceedings of the 17th Allerton Conference on Communications, Control, and Computing,* (1979):480–489.

[7] L. Wang and B. R. Donald. Exact solutions for internuclear vectors and backbone dihedral angles from NH residual dipolar couplings in two media, and their application in a systematic search algorithm for determining protein backbone structure. *Journal of Biomolecular NMR* 29;3(2004):223–242.

[8] L. Wang, R. R. Metta, and B. R. Donald. A polynomial-time algorithm for *de novo* protein backbone structure determination from nuclear magnetic resonance data. *Journal of Computational Biology* 13;7(2006):1267–1288.

Graph cuts have numerous applications in computational structural biology. In this chapter, we introduce a generalized Potts energy model in the Markov Random Field (MRF) framework, and present a multiway graph cut algorithm to find the minimum energy [1]. Then, we present an α-expansion move algorithm to compute visual correspondence with occlusions [2]. Graph cuts provide an algorithm for a wide variety of assignment problems.

Some of these assignment problems first arose in machine vision. We review these vision problems since image processing arises often in computational biology. However, graph cuts also have broader applications, as we shall see. This chapter builds on the MRF concepts introduced in chapter 25.

37.1   Construction of the Energy Function

Markov Random Fields (MRFs)   An image is an array of pixels connected in a grid topology. Suppose that we are given two images. A pixel in one image is said to *correspond* to a pixel in the other if these two pixels are projections along lines of sight from the same physical scene element [2]. In the visual correspondence problem, we can regard the pixels in the second image as *labels*, and our goal is to compute a "label" f_p, which denotes a disparity value, for each pixel p in the primary image. Let f_p denote the disparity at pixel p, and $\mathcal{N}_p$ denote the set of the neighbors of p. In a Markov Random Field (MRF) framework, we have

$$\Pr(f_p|f_q, p \neq q) = \Pr(f_p|f_q, q \in \mathcal{N}_p).$$

Hammersley-Clifford Theorem   Let $f = (f_1, \ldots, f_m)$ denote a *configuration* of all m pixels $\mathcal{P} = \{1, \ldots, m\}$. If we focus on the MRF framework, with a clique potential V, we have

$$\Pr(f) \propto \exp\left(-\sum_{p \in \mathcal{P}} \sum_{q \in \mathcal{N}_p} V_{(p,q)}(f_p, f_q)\right). \tag{37.1}$$

MAP Estimation of MRF Configuration   We must find the optimal configuration f^* such that

$$f^* = \operatorname*{argmax}_{f} \Pr(f|O),$$

where O denotes the observed data. By Bayes' rule, we have

$$f^* = \underset{f}{\operatorname{argmax}} \Pr(O|f) \cdot \Pr(f),$$

where $\Pr(O|f) \cdot \Pr(f)$ is also called the *posterior probability*; $\Pr(f)$ is the prior probability, obtained from equation (37.1), and $\Pr(O|f) = \prod_{p \in \mathcal{P}} g_p(O|f_p)$ is called the *likelihood probability*,[1] which also represents the sensor noise model.

Energy Minimization   We wish to find a configuration f that minimizes the *posterior energy function*:

$$E(f) = -\sum_p \ln g_p(O|f_p) + \sum_{(p,q)} V_{(p,q)}(f_p, f_q). \tag{37.2}$$

The Generalized Potts Model   In a *Generalized Potts Model MRF* (GPM-MRF), a clique potential for any pair of neighboring pixels p and q is defined by

$$V_{(p,q)}(f_p, f_q) = u_{\{p,q\}} \cdot \delta(f_p \neq f_q),$$

where $\delta(f_p \neq f_q) = 1$ if $f_p \neq f_q$, $\delta(f_p \neq f_q) = 0$ otherwise, and $u_{\{p,q\}}$ denotes the penalty for discontinuity at pair (p, q). The posterior energy function of a GPM-MRF is

$$E(f) = -\sum_p \ln g_p(O|f_p) + 2\sum_{(p,q)} u_{\{p,q\}} \cdot \delta(f_p \neq f_q). \tag{37.3}$$

37.2   Optimizing the Energy Function by Graph Cuts

37.2.1   Graph Construction

Given a set of pixels $\mathcal{P} = \{1, \ldots, m\}$ and a set of labels $\mathcal{L} = \{l_1, \ldots, l_k\}$, we construct a graph $G = (V, E)$ such that $V = \mathcal{P} \bigcup \mathcal{L}$ and the edge set E is constructed in the following way. Any two neighboring vertices in $\mathcal{P}$ are connected by an edge, called an *n-link*. Each *n*-link between pixels p and q is assigned a weight:

$$w_{\{p,q\}} = 2u_{\{p,q\}}.$$

A vertex p in $\mathcal{P}$ is connected to a vertex l in $\mathcal{L}$ if $l \in \mathcal{L}_p$. Each edge (p, l) is called *t-link*, which is assigned a weight:

$$w_{\{p,l\}} = \ln g_p(O|l) + K(p) + \sum_{q \in \mathcal{N}_p} w\{p, q\},$$

where $K(p)$ denotes some constant for each pixel p.

---

[1] Here, g is the sensor noise distribution.

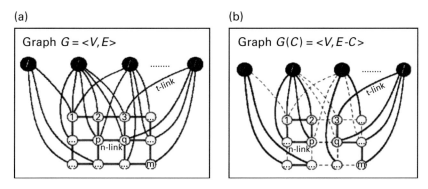

Figure 37.1 (plate 41)
An example of multiway graph cut. Dashed lines in (b) denote the cut edges in the graph [1]. Copyright 1998 IEEE.

Figure 37.1a shows an example of such a constructed graph.

37.2.2   The Multiway Cut Formulation

A subset of edges $C \subset E$ in a graph $G = (V, E)$ is a *multiway cut* if the terminals (label vertices) are separated in the induced graph $G = (V, E - C)$. Figure 37.1 (plate 41) shows a multiway cut example in a graph. The cost of a cut C, denoted by $|C|$, is equal to the sum of weights of all cut edges. The *multiway cut problem* is to find a multiway cut with the minimum cost.

Theorem   *If C is a multiway cut with the minimum cost, then the corresponding configuration f^C (see section 37.3.3) minimizes $E(f)$ in equation (37.3).*

37.2.3   The Multiway Cut Algorithm

A special case of the multiway cut problem, in which there are only two terminals, is a standard min-cut problem and can be solved in polynomial time using the max-flow (Ford-Fulkerson) algorithm. The general multiway cut problem in which there are more than two labels is NP-complete. Currently, efficient approximation algorithms exist that are optimal within a factor of 2. The following algorithm (algorithm 1) is applied to find the minimum cut for a given graph:

Algorithm 1

1: Start with an arbitrary multiway cut C from any pair of terminals.
2: **while** C improves **do**
3:      **for** all pairs of labels l and λ **do**
4:          Find s-t min cut on $G_{\{l,\lambda\}}$.
5:          Update C if a better new cut is found.
6:      **end for**
7: **end while**
8: Return C.

318	37 Graph Cuts for Energy Minimization and Assignment Problems

37.3 Graph Cuts for Computing Visual Correspondence with Occlusions

37.3.1 Notation

Computing Visual Correspondence Suppose that we are given two images. A pixel in one image is said to *correspond* to a pixel in the other if these two pixels are projections along lines of sight from the same physical scene element [2]. In the visual correspondence problem, we can regard the pixels in the second image as *labels*, and our goal is to compute a "label" f_p, which denotes a disparity value, for each pixel p in the primary image. In other words, the visual correspondence problem can be formulated into a GPM-MRF. The reader is referred to section 4 in Boykov et al. [1] for more implementation details.

Occlusion Problem In a visual correspondence problem, some pixels may be visible only in one image, but occluded in the other one. Hence, there is no correspondence for these pixels (or "labels") in the other image.

α-Expansion Move Let $d(a) = (q_x - p_x, q_y - p_y)$ denote the *disparity* of an assignment $a = \langle p, q \rangle$. In an *α-expansion move operation*, some currently active assignments are deleted, and some assignments with disparity α are added.

37.3.2 Energy Function

The basic idea of designing the energy function here is to construct an energy model on assignments so that the graph cut algorithms can be applied. Here each vertex in a graph represents an *assignment* between two corresponding pixels, while in the previous section (37.2), each vertex in the graph denotes a pixel in the image.

Given a configuration f of pixel assignments (or correspondences) between two images, the energy function incorporating the occlusion correspondences is defined by

$$E(f) = E_{data}(f) + E_{occ}(f) + E_{smooth}(f), \tag{37.4}$$

where E_{data} denotes the intensity difference between corresponding pixels, and E_{occ} denotes a penalty for an occluded pixel, and E_{smooth} denotes a smoothness between two neighboring assignments. More specifically, these three terms are designed as follows (we use the same terminology as in section 3.2 from Kolmogorov and Zabih [2]):

1. $E_{data} = \sum_{a \in A(f)} D(a)$, where $D(a) = (I(p) - I(q))^2$ denotes the disparity of an assignment $a = \langle p, q \rangle$, and $I(p)$ denotes the intensity of pixel p.
2. $E_{occ}(f) = \sum_{\text{all pixels } p} C_p \cdot T(|N_p(f)| = 0)$, where C_p denotes a penalty if pixel p is occluded, and $N_p(f)$ denotes the set of active assignments involving pixel p, and $T(\cdot)$ is 1 if its argument is true, and 0 otherwise.
3. $E_{smooth}(f) = \sum_{\{a_1, a_2\} \in \mathcal{N}} V_{a_1, a_2} \cdot T(f(a_1) \neq f(a_2))$, where $\mathcal{N}$ denotes the neighborhood system.

1. Start with an arbitrary unique configuration f

2. Set success := 0

3. For each disparity α

    3.1. Find $\hat{f} = \arg \min E(f')$ among unique f' within single α-expansion of f

    3.2. If $E(\hat{f}) < E(f)$, set $f := \hat{f}$ and success := 1

4. If success = 1 goto 2

5. Return f

Figure 37.2
The α-expansion move algorithm [2].

37.3.3 The α-Expansion Move Algorithm

Overview of the Algorithm The algorithm to find the configuration with the minimum energy is described in figure 37.2. The key step in the expansion algorithm is to find the configuration $\hat{f}$ with smallest energy within a single α-expansion move. This step is implemented by constructing a weighted graph such that the minimum graph cut is equal to the configuration with the minimum energy. The following section shows how to construct the desired graph and compute the minimum cut.

Graph Construction Let f^0 denote the starting configuration. Let $\widetilde{A} = A^0 \bigcup A^\alpha$ denote the set of all possible assignments within one α-expansion move, where $A^0 = \{a \in A(f^0) \mid d(a) \neq \alpha\}$ and $A^\alpha = \{a \in A \mid d(a) = \alpha\}$. Now we construct a directed graph where nodes are all assignments in $\widetilde{A}$ plus two terminal nodes s and t, and edges are constructed according to the following rules:

1. For every vertex $a \in \widetilde{A}$, edges (s, a) and (a, t) are constructed.
2. Edges (a_1, a_2) and (a_2, a_1) are constructed if $\{a_1, a_2\} \in \mathcal{N}$.
3. Edges exist between assignments $a_1 = \langle p, q \rangle$ and $a_2 = \langle p, r \rangle$, where $q \neq r$.

Figure 37.3 shows an example of a constructed graph. The weights assigned for all edges are shown in table 37.1.

Graph Cuts Given a cut $\mathcal{C} = \mathcal{V}^s, \mathcal{V}^t$ on $\mathcal{G}$, the configuration f^C corresponding to the cut $\mathcal{C}$ is defined by

$$\forall a \in A^0 \quad f_a^C = \begin{cases} 1, & \text{if } a \in \mathcal{V}^s \\ 0, & \text{if } a \in \mathcal{V}^t \end{cases}$$

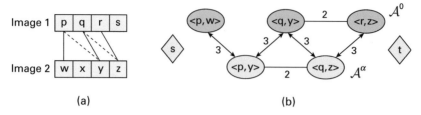

(a)   (b)

Figure 37.3
An example of a constructed graph to find the minimum energy within one α-expansion move. (a) $\mathcal{L} = \{p, q, r, s\}$ and $\mathcal{R} = \{w, x, y, x\}$ denote two original images. Solid lines represent the current active assignments, and dashed lines represent the assignments after an α-expansion move (here $\alpha = 2$). (b) Vertices in the top row represent the assignments in set $\mathcal{A}^0$, and vertices in the bottom row represent the assignments in $\mathcal{A}^\alpha$. The number associated with each edge represents the rule order applied. The edges from rule 1 are not shown here [2]. Copyright 2001 IEEE.

Table 37.1
Assigned weights for all edges [2]

Edge	Weight	For
(s, a)	$D_{occ}(a)$	$a \in \mathcal{A}^0$
(a, t)	$D_{occ}(a)$	$a \in \mathcal{A}^\alpha$
(a, t)	$D(a) + D_{smooth}(a)$	$a \in \mathcal{A}^0$
(s, a)	$D(a)$	$a \in \mathcal{A}^\alpha$
$(a1, a2), (a2, a1)$	$V_{a1,a2}$	$\{a1, a2\} \in \mathcal{N}, a1, a2 \in \tilde{\mathcal{A}}$
$(a1, a2)$	∞	$p \in \mathcal{P}, a1 \in \mathcal{A}^0, a2 \in \mathcal{A}^\alpha, a1, a2 \in N_p(\tilde{f})$
$(a2, a1)$	C_p	$p \in \mathcal{P}, a1 \in \mathcal{A}^0, a2 \in \mathcal{A}^\alpha, a1, a2 \in N_p(\tilde{f})$

$$\forall a \in \mathcal{A}^\alpha \quad f_a^C = \begin{cases} 1, & \text{if } a \in \mathcal{V}^s \\ 0, & \text{if } a \in \mathcal{V}^t \end{cases}$$

$$\forall a \notin \mathcal{A}^0 \quad f_a^C = 0.$$

The following theorem shows that the minimum cut of a constructed graph is equal to the unique configuration with the minimum energy within one α-expansion move.

Theorem [2]   *Let C be the minimum cut on $\mathcal{G}$. Then f^C is the unique configuration that minimizes the energy E within one α-expansion move of f^0.*

The minimum cut of the graph can be computed by using the Ford-Fulkerson algorithm in polynomial time.

Comments   What is the difference between the graph cut algorithm and the other methods for assignment, such as the EM or bipartite matching approach?

Exercise The pixel assignment problem can be generalized to the NMR resonance and NOE assignment problems by first generalizing the grid network to have a more general graph neighborhood topology. What is the best algorithm you can obtain for these NMR problems? Analyze the complexity of your algorithm. Compare to EM and bipartite matching. Will graph cuts be more, or less robust than bipartite matching? Explain. Implement and test your algorithm, using both synthetic and real NMR data.

References

[1] Yuri Boykov, Olga Veksler, Ramin Zabih. Markov Random fields with efficient approximations. *Proceedings of IEEE Conference on Computer Vision and Pattern Recognition* (CVPR) (1998). Santa Barbara, CA.

[2] V. Kolmogorov and R. Zabih. Computing visual correspondence with occlusions using graph cuts. *Proceedings of the Eighth International Conferrence on Computer Vision (ICCV-01) 2001, Vancover, British Columbia.*

38 Classifying the Power of Graph Cuts for Energy Minimization

We now continue our discussion of graph cuts, which were introduced in the previous chapter. In this chapter, we discuss an energy minimization framework using a spatially coherent clustering of feature space, which is a popular approach to image segmentation in the field of computer vision [1]. We then discuss the use of graph cut techniques for energy minimization, and finally give a characterization of the energy functions that can be minimized efficiently using graph cuts [2]. Energy minimization problems are ubiquitous in computational biology, and graph cuts provide an elegant and efficient paradigm that has already proven useful in computer vision, and which shows significant promise in structural biology and biophysics. We attack assignment problems since these energy minimization problems arise naturally in image processing and NMR.

38.1 Feature Space Clustering

Feature Space A feature space is characterized by a set of pixels where with each pixel a set of local properties (called "features") is associated. Features such as intensity, texture, or motion are the commonly studied parameters. Since these properties are local properties, the feature vector of each pixel is a vector of local properties in a high dimensional space.

Feature Space Clustering Since significant features are shared by many pixels, the feature space is clustered and each pixel is labeled with the cluster that contains its feature vector. This is a popular approach for image segmentation problems arising in computer vision.

Spatial Coherence Since a good cluster in feature space is not necessarily coherent in image space, and the correct segmentation in image space may not correspond to a highly distinctive group of feature vectors, it is important that feature space clustering/segmentation algorithms should yield spatially coherent clusters (i.e., corresponding to compact groupings of pixels).

38.2 Energy Minimization Framework for Feature Space Clustering

Feature space clustering can be viewed as the following *pixel labeling* problem in feature space.

38.2.1   The Pixel Labeling Problem

Input   The input image has a set of pixels $\mathcal{P} = \{1, 2, \ldots, n\}$.

$\mathcal{X}$ is the feature space with d dimensions.

Each pixel $p \in \mathcal{P}$ corresponds to a feature vector $x_p \in \mathcal{X}$.

The feature space is summarized by a set of K models, such that each model k describes a cluster in $\mathcal{X}$ and has some model parameter θ_k.

Let $\theta = \{\theta_1, \ldots, \theta_k\}$ be the entire set of clusters.

Output   It is required to compute the parameter set θ and the labeling $f : \mathcal{P} \longrightarrow \{1, 2, \ldots, K\}$, that is, f assigns each pixel p to a model $f_p \in \{1, 2, \ldots, K\}$. Obviously, a cluster can be viewed as either a set of pixels $P_k(f) = \{ p \mid f_p = k \}$, or as a set of points in the feature space $X_k(f) = \{ x_p \mid p \in P_k(f) \}$.

The feature space clustering is done in such a way that the following criteria (quality measure) are satisfied:

• *Good cluster quality in feature space*, that is, low intracluster variation and/or high intercluster variation. Call this measure Q.

• *Spatial coherence in image space*, that is, the labeling f should be spatially coherent.

Let $Q(f, \theta) = $ cost of assigning the label f_p to the pixel p (called *assignment cost*).

Let $V(f_p, f_q) = $ cost of assigning the labels f_p and f_q to adjacent pixels p and q. (This is called *separation cost*.)

Let $\mathcal{N} \subset \mathcal{P} \times \mathcal{P}$ be a *neighborhood system* on pixels that incorporates spatial coherence.

Then, the pixel labeling problem can be formulated as an energy minimization problem where the energy function has the following form:

$$E(f, \theta) = Q(f, \theta) + \sum_{p,q \in \mathcal{N}} V(f_p, f_q). \tag{38.1}$$

Note that here Q is the penalty of assigning a particular pixel to a particular cluster and V imposes the penalty for assigning two different clusters to two neighboring pixels. V is a robust metric in the space of clusters.

The goal is to minimize the overall penalty, that is, to minimize the energy function in equation (38.1).

38.2.2   An EM-Style Energy Minimization Algorithm

We now give an expectation/maximization (EM) algorithm for this energy minimization problem. The energy function in equation (38.1) can be minimized iteratively as follows:

Step 1: E Step   Fix the cluster parameter θ and find the best labeling f.

Step 2: M Step   Fix the labeling f and find the best clusters θ.

Step 3: Repeat until some convergence criterion is satisfied.

Fixing the cluster parameter θ, the energy function looks like

$$E(f) = \sum_{p \in \mathcal{P}} D_p(f_p) + \sum_{p,q \in \mathcal{N}} V(f_p, f_q), \tag{38.2}$$

where $D_p(f_p)$ is the assignment cost of the pixel p and is related to Q by the following equation:

$$Q(f) = \sum_{p \in \mathcal{P}} D_p(f_p). \tag{38.3}$$

Fixing the clusters θ, if for the function Q, equation (38.3) is satisfied for some D_p, then Q is called *linear*. Equation (38.3) is referred to as the *linearity criterion*.

The energy function in equation (38.2) can be minimized efficiently for a wide range of choices of V using the *α-expansion move* algorithm (refer to chapter 37), which is based on the idea of graph cuts.

The M step is done by fixing the labeling f and then optimizing for the cluster parameters θ. Following are different clustering methods that are used here:

Parametric Clustering Method   Here the idea is to *maximize the posterior probability*

$$\Pr(f, \theta | x) \sim \Pr(x | f, \theta) \cdot \Pr(f),$$

where f and θ are assumed to be independent and θ has a uniform prior distribution. The *likelihood term* on f is given by the expression $\Pr(x | f, \theta) = \prod_{p \in \mathcal{P}} \Pr(x_p | \theta_{f_p})$ and the *prior term* is given by $\Pr(f) = \exp\left(-\sum_{p,q \in \mathcal{N}} V(f_p, f_q)\right)$. Taking the negative logarithm, the clustering objective function becomes

$$Q(f, \theta) = \sum_{p \in \mathcal{P}} -\log \Pr(x_p | \theta_{f_p}), \tag{38.4}$$

which meets the linearity criterion. Note that this model of clustering is a generalization of the finite mixture model (FMM) and is justified under the Markov Random Field (MRF) framework.

Nonparametric Clustering Method   Here the idea is to represent a cluster k as a set of points in the feature space X_k.

Let $\Pr(x | X_k)$ be an estimate of the distribution in the cluster k.

Then we can write the following expression for Q in light of equation (38.4):

$$Q(f, \theta) = \sum_{p \in \mathcal{P}} -\log \Pr(x_p | X_{f_p}(f)), \tag{38.5}$$

where (as we have already defined) $X_k(f)$ is the set of features of pixels with the label k under the labeling f.

It is important to note that although this clustering method does not meet the linearity criterion (equation 38.3), it works well in practice with the α-expansion move algorithm.

38.3   Approaches to Incorporating Spatial Coherence

Several approaches have been proposed to incorporate positional information as a feature in the feature space analysis. They can be summarized as follows:

Super Pixel Approach   The idea is to preprocess and compute feature vectors from the preprocessed segments, rather than computing them directly from pixels. This approach fails when superpixels cross the boundaries of object.

Extensions to EM Style Algorithm   The idea is to use an EM style algorithm with additional constraints. This approach suffers from not being able to impose spatial coherence efficiently.

Soft Membership Clustering   Here the feature vectors are labeled with a probability distribution over clusters rather than hardwired to one cluster, an approach justified within the framework of Markov Random Fields. This makes the M step more effective. But the drawback of this approach is that computing soft membership is computationally intractable under MRF.

Use of Hard Membership with Graph Cuts   This is the approach described in Zabih and Kolmogorov [1]. Hard membership implies that each feature vector belongs to exactly one cluster. The advantage with this approach is that it facilitates the use of graph cut–based energy minimization methods, which have tremendous success in energy minimization in the context of pixel labeling problems in stereo vision.

38.4   Classifying Energy Functions that Can Be Minimized Efficiently Using Graph Cuts

Consider the study of computational problems in early vision, which can often be expressed as an energy minimization problem over a very high dimensional space. In general, the energy minimization function can be a nonconvex function, which is often very difficult to minimize. Recently, graph cut–based techniques have been proved to be very useful and can minimize the energy functions to give solutions with promising theoretical bounds. A natural question comes to mind: "How can we classify energy functions that can be efficiently minimized by using the graph cut technique?" The paper [2] answers this fundamental question, by classifying energy functions according to their ability to be represented as a graph such that they can be minimized efficiently. A general-purpose graph construction is also given for each class of energy functions, that minimizes any energy function in that class.

In computational biology, energy minimization is a ubiquitous, yet difficult and expensive task, which is a subroutine in many protocols. Hence, new insights about the properties of energy functions are valuable, and graph cuts provide new tools for their classification.

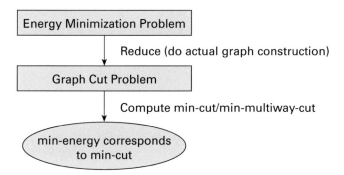

Figure 38.1
Schematic of the graph cut technique. Here, the cuts are labelings and the cost of a cut represents the energy. Credit: Chittu Tripathy and Bruce Donald.

38.4.1 Using Graph Cuts in Energy Minimization

A standard form of energy function is given by

$$E(f) = \sum_{p \in \mathcal{P}} D_p(f_p) + \sum_{p,q \in \mathcal{N}} V(f_p, f_q), \tag{38.6}$$

where $\mathcal{N} \subset \mathcal{P} \times \mathcal{P}$ defines the neighborhood system of interaction, $D_p(f_p)$ denotes the assignment cost function, and $V(f_p, f_q)$ is the separation cost function. This energy function was studied extensively in computer vision and the graph cut technique has been applied successfully to minimize such energy functions.

The general idea behind the graph cut technique is to construct a specialized graph for the energy function to be minimized, so that minimization of energy (locally or globally) corresponds to computing a minimum cut of the graph. Figure 38.1 illustrates the general framework for energy minimization using graph cuts.

38.5 Representation of Energy Functions by Graphs

Graph Representable Functions A function E of n binary variables is *graph representable* if there exists a graph $\mathcal{G} = (\mathcal{V}, \mathcal{E})$ with terminal vertices s and t and a subset of vertices $\mathcal{V}_0 = \{v_1, \ldots, v_n\} \subset \mathcal{V} - \{s, t\}$ such that, for any configuration $x_1, x_2, \ldots, x_n$, the value of the energy $E(x_1, x_2, \ldots, x_n)$ is equal to a constant plus the cost of the minimum s-t-cut among all cuts $C = S, T$ in which $v_i \in S$, if $x_i = 0$, and $v_i \in T$, if $x_i = 1$ ($i \leq i \leq n$). We say that E is *exactly represented* by $\mathcal{G}, \mathcal{V}_o$ if this constant is zero.

Note that the addition of a constant does not change the minimum energy configuration.

This definition leads to the following result. If the energy function E is graph representable by a graph $\mathcal{G}$ and a subset $\mathcal{V}_0$, then it is possible to find the exact minimum of E in polynomial time by computing the minimum s-t-cut on $\mathcal{G}$.

Two important classes of energy functions called $\mathcal{F}^2$ and $\mathcal{F}^3$ in terms of graph representability are discussed in the following sections, which provide general procedures to construct graphs.

38.6   The Class $\mathcal{F}^2$

The class $\mathcal{F}^2$ is a subset of energy functions where each individual term can be expressed as a function of at most two Boolean variables.

Theorem 38.1 ($\mathcal{F}^2$ Theorem)   *Let E be a function of n binary variables from the class $\mathcal{F}^2$, i.e.,*

$$E(x_1, \ldots, x_n) = \sum_i E^i(x_i) + \sum_{i<j} E^{i,j}(x_i, x_j). \tag{38.7}$$

Then E is graph representable if and only if each term $E^{i,j}$ satisfies the following condition, called the regularity *condition:*

$$E^{i,j}(0, 0) + E^{i,j}(1, 1) - E^{i,j}(0, 1) - E^{i,j}(1, 0) \leq 0. \tag{38.8}$$

Note that functions of one Boolean variable are always regular.

38.6.1   Graph Construction for $\mathcal{F}^2$

Construct a graph $\mathcal{G} = (\mathcal{V}, \mathcal{E})$ as follows:

- The vertex set $\mathcal{V} = \{v_1, \ldots, v_n, s, t\} = S \cup T \cup \{s, t\}$, where v_i encodes the Boolean variable x_i, such that $x_i = 0$ if $v_i \in S$, and $x_i = 1$ if $v_i \in T$.
- We write $edge(v, c)$ to mean an edge (s, v) with the weight c if $c > 0$ or an edge (v, t) with the weight $-c$ if $c < 0$.
- For each term $E^i(x_i)$, add an edge $(v_i, E^i(1) - E^i(0))$. Refer to figure 38.2 a and b.
- As figure 38.2 c, d, and e shows, for each term $E^{i,j}(x_i, x_j)$ add the following three edges:
- · $edge(v_i, E^{i,j}(1, 0) - E^{i,j}(0, 0))$,
- · $edge(v_j, E^{i,j}(1, 1) - E^{i,j}(1, 0))$, and
- · (v_i, v_j) with weight $-\pi(E^{i,j}) = E^{i,j}(0, 1) + E^{i,j}(1, 0) - E^{i,j}(0, 0) - E^{i,j}(1, 1)$. The function $\pi(\cdot)$ is called the *functional*, which we formally define in section 38.7. Note that for regular functions of two variables, $\pi(E^{i,j}) \leq 0$.
- Once the graph for each individual term is constructed, *merge* the graphs together to get the final graph. The merging is justified by the *additivity theorem*, which states that the sum of two graph representable functions is graph representable.

Each edge weight is made nonnegative by construction, which is important since we construct the graph to finally compute a min-cut of the graph.

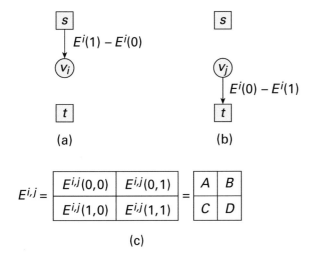

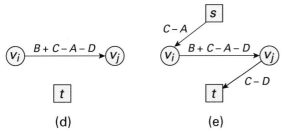

Figure 38.2
Graph cuts for functions $\mathcal{F}^2$. (a) Shows the graphs for E^i where $E^i(1) > E^i(0)$; (b) shows the graphs for E^i where $E^i(1) \leq E^i(0)$; in (c), the table shows the individual terms for the two variable function $E^{i,j}$; in (d) we show the third edge for $E^{i,j}$; (e) represents the complete graph for $E^{i,j}$ when $C > A$ and $C > D$. Note that (e) shows one of the four possibilities [2].

38.6.2 NP-Hardness of General E^2 Functions

Theorem 38.2 *Let E^2 represent an energy function of two variables that is not regular. Then, minimization of functions of the form*

$$E(x_1, \ldots, x_n) = \sum_i E^i(x_i) + \sum_{(i,j) \in \mathcal{N}} E^2(x_i, x_j), \tag{38.9}$$

where E^i are arbitrary functions of one variable and $\mathcal{N} \subset \big\{ (i, j) \;\big|\; 1 \leq i < j \leq n \big\}$, is NP-hard.

38.7   The Class $\mathcal{F}^3$

Theorem 38.3 ($\mathcal{F}^3$ Theorem)   *Let E be a function of n binary variables from the class $\mathcal{F}^3$, that is,*

$$E(x_1, \ldots, x_n) = \sum_i E^i(x_i) + \sum_{i<j} E^{i,j}(x_i, x_j) + \sum_{i<j<k} E^{i,j,k}(x_i, x_j, x_k). \tag{38.10}$$

Then E is graph representable if and only if E is regular.

The regularity condition in case of functions of three Boolean variables will be stated after introducing a few more concepts. We define the *functional* π of an energy function $E(x_1, \ldots, x_n)$ to be a mapping from the set of all functions of Boolean variables to the set of all real numbers given by the following equation:

$$\pi(E(x_1, \ldots, x_n)) = \sum_{x_1, \ldots, x_n \in \{0,1\}} (\Pi_{i=1}^n (-1)^{x_i}) E(x_1, \ldots, x_n). \tag{38.11}$$

For example, $\pi(E) = E(0, 0) + E(1, 1) - E(0, 1) - E(1, 0)$ for a two variable function E. The functional π is linear, *i.e.*, $\pi(E_1 + E_2) = \pi(E_1) + \pi(E_2)$.

For a three-variable function $E^{i,j,k}$ we compute the functional as follows. We take all projections of two variables of $E^{i,j,k}$ (a projection means fixing the value of one variable and then computing $E^{i,j,k}$, which can now be thought of as a function of two variables) with positive values of functional π. If we don't find such a projection, then $E^{i,j,k}$ is regular.

38.7.1   Graph Construction for $\mathcal{F}^3$
Construct a graph $\mathcal{G} = (\mathcal{V}, \mathcal{E})$ as follows:

- The vertex set $\mathcal{V} = \{v_1, \ldots, v_n, s, t\} = S \cup T \cup \{s, t\}$, where v_i encodes the Boolean variable x_i, such that $x_i = 0$ if $v_i \in S$, and $x_i = 1$ if $v_i \in T$.
- We write $edge(v, c)$ to mean an edge (s, v) with the weight c if $c > 0$ or an edge (v, t) with the weight $-c$ if $c < 0$.
- For the terms in the forms $E^i(x_i)$ and $E^{i,j}(x_i, x_j)$, the edges are added as explained in the construction of the graph for $\mathcal{F}^2$. Refer to figure 38.2. We now construct the part of the graph where each term involves three Boolean variables, that is, the terms of the form $E^{i,j,k}(x_i, x_j, x_k)$.
- For each term $E^{i,j,k}(x_i, x_j, x_k)$, add one extra vertex u_{ijk} in addition to the vertices v_i, v_j, v_k, s and t. If $\pi(E^{i,j,k}) \geq 0$, then the following edges are added (see figure 38.3 b):
  - $edge(v_i, E^{i,j,k}(1, 0, 1) - E^{i,j,k}(0, 0, 1))$,
  - $edge(v_j, E^{i,j,k}(1, 1, 0) - E^{i,j,k}(1, 0, 0))$,
  - $edge(v_k, E^{i,j,k}(0, 1, 1) - E^{i,j,k}(0, 1, 0))$,
  - (v_j, v_k) with weight $-\pi(E^{i,j,k}[x_1 = 0])$,
  - (v_k, v_i) with weight $-\pi(E^{i,j,k}[x_2 = 0])$,

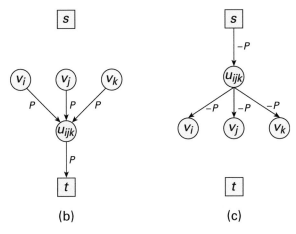

$$E^{i,j,k} = \begin{array}{|c|c|}\hline E^{i,j}(0,0,0) & E^{i,j}(0,0,1) \\\hline E^{i,j}(0,1,0) & E^{i,j}(0,1,1) \\\hline E^{i,j}(1,0,0) & E^{i,j}(1,0,1) \\\hline E^{i,j}(1,1,0) & E^{i,j}(1,1,1) \\\hline\end{array} = \begin{array}{|c|c|}\hline A & B \\\hline C & D \\\hline E & F \\\hline G & H \\\hline\end{array}$$

(a)

(b)   (c)

Figure 38.3
Graph cuts for functions $\mathcal{F}^3$. (a) Shows the terms depending on three variables x_i, x_j, x_k. Here $P = \pi(E^{i,j,k}) = (A + D + F + G) - (B + C + E + H)$. (b) Represents the graph for the case when $P \geq 0$. (c) Shows the graph for the case when $P < 0$ [2].

· (v_i, v_j) with weight $-\pi(E^{i,j,k}[x_3 = 0])$, and

· $(v_i, u_{ijk}), (v_2, u_{ijk}), (v_3, u_{ijk}), (u_{ijk}, t)$ with weight $\pi(E^{i,j,k})$.

If $\pi(E^{i,j,k}) \leq 0$ then we add the following edges (see figure 38.3 c):

· $edge(v_i, E^{i,j,k}(1, 1, 0) - E^{i,j,k}(0, 1, 0))$,

· $edge(v_j, E^{i,j,k}(0, 1, 1) - E^{i,j,k}(0, 0, 1))$,

· $edge(v_k, E^{i,j,k}(1, 0, 1) - E^{i,j,k}(1, 0, 0))$,

· (v_k, v_j) with weight $-\pi(E^{i,j,k}[x_1 = 1])$,

· (v_i, v_k) with weight $-\pi(E^{i,j,k}[x_2 = 1])$,

· (v_j, v_i) with weight $-\pi(E^{i,j,k}[x_3 = 1])$, and

· $(u_{ijk}, v_i), (u_{ijk}, v_2), (u_{ijk}, v_3), (s, u_{ijk})$ with weight $-\pi(E^{i,j,k})$.

• Once the graphs for each individual term is constructed, *merge* the graphs together to get the final graph. The merging is justified by the additivity theorem.

Observe that each edge weight is nonnegative by construction.

38.8   Comments

The graph construction technique used for the class of functions in $\mathcal{F}^2$ does not introduce any additional vertices in $\mathcal{F}^2$ beyond necessity, and in the case of functions in the class $\mathcal{F}^3$, only one extra vertex for every ternary interaction term is introduced. This construction yields much smaller graphs and hence the minimum cut can be computed efficiently.

The graph cut technique using the above graph construction procedure is very powerful when applied to stereo and multicamera scene reconstruction. The usefulness of the technique is strongly supported by the experimental results.

Since $\mathcal{F}^2 \subset \mathcal{F}^3$, the class $\mathcal{F}^3$ includes a larger class of functions that can be minimized using the graph cut framework.

Finally, it will be interesting to see how the graph cut technique can be applied in other areas of computational biology and molecular biophysics.

Exercise   Are the pairwise energy functions commonly used for protein design in class $\mathcal{F}^2$ or $\mathcal{F}^3$? Compare and contrast both classes of functions, to those used in protein design.

Exercise   Can you use graph cuts to model side-chain rotamer assignments, and thereby devise an efficient algorithm for computing the GMEC?

Exercise   Can you use graph cuts to model NMR resonance assignments, and thereby devise an efficient algorithm for structure-based assignment?

References

[1] Ramin Zabih and Vladimir Kolmogorov. Spatially coherent clustering using graph cuts. *Proceedings of IEEE Conference on Computer Vision and Pattern Recognition* (CVPR) 2, (2004):437–444.

[2] Vladimir Kolmogorov and Ramin Zabih. What energy functions can be minimizedvia graph cuts?. *IEEE Transactions on Pattern Analysis and Machine Intelligent*, 26;2(2004):147–159.

39 Protein Unfolding by Using Residual Dipolar Couplings

This chapter discusses ensemble-based computational models for unfolded proteins, and evaluates the effectiveness of these models by comparing the ensemble-averaged residual dipolar couplings (RDCs) and small-angle X-ray scattering (SAXS) data [2, 1].

39.1 Motivation and Overview

Natively unfolded regions have been shown to be common in functional proteins in a vast range of biochemical processes. In contrast to classical structural biology, unfolded proteins must be described by an *ensemble* of intermediate conformers. Many "traditional" NMR experiments, which have been widely used for structure determination in classical structural biology, are still feasible for the study of unstructured (or "intrinsically disordered") proteins in solution, since the NMR experimental data, such as chemical shifts and residual dipolar couplings (RDCs), can still be measured in highly disordered states, and thus used to characterize *average* properties over the ensemble of conformers. See chapter 7.

The methodology applied by Bernado et al. [1, 2] is shown in figure 39.1.

39.2 Ensemble Computation Using Only Local Sampling

We now give an algorithm to compute the conformational ensemble based only on local sampling of backbone dihedral angles, and then compare the computed ensemble-averaged RDCs with experimental data.

The algorithm [1] generates a statistical ensemble of conformations and calculates the average RDCs over the whole ensemble:

1. *Local random sampling of (ϕ, ψ) angles* Build sequential peptide chains using randomly chosen (ϕ, ψ) angles from a database of amino acid–specific conformations, which are obtained from loop regions of high-resolution X-ray structures.

2. *Steric clash checking* For each residue, check whether it has steric clash with another residue in the chain, or with the invariant folded fragment. The (ϕ, ψ) pair with steric overlap is rejected,

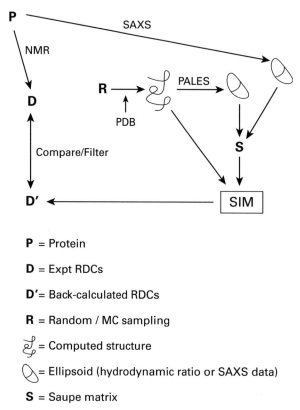

P = Protein

D = Expt RDCs

D′= Back-calculated RDCs

R = Random / MC sampling

= Computed structure

= Ellipsoid (hydrodynamic ratio or SAXS data)

S = Saupe matrix

Figure 39.1
Overview of methodology. The module SIM computes the ensemble average of the back-calculated RDCs, given a set of structures.

and a new (ϕ, ψ) angle is randomly selected until no overlap occurs. In total, $n = 50,000$ feasible conformers are generated.

3. *Ensemble-average RDC computing*   The alignment tensor for each complete conformer is first predicted based on the *hydrodynamic shape*. Then each RDC associated with a corresponding NH vector with respect to this tensor is back-computed (simulated). RDCs from each NH bond are averaged over all $n = 50,000$ conformers.

Figure 39.2 (plate 42) shows comparisons between the ensemble-average (simulated) and experimental RDCs for the αS protein (αS stands for *alpha-synuclein*). The computed ensemble-averaged RDCs fit the experimental data well in the central regions (residues 30–110), but there are significant deviations in the N- and C-terminal regions.

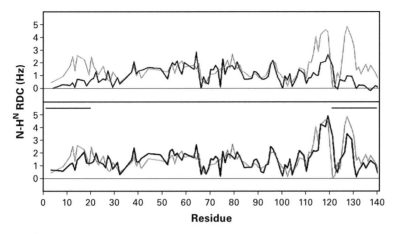

Figure 39.2 (plate 42)
Comparisons between ensemble-averaged RDCs (red) simulated without long-range contacts, and experimental RDCs (blue) for protein αS in (*a*) Pf1 bacteriophage and (*b*) lyotropic media. Simulated data are scaled to maximize fit in the region 22–112. For illustration purposes, the RDC sign is inverted compared to the usual conventions [1]. Reprinted with permission. Copyright 2005 American Chemical Society.

39.3 Ensemble Computation with Both Local Sampling and Long-Range Order

The ensemble computation by incorporating the long-range contacts between the N- and C-terminal domains improves the agreement between simulated and experimental RDCs.

The algorithm is similar to that in section 39.4.2 except that a new step is added to model the long-range contacts. The following shows the modified algorithm:

1. *Local random sampling of* (ϕ, ψ) *angles*   Similar to the corresponding step in section 39.2.

2. *Steric clash checking*   Similar to the corresponding step in section 39.2.

3. *Conformer selection based on long-range contacts*   The primary sequence is divided into several fragments, and two fragment positions are specified. When there is a long-range contact, each generated conformer is accepted only when a C^{β} atom at one specified fragment position is less than 15 Å from a C^{β} atom in the other specified domain.

4. *Ensemble-average RDC computing*   Similar to the corresponding step in section 39.2.

Figure 39.3 shows the differences between experimental and simulated RDCs for protein αS incorporating long-range contacts. In figure 39.3e (plate 43), the distribution of experimental RDCs closely agrees with that of simulated RDCs when the long-range contacts between the N- and C-terminal regions (residues 1–20 and 121–140) are considered.

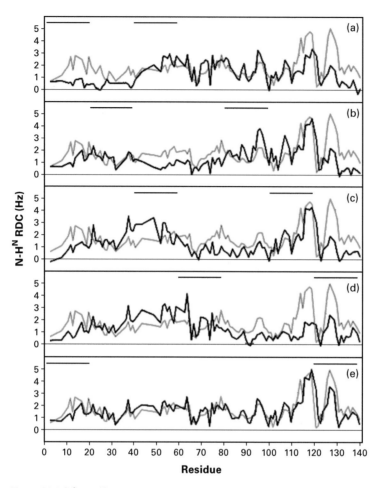

Figure 39.3 (plate 43)
Comparisons between experimental (blue) and simulated RDCs incorporating the long-range contacts (red). Contact regions are indicated by bars. (a) (1–20, 41–60); (b) (21–40, 81–100); (c) (41–60, 101–120); (d) (61–80, 121–140); (e) (1–20, 121–140) [1].

39.4   An Unfolded Protein Structure Model from RDCs and Small-Angle X-Ray Scattering (SAXS) Data

39.4.1   Generation of the Conformation Ensemble

We can generate the conformational ensemble of disordered regions of proteins by using a similar approach as in section 39.2, except that here we need to determine the first peptide unit position in the unfolded domain.

Determination of the First Peptide Unit in the Unfolded Domain   The orientation of the first peptide unit may be inconsistent with the relative orientations of folded domains. Thus, it is necessary to find the optimal orientation of the first peptide plane. In Bernado et al. [2], the first peptide plane is rotated over a grid on S^2, until the alignment tensor frames of the folded and unfolded domains are consistent with each other. The correct orientation of the first peptide plane is used to compute each new ensemble. Compare chapters 13, and 17, sections 17.2 and 17.3.

39.4.2   RDC Computation from the Conformational Ensemble

Prediction of the Alignment Tensor   The alignment tensor for each conformer is calculated by exploiting the similarity between the alignment and the radius of gyration. The shape of the RDC distribution (or a similar property comparable to the radius of gyration), denoted by F, can be regarded as a function of the alignment tensor. We must search for an alignment tensor such that the computed F fits the radius of gyration well. Based on the alignment tensor and the generated structure, the RDCs for each conformer can be computed. Compare chapters 13, 14.

Figure 39.4 (plate 44) shows the comparisons between simulated (ensemble-averaged) and experimental RDCs for protein PX. The RDC distribution, obtained from the ensemble computed from the above scheme, has good agreement with the distribution of experimental RDCs.

39.4.3   Prediction of SAXS Data from the Conformational Ensemble

SAXS data gives the radius of gyration of the protein, and is related to a size and shape of a molecular system. SAXS data may be back-computed as follows. First, side-chains are added to the unfolded domains using the SCCOMP program, and then the SAXS intensities are predicted using the CRYSOL program. After that, the radius of gyration is calculated from the predicted SAXS intensities based on the Debye equation [2].

Figure 39.5 shows the difference between simulated and experimental SAXS data. We see that the size and shape of the ensemble reproduce the SAXS data for the unfolded domains.

Protein Backbone Structure Determination Using Only RDCs   The following sketches the algorithm [3] to compute the protein backbone structure using only RDCs. A variation of this algorithm was used in the protein unfolding work described in the first part of this chapter.

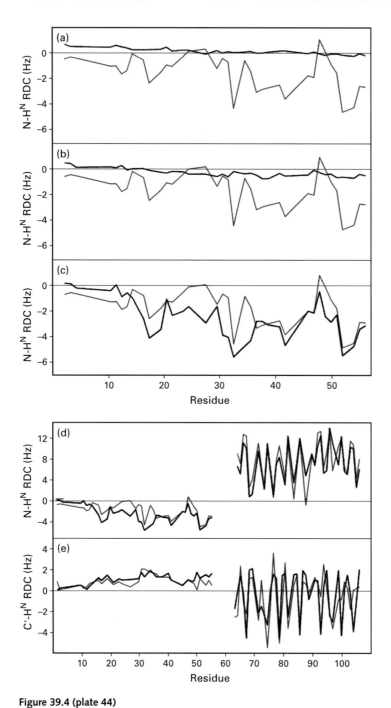

Figure 39.4 (plate 44)

Comparison between simulated (red) and experimental (blue) RDCs for protein PX. PX is numbered from 1 to 109 and contains the 14-residue His tag, such that residues 15–109 correspond to the region 474–568 in the intact C-terminal domain protein. (a) Experimental N-H^N RDCs compared with RDCs simulated in the unfolded domain of PX using random sampling of backbone dihedral angle (ϕ/ψ) space. In this model, (I) no nonbonded interactions were taken

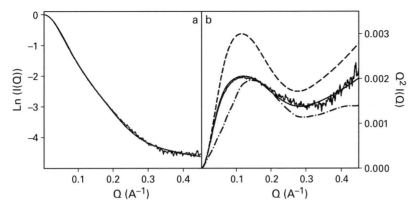

Figure 39.5
Comparisons between simulated and experimental SAXS data for protein PX. (a) Comparison of background corrected experimental SAXS data [intensity I/Q] versus scattering vector $Q(\text{Å}^{-1})$] with the simulated scattering curves averaged over the ensemble. The curves were normalized by dividing all intensities by the value of $I(0)$ extracted from fitting to the Debye relationship. (b) Comparison of Kratky plots where $Q^2|\,I(Q)/\,I(0)$ is plotted against the scattering vector Q. This representation of the data is sensitive to the overall shape of the scattering object and in this case to the relative dimensions of the folded and unfolded domains. Experimental data (gray) are shown in comparison to ensembles calculated by using model III with the native sequence (solid line), a polyglycine sequence (dashed line), and a sequence where the φ/ψ distributions are located in the extended region of the Ramachandran plot (dashed-point line) [2]. Copyright 2005 National Academy of Sciences.

1. *Alignment Tensor Prediction*   The RDC for a known bond vector v_i in the frame of alignment tensor S can be computed by $D_i = D_{\max} v_i^T S v_i$. We must find an alignment tensor such that the following target function is minimized:

$$\chi^2 = \sum ((D_i^{exp} - D_i^{calc})/\sigma_i)^2,$$

where D_i^{exp} and D_i^{calc} represent the experimental and computed RDCs.

2. *Computation of (ϕ, ψ) Angles*   The RDCs in two media can restrict the orientation of each bond vector to a discrete and finite solution space (see chapters 15–18). The paper [3] uses the chiral motif, that is, bond vectors C^α-H^α, C^α-C^β, to filter out ambiguous peptide plane orientations. In cases when some RDCs are missing, the optimal (ϕ, ψ) angles are found using a stochastic search such that all other relevant vector orientations are satisfied.

---

into account except for avoiding overlap between the folded and unfolded domains. This model is essentially a random flight chain model of the polymer and reproduced the expected bell-shaped distribution along the chain. (b) Experimental N-H^N RDCs compared with RDCs simulated in the unfolded domain of PX using random sampling of backbone dihedral angle (φ/ψ) space. In this model (II) nonbonded interactions between residues in the unfolded domain were taken into account by using a simple steric repulsion model. Overlap between the folded and unfolded domains was also avoided. (c) Experimental N-H^N RDCs compared with RDCs simulated in the unfolded domain of PX using random sampling residue-specific (φ/ψ) propensities found in loop regions of a database of folded proteins combined with the simple volume exclusion model (model III). (d) Simulated and experimental N-H^N RDCs from the entire protein by using model III. (e) Simulated and experimental C'-H^N RDCs from the entire protein using model III [2]. Copyright 2005 National Academy of Sciences.

Notes   Since RDCs represent global orientational constraints, the regions where little or no data is available do not propagate to introduce errors in other regions of the backbone.

Exercise   Compare the approach in Hus et al. [3] to RDC-EXACT (chapters 15–18). The algorithm in [3] is called *Meccano*. Explain how Meccano can be extended or modified to obtain a provable algorithm.

Exercise   Compare the algorithm in this chapter to the technique in chapter 7. Contrast the methods, and discuss the similarities and differences.

References

[1] P. Bernado, C. W. Bertoncini, C. Griesinger, M. Zweckstetter, and M. Blackledge. Defining long-range order and local disorder in native alpha-synuclein using residual dipolar couplings. *Journal of the American Chemical Society* 127;51 (2005 Dec 28):17968–17969.

[2] P. Bernado, L. Blanchard, P. Timmins, D. Marion, R. W. Ruigrok, and M. Blackledge. A structural model for unfolded proteins from residual dipolar couplings and small-angle x-ray scattering. *Proceedings of the National Academy of Science U S A* 102;47 (2005 Nov 22):17002–17007.

[3] Jean-Christophe Hus, Dominique Marion, and Martin Blackledge. Determination of protein backbone structure using only residual dipolar couplings. *Journal of the American Chemical Society* 123 (2001): 1541–1542.

40 Structure-Based Protein-Ligand Binding

This chapter discusses uncertainty and dynamics in the protein-ligand docking, and then reviews some methods for flexible protein-ligand docking.

40.1 Uncertainty in Experimentally Derived Structures

40.1.1 Uncertainty in X-Ray Structures
We now discuss possible uncertainties in crystal structures from X-ray diffraction [1].

Interpretation of X-ray Diffraction Data   The raw diffraction data are generally converted to an electron density map by inverting the Fourier transformation after determining the phases. More precisely, the electron density $\rho(x, y, z)$ is written by

$$\rho(x, y, z) = \frac{1}{V} \sum_{h,k,l} |F(h, k, l)| \exp(-2\pi(hx + ky + lz) + i\alpha(h, k, l)),$$

where V is the volume of the unit cell, $F(h, k, l)$ is the structure factor, and $\alpha(h, k, l)$ is the phase angle.

Then a solid atomic model is fit to interpret the density map by iteratively adjusting the model parameters, such as the atomic coordinates. During the interpretation of the density map data, we must consider the following issues:

1. *Missing or Uninterpretable Densities*   During the interpretation of the density map data, the missing or uninterpretable densities need to be corrected according to the crystallographers' experience.

2. *Resolution of the Data*   The resolution of the data is expressed in Å. The resolution of a diffraction pattern, $d_{\min}$, can be obtained by Bragg's law:

$$d_{\min} = \frac{\lambda}{2 \sin \theta_{\max}},$$

where λ denotes the wavelength of the radiation, and θ_{max} denotes the diffraction angle where the diffraction pattern fades away because of the disorder in the crystal. A higher resolution means more reliability of experimental data in terms of accuracy and precision.

3. *Temperature Factor (or B-factor)* The *B-factor* is used to model the effects of static and dynamic disorder in the crystal, and provide a relative indication of the reliability of different parts of the model. The dynamics disorder factor is defined by

$$T = \exp\left(-\frac{B}{4}\left(\frac{2\sin\theta}{\lambda}\right)^2\right) = \exp\left(-\frac{B}{4}\left(\frac{1}{d}\right)^2\right),$$

where B is the mean square displacement of the atomic vibration. If B-factors are high, this means that little or no electron density is available for the atoms in the corresponding side-chain, and thus the coordinates are less reliable.

How well the model predicts the experimental data is usually assessed by the *R-free* value. In addition to the interpretation of the raw X-ray data, we must consider the intrinsic uncertainties in the protein crystal, such as uncertainties in crystal packing, difference in crystalline and solution states, internal motions, and so on. Also, we need to consider the different levels of uncertainties in X-ray structures, such as uncertainties in backbone and side-chains, heteroatoms (water, ligand, and metal), and the ionization state of the protein or ligand.

40.1.2 Uncertainty in NMR Structures
The accuracy of the structures determined by the NMR technique is related to

1. the number and quality of restraints, and
3. the approximate solution generated by computational algorithms.

Generally, we can represent the uncertainties in NMR structure by an ensemble of structures. These issues are discussed in chapters 35 and 42.

40.2 Protein Dynamics

Protein dynamics can be represented as frequencies (timescales) and magnitudes. Some NMR experiments can measure the protein motions within micro- (10^{-6}) to milliseconds (10^{-3}), whereas dynamics within pico- (10^{-12}) to nanoseconds (10^{-9}) are measured by NMR relaxation measurements. So far, there is no uniform mathematical framework for representing various dynamics across all timescales, although this is an active area of research.

The following dynamics, for example, can be measured by current NMR techniques:

1. Backbone dynamics from the NH vector.
2. Backbone dynamics for other backbone vectors.

3. Backbone dynamics from RDCs.

4. Side-chain dynamics.

5. Dynamics from chemical (or conformational) exchanges: slow conformation exchange.

6. Dynamics from H-D exchange.

An engineer or mathematical physicist would be surprised by this use of the term "dynamics." The phenomena measured by NMR are surely *caused* by dynamics. But the readout is more like *stiffness*. For example, local order parameters (S^2) for an NH bond vector are similar to the stiffness of that vector's orientation. See chapter 24.

40.3 Probabilistic Representations of Uncertainty and Dynamics

Because of the uncertainty in measured experimental data and dynamics of protein motions, it may sometimes be difficult to know the precise position of each atom in the protein. A natural representation of atom coordinates is a probabilistic framework, that is, we may in principle assign a probability for each constraint. This would require more complicated algorithms for determining the protein structures. Examples of such algorithms are given in chapter 7.2, and by Nilges and coworkers [4, 5].

40.4 Representation of Protein Flexibility: Ensemble Docking

In computational docking algorithms, the flexibility of protein-ligand docking is considered at three levels, according to their degree of approximation [2]: (1) rigid body docking, in which both the protein and ligand are regarded as rigid solid bodies; (2) semiflexible docking, in which the protein is regarded as rigid, while the ligand is considered flexible; and (3) flexible docking, in which both protein and ligand are considered flexible.

In Knegtel et al. [3], an ensemble of experimental receptor structures is used to model flexible protein-ligand docking. Two related docking schemes are proposed to use the information from conformational variability: *energy-weighted average* and *geometry-weighted average* methods. In the energy-weighted average approach, an average ligand-protein interaction energy for each atom is calculated and considered during docking. In the geometry-weighted average approach, the following steps are executed: (1) Ensembles of protein structures are superimposed on residues near the binding sites; (2) mean and standard deviations for all atom coordinates are calculated; (3) each single structure is divided into ordered and disordered regions; (4) an ensemble of structures is stored for disordered regions, while one average structure is computed for ordered regions; and (5) the scoring function is computed based on the individual ordered regions and ensemble-average disordered regions.

40.5   FDS: Flexible Ligand and Receptor Docking with a Continuum Solvent Model and Soft-Core Energy Function

The following steps are applied in the FDS protocol [6] for protein-ligand docking with flexible ligand and flexible side-chains of receptor:

1. Determine the hydrogen bond motifs or cliques. Each clique is a set of hydrogen-bonded atom pairs from the protein and ligand that simultaneously satisfy the given distance constraints.

2. Generate the structures that satisfy all cliques by using a distance-geometry embedding algorithm, such as DGEOM95.

3. Cluster the cliques using a basic root-mean-square distance (RMSD) to produce representative structures that sample important regions of the active site.

4. The preceding structures are submitted to a Monte Carlo algorithm to find the final binding modes. During the simulation, both the dihedral angles in ligands and side-chains in proteins are sampled, either randomly or according to a rotamer library. A soft-core function for vdW repulsion and a molecular mechanics force field with the generalized born/surface area (GB/SA) continuum model are used.

Exercise   Compare this approach to the K^* algorithm (chapter 12), when side-chain and ligand rotamers are used to model flexibility, but no mutations are allowed. How does K^* allow the pose of the ligand to minimize? Contrast the algorithms, and discuss their similarities and differences.

References

[1] A. M. Davis, S. J. Teague, and G. J. Kleywegt. Application and limitations of X-ray crystallographic data in structure-based ligand and drug design. *Angewandte Chemie International Edition* 42;24 (2003 Jun 23):2718–2736.

[2] I. Halperin, B. Ma, H. Wolfson, and R. Nussinov. Principles of docking: An overview of search algorithms and a guide to scoring functions. *Proteins* 47;4 (2002 Jun 1):409–443.

[3] R. M. Knegtel, I. D. Kuntz, and C. M. Oshiro. Molecular docking to ensembles of protein structures. *Journal of Molecular Biology* (1997 Feb 21)266(2):424–440.

[4] W. Rieping, M. Habeck, and M. Nilges. Inferential structure determination. *Science* 309;5732 (2005 Jul 8): 303–306.

[5] W. Rieping, M. Nilges, and M. Habeck. ISD: A software package for Bayesian NMR structure calculation. *Bioinformatics* 24;8 (2008 Apr 15): 1104–1105. Epub 2008 Feb 28.

[6] R. D. Taylor, P. J. Jewsbury, and J. W. Essex. FDS: flexible ligand and receptor docking with a continuum solvent model and soft-core energy function. *Journal of Computational Chemistry* 24;13 (2003 Oct):1637–1656.

41 Flexible Ligand-Protein Docking

This chapter discusses how to calculate binding energetics from structure [1], and then presents a protocol for flexible docking in solution using metadynamics [2].

41.1 Predicting Binding Energetics from Structure

The relationship between the binding affinity constant, K, and the binding free-energy change, $\Delta G°$, is given by

$$\Delta G° = -RT \ln K,$$

where R is the gas constant, and T is the temperature in Kelvin.

The binding free energy $\Delta G°$ can be written as

$$\Delta G° = \Delta H° - T \Delta S°,$$

where $\Delta H°$ is the enthalpy change and $\Delta S°$ is the entropy change.

To take into account the changes in heat capacity, ΔC_p, that occur during protein-ligand binding reactions in water, the following expression for ΔG^o is used instead:

$$\Delta G° = \Delta H_R° - T \Delta S_R° + \Delta C_p \left(T - T_R - T \ln \frac{T}{T_R} \right),$$

where $\Delta H_R°$ and $\Delta S_R°$ are the enthalpy and entropy changes at the reference temperature T_R and T is the temperature of interest.

The parameters $\Delta H_R°$, $\Delta S_R°$, and ΔC_p can all be calculated by modeling changes in *accessible surface area* (ASA), which is defined as the surface traced out by a sphere, with radius equal to that of a solvent molecule, when rolled over the surface of the molecule of interest. This can also be formulated in configuration space as the *Minkowski sum* of the protein surface and the sphere, namely, $P \oplus B_r = \{p + b \mid p \in P, b \in B_r\}$, where P is the protein and B_r is a ball of radius r. This is a classical problem in computational geometry (see [3]).

Both ΔC_p and ΔH_R° can be computed from the changes in ASA:

$$\Delta C_p = 1.88 \Delta A_{ap} - 1.09 \Delta A_{pol} \, ,$$

and

$$\Delta H^{\circ} = -35.3 \Delta A_{ap} + 131 \Delta A_{pol} \, ,$$

where ΔA_{ap} and ΔA_{pol} denote the changes in apolar and polar ASA upon binding.

The parameter ΔS° consists of the following three terms:

$$\Delta S^{\circ} = \Delta S_{solv}^{\circ} + \Delta S_{conf}^{\circ} + \Delta S_{mix}^{\circ} \, ,$$

where ΔS_{solv}° denotes the contributions from restructuring of solvent, and ΔS_{conf}° denotes the contributions from changes in conformational degrees of freedom of the backbone and side-chain groups, and ΔS_{mix}° denotes the contributions from changes in translational, rotational, and vibrational degrees of freedom upon binding. All ΔS_{solv}°, ΔS_{conf}°, and ΔS_{mix}° can be indirectly calculated from ASA. See section 2 in Murphy [1] for more details.

41.2 Flexible Docking in Solution Using Metadynamics

41.2.1 Overview of Metadynamics

A metadynamics is "a dynamics in the space of collective coordinates that are evolved with a standard restrained molecular dynamics (MD) supplemented by a history-dependent potential" [2]. More precisely, each metadynamics run is an MD run in which a set of harmonic restraints on collective variables $S_{\alpha}(r)$ is imposed, and evolved using the following modified Lagrangian form:

$$H + \sum_{\alpha} \left[\frac{1}{2} M_{\alpha} \left(\frac{ds_{\alpha}}{dt} \right)^2 + \frac{1}{2} k_{\alpha} (s_{\alpha} - S_{\alpha}(r))^2 \right] + V_G(s, t) \, ,$$

where s_{α} is an auxiliary variable representing the value of the restraints, M_{α} is the fictitious mass, and k_{α} is the coupling constant. The potential $V_G(s, t)$ is constructed as a sum of Gaussians centered on the values of $s_{\alpha}(t)$:

$$V_G(s, t) = w \sum_{t'=T, 2T, \cdots t' < t} \exp \left\{ -\frac{[s_{\alpha} - s_{\alpha}(t')]^2}{2\delta_s^2} \right\} \, ,$$

where w and δ_s are the height and the width of the Gaussians and T is the time interval in between two Gaussians added.

The metadynamics method has the following two features:

1. The reduction of search dimensionality by choosing the collective variables S_{α} of the metadynamics.

2. The possibility of the *free energy surface* (FES) calculation from the summed Gaussians.

41.2.2 Application of Metadynamics in Flexible Docking
The following metavariables are used in the metadynamics docking:

1. The angle between the line connecting the centroids of receptor and ligand, and the principal axis of inertia of the ligand.
2. The ligand-active site distance or the ligand-centroid of the receptor distance.

These two metavariables are illustrated in figure 41.1. Other variables, such as the number of ligand-enzyme hydrogen bonds, hydrophobic contacts, and coordinating water molecules, were

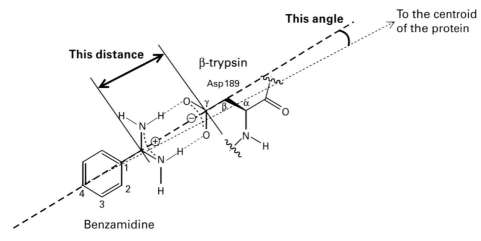

Figure 41.1
Two metavariables in the metadynamics docking. Credit: John MacMaster and Bruce Donald.

Parameters for the distance metacoordinate:

Gaussian height	w	= 0.48 kcal/mole
Gaussian width	s	= 0.4 Angstrom
metadynamics step length	T	= 3 ps
coupling constant	k	= 600 kJ/(mol Angstrom^2)
(ficticious) mass	M	= 150 uma

Parameters for the angle metacoordinate:

Gaussian height	w	= 0.48 kcal/mole
Gaussian width	s	= 0.1 rad
metadynamics step length	T	= 3 ps
coupling constant	k	= 1600 kJ/(mol rad^2)
(ficticious) mass	M	= 400 kJ fs^2/(mol rad^2)

Figure 41.2
The choices of the metadynamics parameters. Credit: John MacMaster and Bruce Donald.

Table 41.1
The overview of results

Goal	Method of Assessment (usually) Compare Prediction to Observation	Simulated Process (time)	Protein-Ligand System Used				
			A — Undocking (3 ns)	A — Docking (3 ns)	B — Undocking (6 ns)	C — Undocking (7.2 ns)	D — Unspecified (8 ns)
#1	Metacoordinates of deepest FES well vs. X-ray structure ligand		"close to"	"slightly shifted"	X	"around"	"around"
	RMSD of heavy atoms within 6 A of ligand		NONE reported	0.3 Angstroms	X	0.4 Angstroms	0.4 Angstrom
#2	Prediction vs. the literature value		−6 vs. −6.5	−6 vs. −6.5	−5 vs. −5.8	−8 vs. −7.1	−20 vs. < −11
#3	Prediction vs. the literature value		X	X	+1 vs +0.7	X	X
#4	Compare to FES from 2D umbrella sampling		"satisfactory" 3ns vs. 15ns	X	X	X	X

2.67 ns simulation = 60 CPU hours on Pentium IV workstation → 22 CPU hours/ns simulated

A: β-trypsin/benzamidine;   B: β-trypsin/chlorobenzamidine;   C: immunoglobulin McPC-603/phosphocholine (PC);   D: cyclin-dependent kinase 2 (CDK2)/staurosporine). Credit: John MacMaster and Bruce Donald.

discarded because either they are claimed by [2] to lack general applicability or they were not able to distinguish between different basins.

The following four protocols were used to test the metadynamics docking method:

1. *Finding the Docking Geometry*   Compare the coordinates of the ligand in the global minimum of the FES with the crystallographic structure. Calculate the average root-mean-square deviation RMSD of the heavy atoms in the ligand and protein between the experimental geometry and the configuration from the metadynamics simulations.

2. *The $\Delta G_{binding}$ Calculation*   The $\Delta G_{binding}$ is computed from the difference between the free energy of the minimum and the deeper free energy of the unbound state.

3. *The $\Delta \Delta G_{binding}$ Calculation.*

4. *The Calculation of the Whole Docking FES*   The whole docking FES is calculated by a bidimensional umbrella sampling [4] using the same angle and distance variables.

41.2.3   Results

In Gervasio et al. [2], the following four docking cases are used to test the metadynamics docking method: β-trypsin/benzamidine, β-trypsin/chlorobenzamidine, immunoglobulin McPC-603/phosphocholine (PC), and cyclin-dependent kinase 2 (CDK2)/staurosporine. The choices of the metadynamcis parameters are shown in figure 41.2. The overview of results is shown in table 41.1.

References

[1] K. P. Murphy. Predicting binding energetics from structure: looking beyond DeltaG. *Med Res Rev,* 1999 Jul; 19(4):333–9.

[2] F. L. Gervasio, A. Laio, and M. Parrinello. Flexible docking in solution using metadynamics. *Journal of the American Chemical Society* 127;8 (2005 Mar 2):2600–2607.

[3] T. Lozano-Perez. Spatial planning: A configuration space approach. *IEEE Transactions on Computers* C-32;2 (February 1983):108–120.

[4] Philip W., Shankar Kumar, and Maximiliano Vasquez Payne. Method for free-energy calculations using iterative techniques. *Journal of Computational Chemistry* 17;10 (1998):1269–1275.

42 Analyzing Protein Structures Using an Ensemble Representation

This chapter discusses the mathematical basis and consequences of comparing an ensemble of unfolded-state molecules versus a given reference structure, and then presents the biological meaning of the structure comparison of unfolded (or intrinsically disordered) proteins at the ensemble level [1].

We address an important problem: Suppose an ensemble of structures is computed for a protein in an unfolded or disordered state (chapters 7 and 39), or even for a unimodal native state (chapters 15–18). Given a reference structure (for example, a crystal structure), or even multiple structures of exchanging states (chapter 27), how do we compare these reference structures to the ensemble? The answer will be surprising to many readers.

42.1 Mathematical Results

42.1.1 Terminology

Let $\widetilde{A}$ and $\widetilde{B}$ be the distance matrices of structures A and B, respectively. The *distance root-mean-square deviation* between distance matrices $\widetilde{A}$ and $\widetilde{B}$ is defined by

$$dRMS(\widetilde{A}, \widetilde{B}) = \sqrt{\frac{1}{n(n-1)} \sum_{i=1}^{n} \sum_{j=1}^{n} (A_{ij} - B_{ij})^2} , \tag{42.1}$$

where $A_{ij} = \|\vec{r}_i - \vec{r}_j\|$ denotes the Euclidean distance between atoms i and j in structure A.

The *Cartesian coordinate-based root-mean-square deviation* between structures A and B is defined by

$$RMSD(A, B) = \sqrt{\frac{1}{n} \sum_{i=1}^{n} \|\vec{r}_i^A - \vec{r}_i^B\|^2} , \tag{42.2}$$

where $\vec{r}_i^A$ denotes the Cartesian coordinates of the ith atom in structure A, and $\| \cdot \|$ denotes the Euclidean l_2-norm, that is, $\|\vec{a}\| = \|(a_1, a_2, \cdots, a_J)\| = \sqrt{\sum_{i=1}^{J} a_i^2}$.

42.1.2 Results

Given an ensemble of structures $A = \{A^1, A^2, \cdots, A^k\}$, $k = 1, \cdots, N$, and a reference structure B, we have the following mathematical results:

$$dRMS(\langle \tilde{A} \rangle_N, \tilde{B}) \leq \langle dRMS(\tilde{A}^k, \tilde{B}) \rangle_N, \tag{42.3}$$

$$RMSD(\langle A \rangle_N, B) \leq \langle RMSD(A^k, B) \rangle_N, \tag{42.4}$$

where A^k denotes the kth structure in the ensemble A, and $\tilde{A}$ denotes the distance matrix of structure A (the same definitions for $\tilde{A}^k$, $\tilde{B}$), and $\langle \cdot \rangle_N$ stands for the ensemble average over all N structures in the ensemble.

42.1.3 Brief Proof

Here we give an intuition of a proof sketch. See the Results section in Zagrovic and Pande [1] for a rigorous proof.

A *metric space* is a set X with a distance function $d : X \times X \to \mathbb{R}$, such that for every $x, y, z \in X$,

1. $d(x, y) \geq 0$ with equality if and only if $x = y$ (nonnegativity);
2. $d(x, y) = d(y, x)$ (symmetry);
3. $d(x, z) \leq d(x, y) + d(y, z)$ (triangle inequality).

By definition of dRMS, we can easily construct a *metric space* with distance function:

$$dRMS : \mathbb{R}^{3n} \times \mathbb{R}^{3n} \to \mathbb{R}.$$

Now consider a simple case in this metric space, as shown in figure 42.1. Suppose we have an ensemble consisting of only two points, a and c, and a reference point b. Let m be the midpoint between a and c, then m stands for the mean of the ensemble. Extend bm to b' such that $|bm| = |b'm|$. By triangle inequality, we obtain $|b'c| + |bc| \geq |bb'|$. Since $|bb'| = |b'm| + |bm| = 2|bm|$

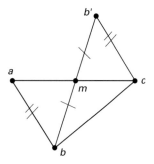

Figure 42.1
An illustration of the intuition. Credit: Michael Zeng and Bruce Donald.

and $|b'c| = |ab|$, we have $|bm| \leq (|ab| + |bc|)/2$. Since $|bm|$ stands for the distance between the mean of the ensemble and the reference—that is, $dRMS(\langle \widetilde{A} \rangle_N, \widetilde{B})$—and $(|ab| + |bc|)/2$ stands for the average over all distances between every point in the ensemble and the reference point—that is, $\langle dRMS(\widetilde{A}^k, \widetilde{B}) \rangle_N$—we thus have

$$dRMS(\langle \widetilde{A} \rangle_N, \widetilde{B}) \leq \langle dRMS(\widetilde{A}^k, \widetilde{B}) \rangle_N.$$

It is easy to extend this intuition to a general case with ensemble size of more than 2.

42.2 Biological Significance

It has been observed that the mean unfolded-state $C^\alpha - C^\alpha$ distance matrices of an α-helix are close to its corresponding native-state distance matrix. To exclude the effect of mathematical averaging, two tests of the α-helical protein villin were performed in Zagrovic and Pande [1]:

Test 1   Use the members of the unfolded-state ensemble as reference structures instead of the native structure, and compare the dRMS (or RMSD) distance between the ensemble mean and each reference structure.

Results   The native-state distance matrix is closer to the mean unfolded-state matrix than that of any member in the ensemble.

Test 2   Use native structures of other, unrelated proteins as reference structures, and perform a similar comparison.

Results   The mean unfolded-state distance matrix is more similar to the corresponding native structure of the unfolded protein than any other native protein.

Reference

[1] B. Zagrovic and V. S. Pande. How does averaging affect protein structure comparison on the ensemble level? *Biophysical Journal* 87;4 (2004 Oct):2240–2246.

43 NMR Resonance Assignment Assisted by Mass Spectrometry

This chapter discusses a novel algorithm for NMR resonance assignment, which is assisted by mass spectrometry [1, 2]. It is quite interesting to combine these two biophysical techniques in one algorithm.

43.1 Motivation

Nuclear magnetic resonance (NMR) is a useful tool for determining the structure of biomolecules, especially for those molecules that are difficult to crystallize for diffraction studies. A key step in structure determination by NMR is the resonance assignment of chemical shifts. The traditional NMR resonance assignment protocol uses triple resonance experiments, such as HNCA, to help identify the connectivities between amino acids. However, for some mammalian proteins that are difficult to express in doubly labeled forms, we may need novel structure determination strategies that depend on only the basic $^{1}H-^{15}N$ HSQC experiment, which requires only ^{15}N isotopic labeling. Prestegard and co-workers [1, 2] propose a new approach using integration of NMR and mass spectrometry (MS) for protein resonance assignment.

43.2 Mass Spectrometry–Assisted NMR Assignment

43.2.1 Principle of the Approach

The MS-assisted NMR assignment approach exploits the fact that both NMR and MS can monitor rates of exchange of amide protons (HX rate) with water deuterons. Based on this observation, we can connect crosspeaks in NMR data and fragment masses in MS data by correlating the HX rates to help the sequential assignment.

43.2.2 Extracting HX Rates by HSQC

The process of extracting HX rates by HSQC is shown in figure 43.1a. We can extract the amide exchange rates by monitoring the loss of individual peak intensities as a function of time after dissolving a protein in deuterated water. We fit the intensity data to the equation $I(t) = I_0(\exp(-kt) + const)$, where k is the residue-specific HX rate and $I(t)$ is the intensity of an

(a)

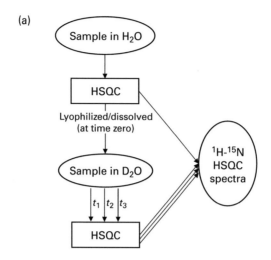

(b)

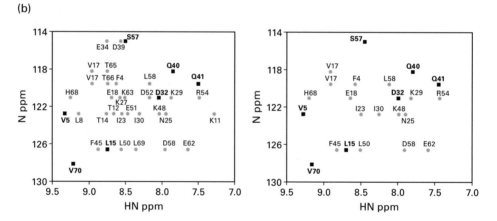

Figure 43.1
Extracting ubiquitin HX rates by [^{1}H, ^{15}N] HSQC. (a) Data in H_2O collected with 64 t_1 increments in 10 minutes. The sample was then lyophilized overnight and brought back to its original volume with 99.9% D_2O at pH 6.0, and immediately returned to the spectrometer for rapid collection of a series of Hadamard spectra. The positions of excited ^{15}N frequencies are shown in dark black. (Residue I36 at 6.1 ppm, 123.4 ppm is not included in the spectra.) (b) First point after 1 minutes 22 seconds in D20 collected with four scans in 42 seconds. [1, 2]. Reprinted with permission. Copyright 2004 American Chemical Society.

HSQC peak at time t. Figure 43.1b shows an example of the loss of peak intensities after dissolving the protein ubiquitin in deuterated water for 1:22 minutes.

43.2.3   Extracting HX Rates by MS

The procedure for extracting HX rates by MS is described in figure 43.2, top. The HX rate measured at each time point is computed by the difference between the centroid of the isotopic peak cluster for the deuterated sample and the centroid of the undeuterated reference. Figure 43.2, bottom shows an example of mass increase after deuterium incorporation.

43.2.4   Correlating HX Rates between NMR and MS

In the following, H denotes a proton, and D denotes a deuteron. The time courses of the exchange from NMR data are calculated by summing the distributions expected for each amino acid. More precisely, the deuteron distribution $d(t)$ is computed by

$$d(t) = \begin{cases} 1, & \text{if half-life of the amide H/D exchange is shorter than 1 minute;} \\ 0, & \text{if half-life of the amide H/D exchange is longer than 1 week;} \\ 1 - \exp(-kt), & \text{otherwise.} \end{cases}$$

where k is the amide proton exchange rate and t is the time interval for exchange.

In order to correlate HX rates from NMR and MS, we first plot the time courses of exchange $D(t) = \sum_{i=1}^{n}(1 - \exp(-k_i t))$, where n is the number of data points in a fragment. Next, we superimpose experimental points from the MS data. A correlation example of deuterium incorporation between the NMR and MS data is shown in figure 43.3.

43.2.5   MS-Assisted Assignment

The following *generate-and-test* procedure is used in Feng et al. [1] to demonstrate how MS helps NMR resonance assignment:

1. Perform an a priori NMR assignment.

2. Predict the exchange data for all residues from HSQC.

3. Compare the exchange data from HSQC with the data from MS.

4. Score the comparison using the function $score = \exp(-\sum(D_{expt} - D_{calc})^2/(n\sigma^2))$, where n is the number of data points, and D_{expt} and D_{calc} are, respectively, deuterium levels from the MS experiment data and the calculation from the NMR data.

Figure 43.4 shows an example of sequential assignment scores for the protein ubiquitin.

43.3   MS-Assisted NMR Assignment in Reductively ^{13}C-Methylated Proteins

The approach in Macnaughtan et al. [2] depends on different rates of *reductive methylation* at each primary amine site. After we run NMR and MS experiments, we can obtain normalized percentages of ^{13}C incorporation from HSQC and MS data, as shown in figure 43.5.

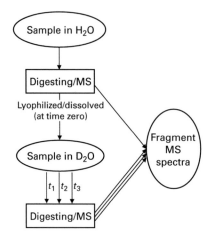

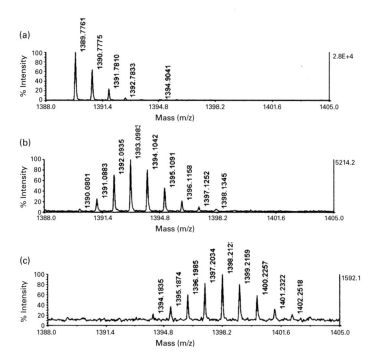

Figure 43.2

Extracting HX rates by MS. Lyophilized ubiquitin was dissolved in D_2O buffered with Na_2HPO_4 at pH 6.1 and incubated for varying lengths of time at room temperature before quenching and digesting the sample. The mass spectrum shows the region around the peptide of average mass 1,390.4 Da (Residues 346–48: AGKQLEDGRTLSD). The undeuterated spectrum is shown in panel *a* as a reference. Panels *b* and *c* are for exchange times of 1 minute and 4 hours [1, 2]. Reprinted with permission. Copyright 2004 American Chemical Society.

(a)

(b)

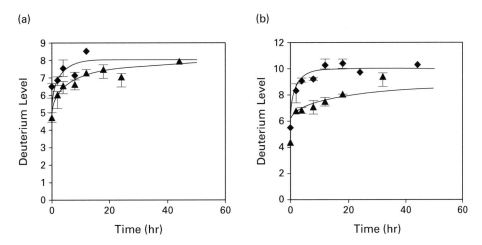

Figure 43.3
The correlation of deuterium incorporation between the NMR and MS data. Correlation of deuterium incorporation from MS data with levels predicted from NMR rate constants for two pairs of peptides from ubiquitin. (*a*) compares data for the 1,021-Da peptide (▲) and the 1,096-Da peptide (◆). (*b*) compares the 1,347-Da peptide (◆) and the 1,176-Da peptide (▲). [1, 2]. Reprinted with permission. Copyright 2004 American Chemical Society.

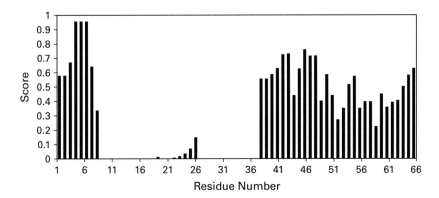

Figure 43.4
The sequential assignment scores for peptide 1,175.5 of ubiquitin. (Residues #5–15). The highest scores are seen near the proper placement position (4–6) [1, 2]. Reprinted with permission. Copyright 2004 American Chemical Society.

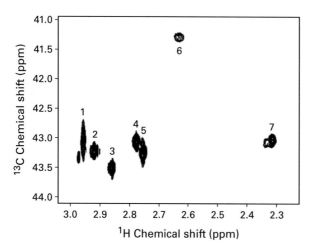

Table 1. Normalized Peak Volumes from $^1H-^{13}C$ HSQC Spectra of the Four Partially ^{13}C-Labeled Samples

NMR peak	0.5:1 sample	1:1 sample	2:1 sample	5:1 sample
1	9.9	18.0	33.7	71.4
2	8.8	12.8	25.3	61.1
3	11.6	17.3	32.0	70.3
4	5.2	7.5	15.4	38.6
5	8.8	11.9	26.1	60.0
6			10.7	30.2
7	19.9	21.5	34.7	78.3

[a]The molar ratio of ^{13}C-formaldehyde to primary amine used in the ^{13}C reductive methylation reaction is used to name each sample.

(a)

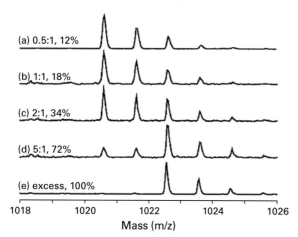

Table 2. Normalized Percentages of ^{13}C Incorporation for Each Peptide Containing Either One or Two Dimethylamines

lysine/amine in peptide	0.5:1 sample	1:1 sample	2:1 sample	5:1 sample
N-terminal amine, K1	15.4	14.1	17.9	44.8
K13	12.1	17.7	33.6	71.5
K33	10.5	11.5	27.6	64.5
K96, K97		13.2	23.0	52.1
K97	15.5	19.8	36.0	81.8
K116	5.5	11.0	24.2	56.9
K96 (back-calculated)		4.7	11.2	23.7

(b)

Figure 43.5
The normalized percentages of 13C incorporation. (a) Left: $^1H-^{13}C$ HSQC of ^{13}C-dimethylated lysozyme at pH 8.1 acquired on a 600-MHz spectrometer in approximately 5 minutes. Right: Normalized peak volumes from $^1H-^{13}C$ HSQC spectra of the four partially ^{13}C-labeled samples. The molar ratio of ^{13}C-formaldehyde to primary amine used in the ^{13}C reductive methylation reaction is used to name each sample. (b) Left: MS isotope profiles for the tryptic peptide containing a single dimethyllysine (Residues #6-14 CELAAAMKR) at various reaction stoichiometries. The reaction conditions in moles of ^{13}C-formaldehyde relative to the moles of reactive amines are listed along with the normalized percentages of ^{13}C incorporation calculated from the profiles. Right: Normalized percentages of ^{13}C incorporation for each peptide containing either one or two dimethylamines [1, 2]. Reprinted with permission. Copyright 2005 American Chemical Society.

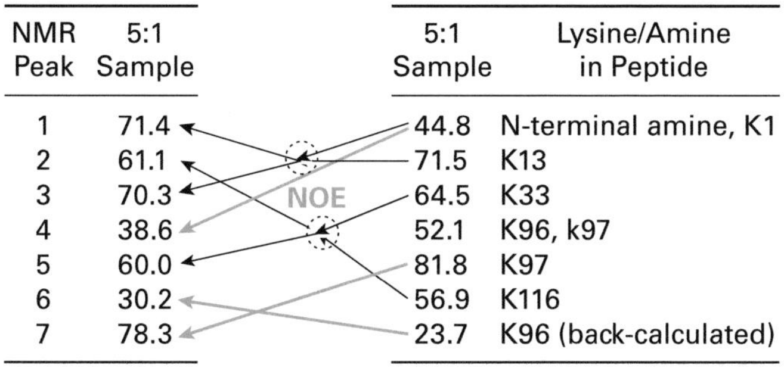

NMR Peak	5:1 Sample		5:1 Sample	Lysine/Amine in Peptide
1	71.4		44.8	N-terminal amine, K1
2	61.1		71.5	K13
3	70.3	NOE	64.5	K33
4	38.6		52.1	K96, k97
5	60.0		81.8	K97
6	30.2		56.9	K116
7	78.3		23.7	K96 (back-calculated)

Figure 43.6
The assignment of NMR peaks (the "5:1 sample" means that the molar ratio of ^{13}C-formaldehyde to primary amine in the sample is 5 : 1) [1, 2]. Reprinted with permission. Copyright 2005 American Chemical Society.

The NMR peak assignment can be performed by comparing the NMR and MS percentages of ^{13}C incorporation, as illustrated in figure 43.6.

Exercise Figure 43.6 appears to define a matching problem similar to bipartite matching. Explain how the Hungarian algorithm, EM, or graph cuts could be applied to solve this problem.

References

[1] L. Feng, R. Orlando, and J. H. Prestegard. Mass spectrometry assisted assignment of NMR resonances in ^{15}N labeled proteins. *Journal of the American Chemical Society* 126(2004):14377–14379.

[2] Megan A. Macnaughtan, Austin M. Kane, and James H. Prestegard. Mass Spectrometry Assisted Assignment of NMR Resonances in Reductively ^{13}C-Methylated Proteins. *Journal of the American Chemical Society* 127;50(2005):17626–17627.

44   Autolink: An Algorithm for Automated NMR Resonance Assignment

This chapter discusses Autolink, an heuristic stochastic algorithm for automated NMR resonance assignment [1]. This advanced lecture is designed to expose the reader to the large number of parameters and thresholds that must typically be chosen in current NMR data analysis algorithms. Choosing these parameters is nontrivial, and modern machine-learning techniques, such as support vector machines and graphical models, offer great promise for their systematic and automatic incorporation.

44.1   Algorithm Overview

Spin System   A *spin system* is defined as a group of coupled resonances, measured as crosspeaks in the NMR spectra.

Link   There is a *link* between two spin systems if their corresponding residues are adjacent in the protein sequence.

Fragment   A *fragment* is a set of two or more linked spin systems.

The Assignment Problem   In an NMR resonance assignment problem, we are given spin systems of a protein sequence, and must find an optimal mapping from spin systems to residues of the sequence.

Overview of Autolink   Figure 44.1 shows an overview of Autolink. The input data are spin systems obtained from CARA [2]. All possible spin system pairings are considered and a "fitness" score is computed for each pair. A pair of spin systems and their fitness score form a "link hypothesis." Each fitness score consists of several subscores representing specific resonance comparisons, such as the comparison of C_1^α shift of one spin system with the C_{i-1}^α shift of another spin system or the comparison of the C_1^β and C_{i-1}^β shifts of spin systems. All link hypotheses form a list of potential spin system links, called the "priority list," which is regarded as the input of the next

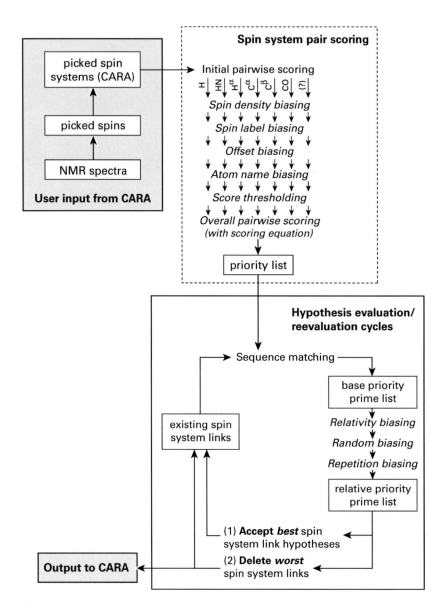

Figure 44.1
An overview of Autolink [1]. Reprinted with permission from Elsevier.

step, "hypothesis evaluation/reevaluation" cycles. In hypothesis evaluation/reevaluation cycles, some link hypotheses are accepted for the assignment of their spin systems to some regions of the protein sequence, and some link hypotheses are rejected from the candidate list. Next, the link hypotheses fitness scores are modified by factors that depend on the acceptance or rejection of other link hypotheses. The process repeats until convergence, that is, until none of the remaining link hypotheses can be accepted.

44.2 Spin System Pair Scoring

The spin system pairing fitness score is a measurement of how likely each spin system pairing corresponds to adjacent spin systems within the protein. As shown in figure 44.2 (plate 45), each spin system pair score consists of several subscores according to different user-selected types, such as C^α, C^β, CO, H(NOE). Here, CO means C'. Each subscore is calculated by

$$\sum_{s1}\sum_{s2} f(\Delta),$$

where $\sum_{s1}$ and $\sum_{s2}$ sum over all the spin comparisons, and $f(\Delta)$ is a user-defined function describing a mathematical comparison of spin systems.

In addition, each spin pair score is biased according to several user-controlled functions to interpret spin density, overlap, the labeling of each spin, the atom name in each atom label, and the offset in spin label. All these biases will be described in the following section.

44.2.1 Spin Density Bias
The spin density bias is used to account for the fact that uncommon chemical shifts are generally more valuable for determining spin systems linkages than those spins with relatively common ppm values. Spin pair scores are modified according to the following formula:

$$rel\_score' = (1 - sdc) \times rel\_score + \frac{sdc \times score_{1 \to 2}}{density_{1 \to \Sigma}},$$

where rel_score is the base score of the pair of spins 1 and 2, rel_score' is the density-compensated score, sdc is a user-defined input parameter, and $density_{1 \to \Sigma}$ is called the *spin density*, which denotes the average base score of spin 1 with all spins.

44.2.2 Assigned Spin Bias
The assigned spin bias is used to account for the assignment of ambiguous spins. It is considered by the following formula:

$$rel\_score' = (1 - asb) \times rel\_score_{unlabeled} + asb \times rel\_score_{labeled},$$

where rel_score' is the modified score, asb is a user-defined input parameter, $rel\_score_{unlabeled}$ is the spin pair score without spin labeling, and $rel\_score_{labeled}$ is the spin pair score considering labeling.

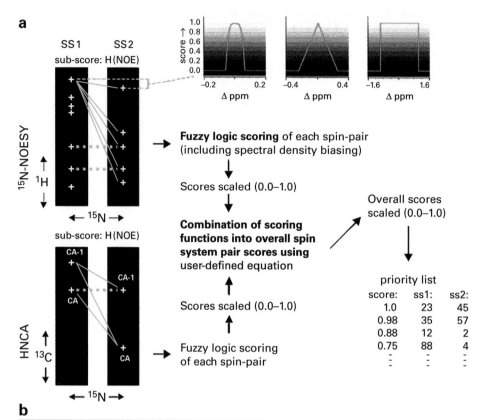

b

Equation	Interpretation
$C\alpha$	Consider only spin system pairs with compatible $C\alpha$ (ignore other spins).
$H \times C\alpha$	Consider only spin system pairs with compatible $C\alpha$s and H xpeaks (NOESY) in common.
$H \times C\alpha \times C\beta$	Consider only spin system pairs with compatible $C\alpha$s, compatible $C\beta$s, and H xpeaks in common.
$C\alpha \times (H + C\beta)$	Consider only spin system pairs with compatible $C\alpha$s and either compatible $C\beta$s or H xpeaks in common.
$H\&\&C\alpha$	Consider only spin system pairs with compatible $C\alpha$s and H xpeaks in common.
$H\|\|C\alpha$	Consider only spin system pairs with compatible $C\alpha$s or H xpeaks in common.
$H\&\|C\alpha$	Consider only spin system pairs with compatible $C\alpha$s or H xpeaks in common, but especially favor those with both.
$C\alpha^2 \times (H + C\beta + CO)$	Consider only spin system pairs with stringently compatible $C\alpha$s and at least one of the following: H xpeaks in common, compatable $C\beta$s, or compatible COs.

Figure 44.2 (plate 45)

The spin system pairing scoring. (*a*) Diagram of spin-system pair scoring. The spins of both spin systems (represented here by crosspeaks in spectra) are compared using a user-defined scoring function. Spin comparisons are signified by lines connecting relevant points in the NMR spectra. Model subscoring functions are plotted in gold at the top of the figure. The subscores are calculated by adding together all of the relevant comparison scores of each spin of one spin system (ss1) with each spin of another spin system (ss2). After the individual subscore types ($C\alpha$ and H(NOE) are shown here) are calculated, they are then combined to form an overall spin-system pair score. The manner in which they are combined is determined by a user-defined scoring equation. The overall scores are recorded in the "priority list." (*b*) Examples of scoring equations and literal interpretation. The left column shows examples of valid scoring equations demonstrating the quasi-operators "&&," "i," and "&|." "xpeak" = crosspeak. The right column describes the operations of the scoring equations [1]. Reprinted with permission from Elsevier.

44.2.3 Offset Bias

The offset bias is used to interpret the uncertainty in the intra- or interresidue correlation. It modifies the spin pair score to the following formula:

$$rel\_score' = (1 - ob) \times rel\_score_{labeled} + ob \times rel\_score_{labeled+offset},$$

where rel_score' is the new labeled score component, ob is a user-defined control parameter, $rel\_score_{labeled}$ is the spin pair score considering the presence of labels but ignoring offsets, and $rel\_score_{labeled+offset}$ is the spin pair score considering offsets.

44.2.4 Atomic Assignment Bias

The atomic assignment bias is used to consider the ambiguity in the atomic assignment (i.e., the "C^α" in "C^α_{i-1}"). It modifies the spin pair score to the following equation:

$$rel\_score' = (1 - aab) \times rel\_score_{labeled} + aab \times rel\_score_{labeled+same},$$

where rel_score' is the new labeled score component, aab is a user-defined control parameter, $rel\_score_{labeled}$ is the spin pair score considering the presence of labels but ignoring the atomic assignment, and $rel\_score_{labeled+same}$ is the spin pair score considering the atomic assignment.

44.2.5 Overall Spin System Pair Scoring

All the preceding subscores are combined to form an overall score for each spin system pair based on a heuristic, fuzzy-logic approach (see section 2.3.6 in Masse and Keller [1] for more details). Next, these combined scores are sorted in terms of their values and stored in the priority list.

44.3 Hypothesis Evaluation/Reevaluation Cycles

In the hypothesis evaluation/reevaluation cycles, a "base priority prime list" is first calculated based on the existing spin system residue assignment. Then a "relative priority prime list" is formed by modifying the base list with biasing potentials (relative biasing, random biasing, and repetition biasing). After that, some best spin system link hypotheses are accepted and some worst spin system links are deleted. Such a process is repeated until convergence, as shown in figure 44.1. There is no guarantee that the algorithm converages, nor on the time complexity, soundness, or solution accuracy.

44.3.1 Calculation of the Base Priority Prime List

The base priority prime list is created by considering how well the fragment from the acceptance of each hypothesis would match positions of the protein sequence. The fitness of each fragment is calculated by

$$fragment\,Score = [score_{ref}(ss_1, res_1) \times \cdots \times score_{ref}(ss_n, res_n)]^{1/n}, \tag{44.1}$$

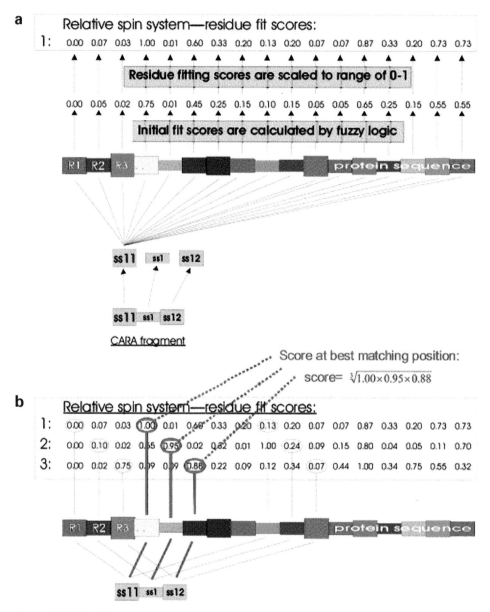

Figure 44.3 (plate 46)
The computation of fitness of a fragment. Matching of spin system fragments to the protein sequence proceeds in two stages, (*a*) individual spin system–sequence matching, and (*b*) fragment-sequence matching. (*a*) Each spin system of each fragment is first fit, using fuzzy logic, to each residue position of the protein sequence. In this diagram, comparison of a single spin system (ss11) to each position of the protein sequence is signified by solid black lines. The fit scores for each spin system are then scaled to a range of 0–1. The final fitting scores for each spin system to the protein sequence are shown here in rows 1, 2, and 3, in the ''Relative spin system residue fit scores'' boxes. (*b*) Fragment-sequence fitting. Fragment scores are calculated as the geometric average of the single spin system relative fits for consecutive positions within the protein sequence [1]. Reprinted with permission from Elsevier.

where n is the number of spin systems in the fragment, ss_i are the spin systems of the hypothetical fragment, res_i are the residue points in the protein sequence, and $score_{ref}(ss_x, res_y)$ is the *relative* score of spin system x against residue position y in the protein sequence. An example in figure 44.3 (plate 46) illustrates the computation of fitness of a fragment assigned to some region of the protein sequence.

44.3.2 Calculation of the Relative Priority Prime List

Score Delta Bias   Each spin system pair score is biased according to the next best hypothesis in the list, as shown in the following formula:

$$score'_{A \to B} = (1 - sdb) \times score_{A \to B} + sdb \times [score_{A \to B} \times (score_{A \to B} - score_{A \to ?})],$$

where $score_{A \to B}$ is the score of spin system "A" $\to$ spin system "B" before score delta biasing, $score_{A \to ?}$ is the next best score for spin system "A", $score_{? \to B}$ is the next best score for spin system B, and sdb is a user-defined control parameter between 0 and 1.

Repeat Bias   The repeat bias is used to avoid loops of the Autolink program. Autolink uses a counter to remember how many times a particular spin system has been linked to another spin system. The repeat bias is defined by

$$score' = score \times rb^{\#of\,Repeat},$$

where rb is a user-defined input parameter between 0 and 1, and $\#of\,Repeat$ is the number of times the spin system's link partner has been detected.

Random Factor Bias   The random factor bias is defined by

$$score' = (1 - rfb) \times score + rfb \times score \times randomFactor^{exp} \times randAmp,$$

where rfb is a user-defined weighting factor representing the percentage of the base priority prime score multiplied by the random factor, exp is a user-defined exponent, and $randAmp$ is a parameter that controls the magnitude of the random factor's effect.

Evaluation of the Relative Priority Prime List   Autolink divides the linking/unlinking process into several rounds, and a maximum number of added/deleted links is defined for each round. During each round, Autolink evaluates the spin system link hypotheses by computing the base priority prime list and subsequently the relative priority prime list, and then accepts a user-defined number of hypotheses at the top of the list. However, Autolink does not accept links that would preclude a higher ranking link in the list. After each linking round, the priority prime lists are recalculated and the worst scoring links are removed.

Linking/unlinking rounds are repeated until either a user-defined number of rounds is reached, or the number of remaining spin system link hypotheses is below a user-defined threshold.

Spin System Fragment Assignment   Initially, all of the fragments from previous linking/unlinking cycles are unassigned. All fragment are scored by equation (44.1) against the protein sequence and sorted in terms of that score. The fragment with the highest priority and only one fitting sequence position is assigned first. After the first fragment is assigned, each fragment is rescored against the remaining unassigned parts of the protein sequence, and again, the highest priority fragment matching only one sequence position is assigned. Such a procedure is repeated until no fragments can be unambiguously assigned to the protein sequence.

References

[1] J. E. Masse and R. Keller. AutoLink: Automated sequential resonance assignment of biopolymers from NMR data by relative-hypothesis-prioritization-based simulated logic. *Journal of Magnetic Resonance* 174;1(2005 May):133–151.

[2] R. Keller and K. Wuthrich. Computer-aided resonance assignment (CARA). Available at: http://www.nmr.ch.

45 CS-Rosetta: Protein Structure Generation from NMR Chemical Shift Data

This chapter describes a version of the Rosetta protocol for structure prediction [1]. Rosetta has been used in protein structure prediction and design, and in the design of protein:ligand interactions.

45.1 Introduction

Modern technological advances have expedited the determination of protein structures. For example, NMR spectroscopy provides experimentalists with restraints, such as NOEs and RDCs, to calculate the structure of a molecule. Some programs such as CYANA and DYANA take experimentally measured NOEs as input, and use simulated annealing and simplified molecular dynamics to compute a structure. However, the experiments to measure NOEs are time-consuming and can be expensive, and it is also sometimes challenging to assign the NOEs. Indeed, robust algorithms for automated NOE assignment are an active area of research in NMR methods development (chapters 8, 14, 17.4, 18, and 33).

Other scientists have developed techniques that *predict* protein structure, using less experimental information than traditional NMR protocols. One such strategy starts with the protein sequence information and uses empirical molecular mechanics energy functions and computational biophysics, plus information from a library of previously determined structures, together with statistical information about backbone and rotamer conformations. It would be incorrect to say these approaches do *not* use empirical data, since the weights and nuisance parameters in the energy functions and employed by the algorithm have been *fit* to these other, previously determined empirical structures. But they do not use experimental data measured on the *particular* protein for which a structure is being predicted. In this sense, it would be fair to view such approaches from a point of view including statistical parameter estimation or machine learning. Thus, the scoring function and even the algorithm is constructed using a *training set* of data and structures, and is then demonstrated on a (usually disjoint) *testing set* of structure prediction problems. In these early days of computational structural biology, issues of convergence, statistical bias, undersampling, and overfitting are also starting to receive attention. Finally, Rosetta relies critically on a library of fragments from a curated database of protein structures. Rosetta may be viewed as a stochastic algorithm combining local homology fitting, energy minimization, and random sampling. This having been said, the Rosetta suite of software is large and complex; this chapter only touches on a few of its concepts.

The Rosetta protocol employs classical Monte Carlo sampling plus a database search strategy to attempt to produce a likely protein fold, without using empirical restraint data (e.g., NOEs) for the protein. Rosetta has been one of the most successful blind-prediction programs, and the authors of CS-Rosetta augmented the basic Rosetta program to use NMR chemical shifts (CS) in an attempt to increase accuracy.

45.1.1 Rosetta

Rosetta is a protein structure prediction program, that starts with the primary sequence and searches conformation space to find a global fold. Given the sequence of a protein, Rosetta predicts protein structure via database search and comparison, combined with modeling energy minimization. Rosetta uses a Monte Carlo simulated annealing procedure. In this chapter we will distinguish vanilla *Rosetta* (which does not use experimental data) from *CS-Rosetta* (which uses experimentally measured NMR chemical shifts). "CS" stands for "chemical shifts."

Monte Carlo Procedure   *Monte Carlo* is a random-walk-based exploration of the search space. Rosetta uses *simulated annealing*, an algorithm that randomly explores neighboring conformations using a simulated Boltzmann distribution until it converges to some point in the search space. The simulated annealing algorithm in general has three phases: *generation*, *acceptance*, and *update*. The *generation* phase is the phase in which a new configuration is randomly selected from neighboring points in the search space. Some values of the configuration are modified slightly, and the new configuration is returned. The *acceptance* phase accepts a configuration generated by the generation phase if it is of lower overall energy; if a configuration is of *higher* energy than the current configuration, the move is accepted probabilistically depending on the energy difference and a parameter, T, commonly "temperature." In the *update* stage, T is modified based on the current configuration's energy and as some function of time. A function mapping configurations and times to values for T is called an *annealing schedule*. There exists some annealing schedule, such that the search will converge as time goes to infinity. However, finding this schedule is NP-hard.

Correctness and Time Bounds   Probabilistically, the Monte Carlo procedure is expected to identify the globally optimal solution as rapidly as brute force search: If there are n total possible states E(solution = optimal) = $1/n$, and E(number of tries to find optimal solution) = $1/(1/n) = n$. In practice, brute force search often suffers from the curse of dimensionality, and with simulated annealing the possibility of identifying the optimal solution without covering the full combinatoric search space becomes nonnegligible. Fabio Romeo and Alberto Sangiovanni-Vincentelli proved that the probability of identifying the optimal solution goes to 1 as time goes to infinity, but this proof offers little insight into the behavior of the algorithm given finite time constraints [2]. The model of finite time and space, and asking what can an algorithm compute in a finite time bound whose value is some function of the input size (e.g., polynomial or exponential), models a computer's efficiency to solve the problem.

General Applicability   Simulated annealing (SA) has the potential to find the optimal solution in subexponential time. However, it's not clear if it will run faster than brute force search. It's not even clear how close the answer returned by SA is to the right answer. Understandably, for polynomial time solvable problems, such as classical search and sort, this is an unreasonably loose bound on performance, and deterministic algorithms are preferred. In NP-complete cases, however, ease of programming and potential success rate continue to make simulated annealing a popular strategy.

45.1.2   CS-Rosetta

Shen et al. [1] use SPARTA, also created by their lab, to build a fragment library from which Rosetta draws for fragment replacement. After the fragment library is generated for the protein sequence, traditional Rosetta is run to generate candidate structures. In summary, the CS-Rosetta structure determination proceeds as follows:

1. *Fragment Library Generation*   CS-Rosetta employs SPARTA, which is an algorithm for chemical shift back-calculation trained on 200 proteins, to select from the 5,665-protein structural database a library of protein fragments that best fit the chemical shifts observed from experimental data, fit via molecular fragment replacement (MFR; see chapter 17.2) [3]. This fragment library is fed into traditional Rosetta.

2. *Fragment Replacement*   Fragment replacement begins with an uncoiled peptide strand. Random 9-residue windows are selected by Rosetta, and the fragment library from the previous step is searched for the 25 best matches ranked by sequence similarity and conformational energy. A random match is selected from the 25 and the backbone angles of the protein are modified to match the fragment. This proceeds until all residues have been assigned, and up to 2,000 iterations past that point. As in all other parts of the procedure, the change is accepted if the resulting structure has lower energy, and accepted according to the simulated annealing Metropolis criterion if higher.

3. *Side-Chain Minimization*   Side-chains are computed in CS-Rosetta (and traditional Rosetta) by randomly selecting a residue, randomly assigning a rotamer, and keeping that assignment according to the simulated annealing Metropolis criterion.

4. *All-Atom Minimization*   After sufficient iterations of side-chain minimization have been completed, CS-Rosetta randomly selects an atom and perturbs it randomly, and minimizes the whole protein to accommodate the perturbation. Again, SA procedures are observed to encourage the program to travel into the global and not the local minimum.

5. *Termination*   The program was assumed to have converged when the next lowest energy conformation generated was less than 2 Å RMSD from the current best, and the structure accepted. 10,000 to 20,000 all-atom models were generated for each protein.

Computational Experiments Performed   Shen et al. ran CS-Rosetta on 16 small proteins whose structure was known at the time of the experiment and performed 9 additional blind structure predictions. They also compared accuracy of SPARTA to DC, an alternative algorithm, and compared the accuracy of models generated.

45.2   Results

Even though SPARTA was only 10% more accurate than DC, the fragments selected were significantly more accurate for Rosetta's MFR procedure. When using DC instead of SPARTA, the backbone coordinate error obtained for fragments selected from the structure database is 25%–65% higher than those for the respective SPARTA-assigned database. Rosetta does not account for chemical shift data when performing side-chain prediction and minimization, which resulted in some deviation from the native structure, and, interestingly, the "goodness" of the structure. For the 9 unknown structures, CS-Rosetta performed well in the ordered regions, deviating no

Table 45.1
Accuracy of CS-Rosetta structures for 16 proteins

Protein Name	PDB ID	N_α/N_β	N_{res}	N_{cs}	$RMSD_{bb}$, Å	$RMSD_{all}$, Å
GB3	2OED	14/26	56	332	0.69	1.40
CspA	1MJC	0/33	70	405	1.57	2.19
Calbindin	4ICB	47/0	75	435	1.20	2.01
Ubiquitin	1D3Z	18/25	76	426	0.75	1.35
XcR50	1TTZ	28/16	76	352	1.53	2.30
DinI	1GHH	36/21	81	463	1.76	2.29
HPr	1POH	29/23	85	419	1.01	1.79
MrR16	1YWX	23/35	88	514	1.52	2.28
TM1112	1O5U	10/52	89	524	1.51	2.22
PHS018	2GLW	20/41	92	531	1.28	2.08
HR2106	2HZ5	37/25	96	470	1.65	2.42
TM1442	1SBO	41/23	110	647	1.09	1.88
Vc0424	1NXI	55/25	114	679	1.72	2.51
SpoOF	1SRR	55/25	121	590	1.24	2.02
Profilin	1PRQ	41/41	125	595	1.71	2.34
Apo_lfabp	1LFO	15/70	129	688	1.64	2.18

Note: Running CS-Rosetta on the 16 known structures, predicted structures were no more than 2.5 Å RMSD from the native conformation. Reprinted from [1].

more than 2 Å from the native structure, and no more than 1 Å RMSD for 6 of the 9 proteins. In the disordered regions, CS-Rosetta was less able to produce native conformations. This is explained by the nature of Rosetta's scoring function, which favors hydrogen bonds and therefore extended the protein's secondary structure a few residues beyond the native structure. This could be alleviated by flagging the loops of the protein, which can be identified via Random Coil Index (RCI), and this improved accuracy further in the ordered sections. Additionally, the protocol DP (DP score) was used to score experimentally derived structures and CS-Rosetta predicted structures, and DP scored the experimentally derived structures 10–40% higher, mostly on account of differences in core side-chain packing. Recall that DP score ("Discriminating Power") was introduced in chapter 33.

Chemical shift data, measured in parts per million, are among the most exquisitely sensitive NMR data to the local electronic environment of a nucleus. The results of CS-Rosetta (table 45.1, figures 45.1, 45.2, plates 47, 48) show that there is a computational protocol to determine

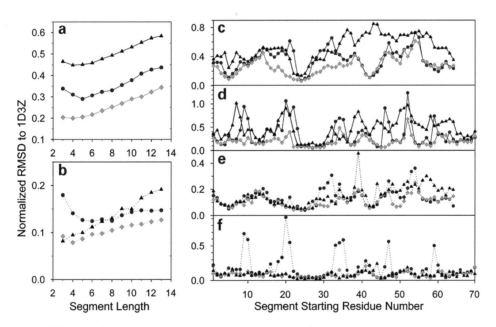

Figure 45.1 (plate 47)
Plots of normalized accuracy of database fragments selected for ubiquitin. For each ubiquitin segment, 200 fragment candidates of the same length were selected using either the standard Rosetta procedure (filled triangles), or an MFR search of the 5,665-protein structural database, assigned by the programs DC (filled circles) or SPARTA (filled diamonds). For all panels, coordinate RMSDS (N, C^{α}, and C^{β}) between query segment and selected fragments are normalized with respect to randomly selected fragments. (a and b) Average (a) and lowest (b) normalized RMSD of 200 selected fragments, as a function of fragment size, relative to the X-ray coordinates of the corresponding ubiquitin segment, averaged over all (overlapped) consecutive segments. (c and d) Average normalized RMSD of 200 nine-residue (c) and three-residue (d) fragments relative to the X-ray coordinates, as a function of position in the ubiquitin sequence. (e and f) Lowest normalized RMSD of any of these selected nine-residue (e) or three-residue (f) fragments. Reprinted from [1]. Copyright 2008 National Academy of Sciences.

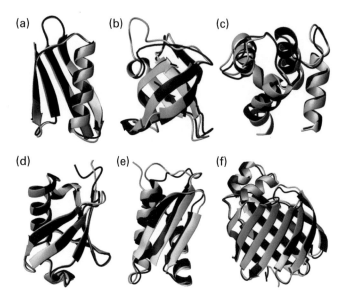

Figure 45.2 (plate 48)
Six structures generated by CS-Rosetta; the ordered secondary structure is very similar but the loops differ significantly between experimental and predicted structures. Backbone ribbon representations of the lowest-energy CS-Rosetta structure (red) superimposed on the experimental X-ray/NMR structures (blue), with superposition optimized for ordered residues. (a) GB3, (b) CspA, (c) calbindin, (d) ubiquitin, (e) DinI, (f) Apo_lafbp. Reprinted from [1].

structures from these data alone. This is significant, since CS data is currently the cheapest NMR data to record. Note that the CS-Rosetta protocol requires the resonances (chemical shifts) to be assigned.

References

[1] Yang Shen, Oliver Lange, Frank Delaglio, Paolo Rossi, James M. Aramini, Gaohua Liu, et al. Consistent blind protein structure generation from NMR chemical shift data. *Proceedings of the National Academy of Sciences* 105;12 (2008): 4685–4690.

[2] D. Mitra, F. Romeo, and A. S. Vincentelli. Convergence and finite-time behavior of simulated annealing. *Advances in Applied Probability* (Sept 1986).

[3] G. Kontaxis, F. Delaglio and A. Bax. Molecular fragment replacement approach to protein structure determination by chemical shift and dipolar homology database mining. *Methods in Enzymology* 394 (2005):42–78.

46 Enzyme Redesign by SVM

This chapter discusses a support vector machine (SVM)–based method to predict the substrate specificity for wild type and mutants in a given protein sequence family [1]. SVMs have wide application in computational biology, and often good results may be obtained.

46.1 Overview

In Rausch et al. [1], a new support vector machine (SVM)–based approach was proposed to predict the substrate specificity of subtypes (adenylation domain, or A domain) of a given protein sequence family (nonribosomal peptide synthetases, or NRPS). Based on the physicochemical properties of the amino acids, the residues of NRPS were first encoded into vectors in a high dimensional feature space. A database of A domains with known specificities compiled from the literature was used to train the SVM classifier. The classifier was then used to predict the specificity of a new A domain. The machine learning approach can be generalized to the prediction of functional subspecificities of other classes of enzymes that "share a conserved structure but catalyze different substrates."

46.2 Data Representation

A database of A domain sequences was collected from the protein databases and literature, and each sequence requires the occurrence of a complete NRPS module with at least one condensation domain, one A domain, and one peptidyl carrier domain. Specificity annotations of A domains are obtained from the published literature. In an A domain sequence with known specificity, the 34 residues within 8 Å distance from the substrate are extracted. Each residue is classified using 12 features, including the number of hydrogen bond donors, polarity, volume, secondary structure preferences, hydrophobicity, and isoelectric point. Thus, each A domain is represented by a vector with $34 \times 12 = 408$ total features.

46.3   The Support Vector Machine (SVM) Approach

The SVM provides a powerful tool for data classification in a high-dimensional feature space. Burges provides a good introduction and survey on SVM [2].

The classical SVMs are "inductive," where all training data are labeled. In many biological datasets, however, the amount of labeled data is small, and a large amount of unlabeled data is available. To address the problem of learning from both labeled and unlabeled data, the "transductive" SVM (TSVM) was developed. TSVMs assume that the missing labels of unlabled data points are consistent with those of nearby points within the same cluster.

To evaluate the accuracy of a classifying algorithm, the following measurements are used:

1. Error rate $= \mathrm{err} = (FP + FN)/(FP + FN + TP + TN)$;

2. Recall $=$ sensitivity $S_n = TP/(TP + FN)$;

3. Precision $=$ specificity $S_p = TP/(TP + FP)$;

4. Matthews correlation coefficient MCC,

$$= \sqrt{\frac{(TP \cdot TN) - (FN \cdot FP)}{(TP + FN)(TP + FP)(TN + FN)(TN + FP)}},$$

where TP, FP, TN, and FN denote the number of true positive, false positive, true negative, and false negative predictions, respectively.

When using a single SVM model to predict the specificity, it is possible for all single SVMs to return a "negative" result—that is, no final prediction will be possible. In such a case, the single

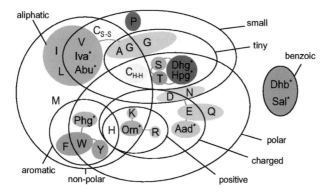

Figure 46.1 (plate 49)
Addressing the side specificity problem by clustering specificities for amino acids with similar physicochemical properties. The colored sets show how similar amino acids have been clustered to composite specificities of A domains. To get larger clusters, several smaller clusters were joined, as indicated by red lines connecting colored sets. An asterisk indicates rare nonproteinogenic amino acids [1]. Reprinted with permission.

Table 46.1
Results of cross-validating the different SVMs by leave-one-out (LOO)

Specificity of SVM	Positive training points	Kernel type	Leave-one-out cross-validation				Quality of SVM
			Error	S_n	S_p	MCC	
Large clusters	282 Labeled and 664 unlabeled data points (18 + 646)						
Dhb=Sal	11	l	0.4	100	92	96	++
Asp=Asn=Glu=Gln=Aad	43	r	1.4	100	91	95	++
Pro=Pip	20	r	0.7	90	100	95	++
Cys	17	r	0.7	100	89	94	++
Ser=Thr=Dhpg=Dpg=Hpg	50	r	2.5	96	91	92	++
Gly=Ala=Val=Leu=Ile=Abu=Iva	92	r	4.3	95	93	90	+
Orn=Lys=Arg	16	l	0.7	88	88	87	+
Phe=Trp=Phg=Tyr=Bht	33	r	3.2	88	85	85	0
Small clusters	273 Labeled and 673 unlabeled data points (27 + 646)						
Dhb=Sal	11	l	0	100	100	100	++
Aad	7	l	0	100	100	100	++
Glu=Gln	15	l	0	100	100	100	++
Dhpg=Dpg=Hpg	20	l	0.4	100	95	97	++
Ser	13	l	0.4	92	100	96	++
Cys	17	l	0.7	100	89	94	++
Thr	16	l	0.7	94	94	93	++
Pro	16	r	0.7	94	94	93	++
Asp=Asn	21	l	1.1	90	94	92	++
Val=Leu=Ile=Abu=Iva	60	l	2.9	92	95	91	+
Orn	8	l	0.7	88	95	87	+
Gly=Ala	32	l	3.3	81	90	84	0
Tyr	18	r	2.2	94	77	84	0
Arg	5	l	0.7	80	80	80	0
Phe=Trp	14	l	3.7	57	67	60	0

The more training data that are available the more reliable the trained predictive models are. The "quality of SVM" in the last column, therefore, is a qualitative measure for the MCC. Kernel type l stands for linear kernel and r stands for radial basis function kernel. Error rate, sensitivity (S_n), specificity (S_p) and Mathews correlation coefficient (MCC) are given in percentage. [1]. Reprinted with permission.

SVM model may not be powerful enough to classify the data, and a *multiclass* SVM model [3] may be used to solve the problem.

The Side Specificity Problem   In reality, A domains often have significant side specificities, which lead to alternate peptide products that differ at the corresponding point. This problem is addressed by clustering specificities for amino acids with similar physicochemical properties, as shown in figure 46.1 (plate 49). Two choices have been considered for clustering: grouping specificities into a few large clusters or into more small clusters. Grouping specificities into larger clusters has the advantages of (1) a larger positive dataset for SVM training, (2) a larger spectrum of sequence variations, (3) a larger subspace in the hyperspace, (4) a lower risk of overfitting, and (5) the recognition of similar substrate specificities. On the other hand, grouping specificities into small clusters has the advantage of more precise predictions, but entails a higher risk of overfitting to the training data.

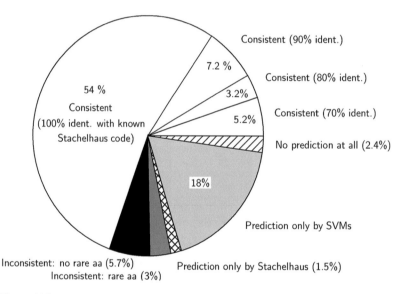

Figure 46.2
The comparisons of the SVM-based method with the sequence-based prediction method described in chapter 12.1 Of the 1,230 adenylation domains (with HMMER automatically extracted from the June 2005 version of UniProt) 70% or 858, obtained consistent predictions by both predictors (white sectors). For most of these consistent predictions (54% of the total, or 666), the Stachelhaus method was based on an exact match with a known "specificity-conferring code," the others had at least a 70% match. To 2.4% or 29 sequences, none of the predictors can assign any specificity (no match >70%, diagonal hatches). An 18% or 217 sequences, could be classified only by the SVMs and not by the Stachelhaus method (light gray sector), 18 A domains (1.5%) could not be classified by the SVMs but by the Stachelhaus method (cross-hatched), and 2 of them are rare specificities. The Stachelhaus predictions for the rest are based mainly on 70% matches to known specificity "codes." For 108 sequences (8.8%) the predictions were inconsistent, but 38 of them (3% of the total, gray sector) had matches to rare amino acids that were not used for training the SVMs. The remaining 70 incompatible predictions were mainly based on < 80% identity matches with known "specificity-conferring codes" (black sector) [1]. Reprinted with permission.

46.4 Results

In implementing TSVM, different kernel functions, such as linear, polynomial, radial basis, and sigmoid functions, were tried. A grid search approach is used to determine the optimal kernel parameters. For each specificity cluster, the sequences belonging to a given specificity are labeled by "+", all other sequences with different but known specificity are labeled by "−", and the uncharacterized sequences are labeled by "0". Table 46.1 shows a tabular overview of the results of the quality assessment of the models. Figure 46.2 shows the comparisons of the SVM-based method with the sequence-based prediction method described in chapter 12 section 12.1.

References

[1] Christian Rausch, Tilmann Weber, Oliver Kohlbacher, Wolfgang Wohlleben, and Daniel H. Huson. Specificity prediction of adenylation domains in nonribosomal peptide synthetases (NRPS) using transductive support vector machines (TSVMs). *Nucleic Acids Research* 33;18 (2005):5799–5808.

[2] C. J. C. Burges. A tutorial on support vector machines for pattern recognition. *Data Mining and Knowledge Discovery* 2 (1998): 121–167.

[3] Bernhard Schlkopf, Koji Tsuda, and Jean-Philippe Vert (eds.). *Kernel Methods in Computational Biology*. Cambridge, MA: MIT Press, 2004.

47 Cross-Rotation Analysis Algorithm

Algorithms in structural biology and molecular biophysics are advancing our long-range goal of understanding biopolymer interactions in systems of significant biochemical as well as pharmacological interest. This chapter gives a particular example, an algorithm to analyze non-crystallographic symmetry in X-ray diffraction data of biopolymers, where one must "recognize" a finite subgroup of $SO(3)$ (the Lie group of 3-dimensional rotations) out of a large set of molecular orientations. The problem may be reduced to clustering in $SO(3)$ modulo a finite group, and solved efficiently by "factoring" into a clustering on the unit circle followed by clustering on the 2-sphere S^2, plus some group-theoretic calculations. This yields a polynomial-time algorithm that is efficient in practice [2], and which recently enabled biological crystallographers to reveal the architecture of a parasite's enzyme [6]. This will help researchers reduce the threat of certain diseases among those with weak immune systems. See chapter 17, section 17.6, for more details about this enzyme structure, and for a comparison of the rotation search algorithms in NMR (e.g., chapter 17, sections 17.2 and 17.3) and crystallography (this chapter and chapter 48).

More generally, many fascinating computational problems arise in the analysis of X-ray diffraction data. X-ray crystallography techniques (see section 40.1.1) yield high-resolution macromolecular structures of biomolecules, and therefore are very important to structural biologists. In this lecture, we introduce a quaternion-based algorithm to identify (and generate missing) cross-rotation peaks to model orientations consistent with proper noncrystallographic symmetry (NCS).

47.1 CRANS

Proper noncrystallographic symmetry (NCS) facilitates orienting and translating copies of the molecular-replacement (MR) model to use initial phases from an homologous model. Traditionally, the presence and degree of NCS is initially identified by a *self-rotation function* [4], while the orientations of each copy of the model are subsequently identified using the *cross-rotation function* [4].

The key observation behind MR using NCS is that symmetries in the protein structure yield symmetries in the data in the Fourier domain. This is because the operations of rotation and the Fourier transform commute.

The *self-rotation function* samples three-dimensional orientation space $SO(3)$. For each sampled orientation $q \in SO(3)$, the diffraction data D is rotated by q and correlated with the unrotated data. This signal $A(D, D(q))$ defines a scalar field on rotation space, with peaks where the rotated data correlates with the unrotated data D. Symmetries will cause these peaks, and hence the peaks can be used to infer the NCS. For example, in C_5 symmetry we will see rotation peaks around a common axis (vector), spaced 72 degrees apart. Here, A denotes the correlation function, and $D(q)$ rotating the data by q.

The *cross-rotation function* is similar. The scalar field computed is instead $A(D, H(q))$, where H is the simulated diffraction from a homology putative model. $H(q)$ denotes the rotation of H by q (similar to $D(q)$).

The term *self-rotation* encodes the notion that the data is correlated with itself (modulo rotation). The term *cross-rotation* encodes the notion that the correlation is "across" the raw data and the back-calculated structure factors, from a rotated putative homolog.

The correct peaks of the cross-rotation function should have rotation-function scores much higher than the incorrect ones, and they must be compatible with the NCS constraints. That is, they must correspond to the correct NCS-consistent orientations. Therefore, one method to identify the correct peaks of the cross-rotation function compatible with n-fold NCS is selecting the n peaks that have the highest rotation-function scores. However, if the model is partial [3], this method does not work well. Furthermore, it could be the case that some of the correct rotations are missing in the cross-rotation peak list. The *CRANS* (cross-rotation analysis) algorithm [2] was designed to overcome these problems. The advantage of CRANS is that it can identify the peaks that are consistent with the NCS constraints, and it can generate the missing peaks.

47.1.1 Methods

The key idea of CRANS is to identify a set of peaks (rotations) whose *rotation differences* are compatible with that of a point group with a single rotation axis. More precisely, let $C = \{r_1, \ldots, r_w\}$ be a set of w peaks, such that r_i is a member in the rotation group of the 3-dimensional Euclidean space. CRANS first computes w^2 rotation differences $d_{i,j} = r^{-i}r^j$ (r^{-i} represents the inverse of r^i). Then, the algorithm performs a *subgroup search* to find all subsets of peaks R such that the *rotation difference set* $D(R) = \{r^{-i}r^j : r^i, r^j \in R\}$ is a subset of the *special orthogonal group of three-dimensional rotations $SO(3)$*, and the rotations in R share a common axis. If some peaks are missing, CRANS then "completes" the partial set of identified NCS-consistent rotations by generating missing rotations using the group properties of quaternions [1, 5]. The algorithm is illustrated in figure 47.1 (plate 50).

47.1.2 Complexity

The time complexity of each step of CRANS is summarized in the following table, in which m is the number of cross-rotation peaks, the NCS is n-fold, and the number of NCS-consistent subgroups is g ($\ll m$).

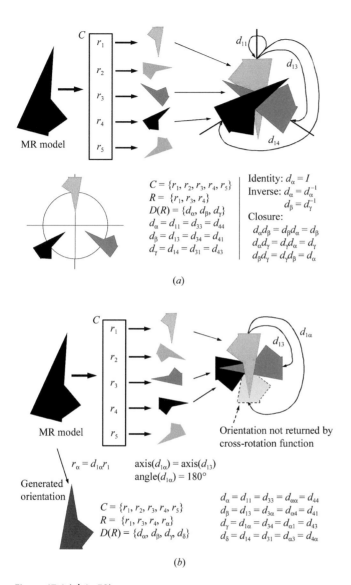

(a)

(b)

Figure 47.1 (plate 50)

Two-dimensional examples for threefold (a) and fourfold (b) NCS. (a) A model is shown with the results of a simplified cross-rotation search C containing only five rotations. Orientations corresponding to rotations r_1 (purple), r_3 (green), and r_4 (blue) form an NCS-consistent rotation set R. For clarity, only a few rotation differences (d_{11}, d_{13}, and d_{14}) are shown in the upper right overlapping-orientation figure. The rotation differences, $D(R)$, form a complete rotation difference set and satisfy the group properties of associativity (not shown), identity, inverse, and closure. (b) A fourfold NCS example is shown using similar notation to that in (a). In this example, only three of the four NCS-consistent rotations (r_1, r_3, and r_4) are contained in the cross-rotation peak list C. The missing rotation $r_\alpha = d_{1\alpha} r_1$ is computed using $d_{1\alpha}$ defined by the axis of the three identified NCS-consistent rotations and the missing angle. The now complete NCS-consistent rotation set R has a complete rotation difference set $D(R)$, which satisfies the subgroup properties.

Step	*Worst Case*	*Expected*
1. Compute all rotation differences $\Delta = \{rs^{-1}\}$	$O(m^2)$	
2. Cluster those differences $\Delta =$ the n symmetry angles (filter)	$O(nm^2)$	
3. Cluster rotation Δ sets sharing a common axis	$O(m^4)$	$O(m^2)$
4. Patch missing	$O(g)$	
Total	$O(m^4 + g)$	$O(nm^2 + g)$

Exercise   Technically, the subgroup search should be called a *coset search*. Explain why.

References

[1] W. Hamilton. (1969). *Elements of Quaternions*, 3rd ed. New York: Chelsea Publishing Co.

[2] R. Lilien, C. Bailey-Kellogg, A. Anderson, and B. Donald. *Acta Crystallographica* D60 (2004):1057–1067.

[3] B.-H. Oh. (1995). *Acta Crystallographica* D51, 140–144.

[4] M. G. Rossmann, and D. M. Blow. *Acta Crystallographica* 15 (1962):24–31.

[5] E. Salamin. *Applications of Quaternions to Computation with Rotations*. Technical Report, Stanford University, AI Lab, Internal working paper, 1979.

[6] R. O'Neil, R. Lilien, B. Donald, R. Stroud, and A. Anderson. Phylogenetic classification of protozoa based on the structure of the linker domain in the bifunctional enzyme, dihydrofolate reductase-thymidylate synthase. *Journal of Biological Chemistry* 278 (2003):52980–52987.

48 Molecular Replacement and NCS in X-ray Crystallography

This chapter presents the application of normal-mode analysis (NMA) in molecular replacement (MR) and crystallographic refinement [1]. We also discuss examples of NCS-constrained exhaustive search using oligomeric models [2], building on chapter 47.

48.1 Background

48.1.1 The Phase Problem

X-ray crystallography is one of the most important approaches for solving structures of biological molecules at nearly atomic resolution. In an X-ray crystallography experiment, the X-ray beam is scattered mainly by the electrons of atoms, forming a diffraction pattern on the collection screen that records the intensity of diffraction (figure 48.1, plate 51). In effect, the diffraction pattern experimentally contains only the *magnitudes* (the Fourier moduli) of the complex structure factors $\mathbf{F_q}$ that correspond to the Fourier transform of the electron density map of the crystal.

If the magnitudes can be supplied with phases, an inverse Fourier transform would allow us to reconstruct the electron density map in the crystal, from which we can obtain rich three-dimensional (3D) structural information about the molecule. Unfortunately, the phase information is exactly the missing piece in an X-ray crystallography experiment, and its reconstruction hence becomes a very famous problem known as the "phase problem."

48.1.2 Molecular Replacement

Since early 1930s many methods have been developed for solving the phase problem. Among these are Fourier refinement, the Patterson map, heavy atoms, direct methods, and more recently, molecular replacement [5]. Based on the assumption that similarity in amino acid sequence may imply similarity in 3D structure, molecular replacement (MR) is essentially a trial-and-error approach for solving crystal structures for which there is a known, homologous structural model through which initial estimates of phases for the unknown structure could be obtained [3].

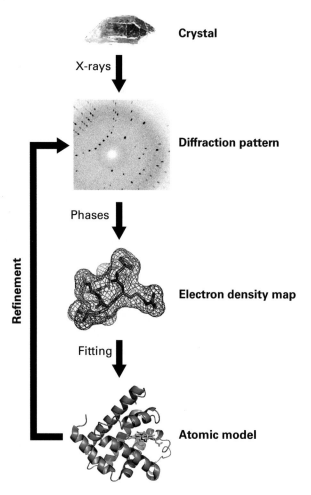

Figure 48.1 (plate 51)
Schematic diagram for X-ray crystallography in structure determination [6].

48.2   NMA in Molecular Replacement

48.2.1   Objectives

The algorithm in Delarue [1] aims to address the following problems:

- Conformational sampling to optimize the success rate of molecular replacement.

- Refinement of a model against X-ray or cryo-EM data in the presence of large-amplitude structural changes.

- Multicopy refinement and an all-in-one-go weight refinement approach for MR.

48.2.2 Normal Modes and Elastic Network Models (ENM)

Normal modes are eigenvectors of the Hessian matrix of the free energy:

$$H_{i,j} = \frac{\partial^2 V}{\partial x_i \partial x_j},$$

where the Tirion potential energy is of the form

$$V = \frac{C}{2} \sum_{i,j} (d_{ij} - d_{ij}^0)$$

under the Elastic Network Model (ENM) assumptions (see chapter 21). Here, d_{ij} is distance between atoms and it has an upper threshold of around 10 Å , and d_{ij}^0 is its equilibrium value. If the system is at equilibrium and the potential energy is purely harmonic, then the motion $\mathbf{r}_i(t)$ of each atom i is a linear combination of normal modes

$$\mathbf{r}_i(t) = \sum_k c_k \sin(\omega_k t + \varphi_k) \mathbf{u}_k^i,$$

where ω_k is the frequency associated with eigenvector λ_k, c_k is the amplitude constant, and φ_k is some phase shift. Note that at a given temperature, the lowest-frequency modes are those that are most likely to reproduce large amplitude movements. An easy way to understand this qualitatively is by analogy to a simple energy function for harmonic oscillators:

$$E = \frac{1}{2} \cdot 4\pi^2 m f^2 \cdot A^2,$$

where f is the frequency of oscillation, m is the mass and A is the amplitude. Since each normal mode is assigned the same energy E, it follows that the amplitude of movement A along a lower frequency f normal mode will be greater than those with higher frequencies. Supporting evidence includes the observation that on average known structural transitions can be reasonably described with two modes that tend to be among the lowest frequency modes, and results from comparing predicted and measured crystallographic B factors.

How Many Normal Modes Do We Need to Represent the Full Transition? The full structural transition can only be described by the complete set of normal modes. However, it has been observed that usually the 10–20 lowest normal modes can describe 90%–95% of the structural transitions.

How to Select Biologically Relevant Modes? Two approaches are available:

- Select modes that are robust in slightly different ENMs.
- Tune the links between atoms in ENMs by multiple sequence alignment.

NMA Application in MR   In X-ray crystallography, NMA has been used to generate model variants by systematically varying the amplitude of a given normal mode in a given range. While 1D and 2D exhaustive scans of a given mode are possible, higher dimensions of scanning are computationally prohibitive. One way around this is a randomized approach. Sampling conformational flexibility within a given RMSD range is made possible by generating random linear combinations of a limited set of modes:

$$\mathbf{r}_i = \mathbf{r}_i^0 + \sum_{k=1}^{m} c_k \mathbf{u}_k^{(i)}, \tag{48.1}$$

where $\mathbf{r}_i^0$ is the initial model, c_k is randomly generated, and m is the size of set. Given this formulation, the structure factors (see section 40.1.1) corresponding to the models deformed by equation (48.1) can be written as

$$\mathbf{F}_{\text{calculated}}(\mathbf{H}) = \sum_{i=1}^{n} f_i \exp\left\{ 2\pi i \mathbf{H} \cdot \left[\mathbf{r}_i^0 + \sum_{k=1}^{m} c_k \mathbf{u}_k^{(i)} \right] \right\}, \tag{48.2}$$

where $\mathbf{H}$ is the coordinates in reciprocal space expressed in Miller indices, n is the total number of atoms perturbed and f_i is the scattering factor of atom i. Then the usual Pearson correlation coefficient $\rho(|\mathbf{F}_{\text{calculated}}|, |\mathbf{F}_{\text{observed}}|)$ can be computed, and c_k's and appropriate coordinates that maximize $\rho(|\mathbf{F}_{\text{calculated}}|, |\mathbf{F}_{\text{observed}}|)$ are identified. Note that this refinement is made after rotation and translation search in MR, in place of the rigid body refinement program.

An alternative to this approach is to simultaneously refine multiple models, linked in one term by adding adjustable weights to each model. Attempts are made to avoid overfitting by fixing models and only varying their weights, and mean-field optimization of weights is then performed using Pearson correlation as a score, starting from uniformly assigned weights. It has been observed that the procedure converges with a dominating weight for the true expected solution for the test example. However, more comprehensive study is needed to analyze the scope and robustness of this approach in refinement.

48.2.3   Summary

In conclusion, normal mode analysis is a powerful tool to generate structural variants that would increase the chance of success in MR. It has the advantage that it can be refined directly by X-ray data and could potentially be made more efficient by a weight refinement regime. However, the weight refinement regime has only succeeded in refinement for a few test cases, and more systematic tests will be needed to assess its applicability.

48.3   NCS-Constrained Exhaustive Search Using Oligomeric Models

The presence of noncrystallographic symmetry (NCS) can be helpful for phasing using molecular replacement (MR). The orientation of NCS axes may be identified by the *self-rotation function*

(SRF), whereas the orientations of each copy of the model are identified by the *cross-rotation function (CRF)* (see chapter 47). Since the cross-rotation function uses only a fraction of the Patterson vectors, its efficiency can be limited.

When the directions of NCS can be derived accurately and oligomeric homologs of the unknown structure are available, the cross-rotation step of MR can be omitted. Functions are developed to build oligomers from monomers and to orient them properly, that is, in agreement with SRF. The oligomers are subsequently used in translation search. Such oligomer models may reduce the dimension of search space, making subsequent exhaustive search using the translation function a plausible option. However, this approach is applicable only in rare or ideal circumstances when protein quaternary structure is proven to be conserved by analysis of multiple homologs, when the unknown protein has a similar oligomeric structure, and when the cross-rotation function step causes problem with the conventional MR.

48.3.1 Methods

The technical procedure in Isupov and Lebedev [2] includes:

1. Orientations of the NCS axes are defined from SRF.

2. Reference oligomer axes are aligned with the NCS axes.

3. A list of allowed oligomer orientations relative to the reference orientation is created for subsequent translation function (TF) trials.

4. Visual inspection and case by case manual analysis.

48.3.2 Examples

Three examples are given in Isupov and Lebedev [2], whose structures were solved following the procedure outlined in section 48.3.1:

1. Oxygenating component of 3,6-diketocamphane monooxygenase from *Pseudomonas putida*.

2. Anti-TRAP from *Bacillus licheniformis*.

3. Thioredoxin peroxidase B from human erythrocytes.

In the third example, the internal organization of the oligomer is varied, requiring generation of a new model for each translation search.

References

[1] Marc Delarue. Dealing with structural variability in molecular replacement and crystallographic refinement through normal-mode analysis. *Acta Crystallographica. Section D, Biological Crystallography* 64;1 (2008 Jan):40–48. Epub 2007.

[2] M. N. Isupov and A. A. Lebedev. NCS-constrained exhaustive search using oligomeric models. *Acta Crystallographica. Section D, Biological Crystallography* D64 (2008):90–98 [doi:10.1107/S0907444907053802].

[3] P. Evans and A. McCoy. An introduction to molecular replacement. *Acta Crystallographica. Section D, Biological Crystallography* D64 (2008):1–10.

[4] Lilien RH, Bailey-Kellogg C, Anderson AC, Donald BR. A Subgroup Algorithm to Identify Cross-Rotation Peaks Consistent with Non-Crystallographic Symmetry. *Acta Crystallographica. Section D, Biological Crystallography* D60 (2004):1057–1067.

[5] David Sayre. X-ray crystallography: The past and present of the phase problem. *Structural Chemistry* 13;1 (February 2002): 81–96.

[6] Article on X-ray crystallography. Wikipedia. Accessed: 2010. URL: http://www.wikipedia.org.

49 Optimization of Surface Charge-Charge Interactions

This lecture presents the application of a genetic algorithm in optimizing solvent-exposed charge-charge interaction of Fyn SH3 domain. Based on experimental results, the designed variants have demonstrated an enhancement in thermostability compared to the wild-type protein [1].

49.1 Algorithm Input

The input to the genetic algorithm is a group of artificial (computational) "chromosomes" with length equal to the number of surface residues, which are subjected to optimization in the protein of interest. The chromosomes are *feature vectors*, and should not be confused with biological chromosomes made of DNA, histones, and chromatin. The term "chromosome" in genetic algorithms is chosen as a trope to biological inspiration. The chromosome is encoded with a sequence of "+", "−", and "o", each of which represents the generic type of its corresponding residues. For example, Lys and Arg are classified as generic positive groups (+), Asp and Glu belong to generic negative groups (−), and His, N-terminal, C-terminal are noncharged (o). Therefore, a chromosome in the input is a specific charge distribution among surface residues, which will be optimized. For example,

$$\text{ArgGluHisAsp} = + - o -$$

49.2 Genetic Algorithm

A genetic algorithm (GA) [2] runs iteratively on input chromosomes until certain exit conditions are met. Each cycle of genetic algorithm consists of three phases: *chromosome scoring*, *parental chromosome selection and crossover*, and *child chromosome mutation*, as shown in figure 49.1. In the phase of chromosome scoring, GA computes the energy score of each chromosome based on energy models. In the phase of *parental chromosome selection and crossover*, chromosomes with favorable scores are selected. Among these chromosomes, one part of them are passed directly to the next phase, and the rest are paired up and subjected to random crossover to generate two new child chromosomes. In the last phase, *child chromosome mutation*, some of the child

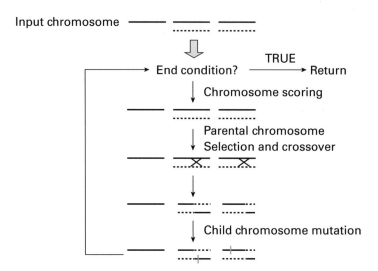

Figure 49.1
Overview of genetic algorithm.

chromosomes are randomly mutated at single positions. In the following subsections, more details will be presented for each phase of the algorithm. Genetic algorithms encode a heuristic search strategy, with no guarantees.

49.2.1 Chromosome Scoring

Each chromosome is assigned an energy score Z, which is calculated as follows:

$$Z = E_{TK} + E_P \cdot \delta_{WT}. \tag{49.1}$$

E_{TK} is the charge-charge interaction energy equal to the contribution of surface electrostatic interaction to Gibbs free energy of protein unfolding. This energy is approximated based on the following equation:

$$E_{TK} \approx \frac{1}{2} \sum_{i=1}^{n} \langle W_i \rangle = \sum_{\chi} \left[\sum_{i,j=1}^{n} \frac{1}{2} E_{ij} (q_i + x_i)(q_j + x_j) \right] \rho_N(\chi).$$

The equation sums over $\langle W_i \rangle$, the charge-charge interaction energy of group i with the rest of the ionizable groups in the protein over all charged groups. The sum can also be viewed as the arithmetic sum of total electrostatic interactions over all interacting charged groups in each protonation state of the protein. E_{ij}, the charge-charge interaction of a unit charge placed at position i and j, is calculated from the TK-SA model [3], which simplifies a globular protein to approximate it as a sphere with annotated solvent-accessible areas.

If a chromosome is different from the chromosome encoded from wild type, a penalty in the unfolding energy will be paid. This penalty is calculated in the second term of equation (49.1), in which E_P is the penalty energy per difference, and σ_{WT} is the number of differences. By including the penalty in the Z score, a chromosome having a higher similarity to the wild type will be assigned a more favorable score, which indirectly maintains the wild-type structure and function of the protein's variants.

Time Complexity For each chromosome, E_{TK} is computed in $O(n^2)$ time, where n is the length of the chromosome. $E_P \cdot \sigma_{WT}$ is calculated in $O(m)$ time, where m is equal to σ_{WT}. If there are p chromosomes in the input, the total running time for chromosome scoring will be $O(p(n^2 + m))$.

49.2.2 Parental Chromosome Selection and Crossover

In the second phase, chromosomes with a Z score that satisfies a certain threshold are selected as parental chromosomes, which form the parental population. Chromosomes with the best Z scores are passed directly to the next generation. The rest of the chromosomes in the parental population are paired up randomly and subjected to random crossover to generate child chromosomes. By the selection process, the overall score of the new generation is expected to be no worse than the input populations.

Time Complexity It takes $O(p)$ time to construct parental population, where p is the number of chromosomes in the input population. Chromosome pair-up and crossover takes $O(p^2)$ time, since the maximum size of parental population is p. Therefore, the total time complexity for chromosome selection and crossover phase is $O(p^2)$.

49.2.3 Child Chromosome Mutation

The newly generated chromosomes are subjected to random point mutations. A random number is assigned to determine which chromosome and which position on the chromosome will be mutated. The child chromosome population that passes this phase will enter the genetic algorithm as input for the next cycle of optimization.

Time Complexity The time complexity for this phase is $O(p)$.

There is no bound on how many phases (iterations) are required for a GA to converge or find an optimal solution. There is no guarantee of the quality or accuracy of the computed solution by a GA.

49.3 Computational and Experimental Validations

In the paper [1], Fyn5 is the output of the genetic algorithm with optimized surface charge-charge interaction. Five residues—Glu11, Asp16, His21, Asn30, and Glu46—are modified into Lys. Another three variants, named Fyn1, Fyn2, and Fyn3, are constructed with one, two, and three

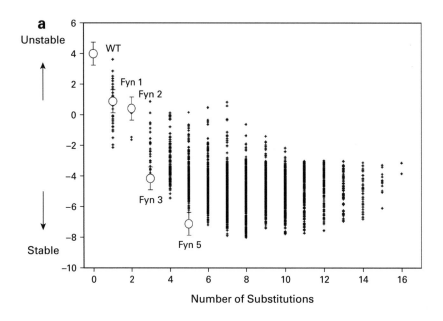

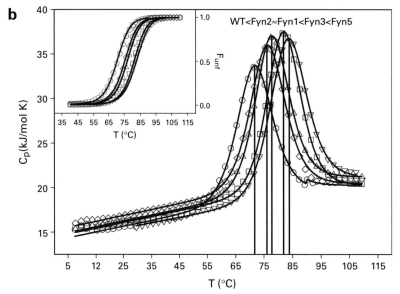

Figure 49.2
(a) Computed unfolding energy of variants and wild type. (b) DSC and CS spectrum of variants and wild type [1].

mutations for the study of cumulative effect of the mutations. Based on the prediction, the rank of protein stability is computed to be: Fyn5 > Fyn3 > Fyn2 > Fyn1 > wild type. Finally, both computational and experimental approaches were applied to validate the predicted stability among these variants.

49.3.1 Computational Validations

The computed unfolding energies of the variants are shown in figure 49.2a. The stability increases from wild type to Fny5 variant.

49.3.2 Experimental Validation

The four variants and the wild type were tested by differential scanning calorimetry (DSC), and circular dichroism spectroscopy (CD), which is shown in figure 49.2b. Both spectra generally demonstrated an increase in denaturing temperature from wild type to Fny5 variants, that is, the stability was increased from wild type to Fny5. However, ambiguity was observed for the Fny1 and Fny2 variants.

References

[1] Katrina L. Schweiker, Arash Zarrine-Afsar, Alan R. Davidson, and George I. Makhatadze. Computational design of the Fyn SH3 domain with increased stability through optimization of surface charge-charge interactions. *Protein Science* 16 (2007):2694–2702.

[2] Beatriz Ibarra-Molero and Jose M. Sanchez-Ruiz. Genetic algorithm to design stabilizing surface-charge distributions in proteins. *Journal of Physical Chemistry* 106;26 (2002):6609–6613.

[3] Beatriz Ibarra-Molero, Vakhtang V. Loladze, George I. Makhatadze, and Jose M. Sanchez-Ruiz. Thermal versus guanidine-induced unfolding of ubiquitin. An analysis in terms of the contributions from charge-charge interactions to protein stability. *Biochemistry* 38 (1999):8138–8149.

50 Computational Topology and Protein Structure

50.1 Topology

Protein biochemists sometimes use the word "geometry" to describe the local geometric structure of a protein, and the word "topology" to describe more global, shapelike features. Strictly speaking, this is an abuse of notation, since the term *topology* refers to properties that are *invariant* under any embedding into any space. Nevertheless, the rigorous mathematical notions of topological invariance are quite useful to analyze protein structure.

The term *homology* has precise meanings in both biology and mathematics. While the meanings are related, they are different. In this chapter, we use the term in its mathematical sense. We will start with some basic definitions, to build intuition.

Suppose we are given the structure coordinates of a protein. Such a protein can be represented by a *triangulation,* in which the atomic positions are taken to be a set of *vertices* P, and these vertices are connected by *edges*. Next, the edges form triangles. The triangles then make up the faces of tetrahedra. Alternatively, a *Delaunay triangulation* of the set of points P representing atomic nuclei could be constructed. Or, the *surface* of the protein might be triangulated. In all three cases, the basic object of study is a *triangulation*, which is used as an approximation to the protein shape and structure. The Delaunay triangulation is the geometric dual of the *Voronoi diagram.* Triangulations arise in a wide variety of settings in computational science, and so we will discuss them somewhat generally. In computational structural biology, as well as other applications, the value proposition is that the detailed geometric and topological properties of the triangulation can be studied to understand the mechanism and function of the biological macromolecule.

In a triangulation (which mathematicians call a *complex*), triangles have an *orientation* (intuitively, clockwise or counterclockwise), which defines their *sign* ($+1$ or -1). This notion can be generalized to an orientation for tetrahedra. Moreover, the triangles and tetrahedra in a complex can either be filled (solid), or empty (holes). Homology groups are sometimes introduced as a measure of "counting these holes" (where their signs can be positive or negative) in each dimension. Although this intuition is reasonable, it is highly simplified, and does not describe the algebraic power of homology groups to compute and compare the topology of complicated geometric objects such as proteins.

We describe an algorithm for computing the *homology type* of a triangulation. By *triangulation* we mean a finite simplicial complex; its *homology type* is given by its homology groups (with integer coefficients). The algorithm could be used in structural biology to tell whether two finite-element meshes or Bézier-spline surfaces representing solvent accessible protein surfaces are of the same "topological type." Finally, if a triangulation is defined by a network of NOEs, we might ask, using topological methods, whether it can be embedded in $\mathbb{R}^3$. Homology computation is also a purely combinatorial problem of considerable intrinsic interest.

50.2 Homology

In this chapter, we study computing all the homology groups of a triangulation. In protein structure analysis and computer-aided design (CAD), many geometric objects are triangulations. For example, finite-element methods use 2D and 3D triangulations of solids, called finite-element meshes. Triangular parametric surfaces such as rational Bézier surfaces are widely used in geometric modeling systems [41]. A collection of these objects is, topologically, a triangulation (a 3D finite-element tetrahedralization is simply a 3D triangulation). Furthermore, algorithms exist for triangulating polygons and tetrahedra, and the problem of triangulating algebraic varieties has been studied by Hironaka, McCallum, and others [14, 31].

Intuitively, a p-simplex is a convex hull of $p + 1$ linearly (affinely) independent points, called the *vertices* of the simplex. A set of these simplices, with appropriate adjacency conditions, is called a *simplicial complex*. In this chapter we will adopt the following:

Definition 50.1 *A triangulation is a finite simplicial complex.*

Thus, a triangular parametric surface will be regarded as a homeomorph of a triangulation. Since homology type is a topological invariant, it is the same for all homeomorphs of a triangulation, and for all other triangulations of the same geometric object.

In structural biology and NOE network analysis, we might construct triangulations (either by machine or by hand), and then modify them incrementally in the protein design or NOE assignment process. These modifications can involve adding new triangles and tetrahedra, identifying edges and vertices, and so on. We wish to ask, when are two triangulations "topologically equivalent"? For example, we may wish to compute whether two protein designs have the same "topological type." Alternatively, after modifying a design, we wish to know whether its "topology" has been altered. Indeed, the topological type of a design may be taken as a design specification, or as an invariant, which no modeling operation should violate. Finally, we are interested in the physical realizability of a geometric protein design. One basic question we may ask is, "Can the design be embedded in Euclidean space ($\mathbb{R}^3$)?" In general these questions are very hard [13]. One way to attack this problem is to search for computable topological invariants. For example, one topological invariant we might consider is the *fundamental group*. Given a simplicial complex it is possible to "effectively compute" its fundamental group, in the sense that one may compute a presentation for that group. However, deciding the isomorphism of two groups from their presentations is uncomputable. Indeed, by showing that any finitely presented group can be the fundamental group of a compact 4-manifold [36], it has been shown that there exists no algorithm for deciding the topological equivalence (homeomorphism) of two compact, orientable, triangulable 4-manifolds [29].[1] Fast algorithms would also be useful for computing cohomology groups. One motivation is that cohomology group computation is a vital subroutine called in the construction of Postnikov complexes, which calculates the higher homotopy groups.

We develop the basic concepts of triangulations, and the basic algorithmic structure of the resulting matrices representing the boundary homomorphisms.

50.3 Simplicial Complexes

A triangulation or tetrahedralization of a solid is a geometric realization of a simplicial complex. This object contains 0-simplices (vertices), 1-simplices (edges), 2-simplices (triangles), 3-simplices (tetrahedra), and so on. The set of these simplices is called a *simplicial complex*. The intersection of any two simplices must be another simplex of lower dimension. Clearly, we can represent a p-simplex by its vertices $[v_0, \ldots, v_p]$. We wish to treat a simplicial complex as a purely combinatorial object. We do this by treating the simplices of a complex as combinatorial objects "independent of their coordinates." We define the following:

Definition 50.2 *An* abstract simplicial complex K *consists of a set* $\{\, v \,\}$ *of vertices and a set* $\{\, s \,\}$ *of finite non-empty subsets of* $\{\, v \,\}$ *called* simplices *such that*

1. Any set consisting of exactly one vertex is a simplex.

2. Any nonempty subset of a simplex is a simplex.

Simplices *is plural; the singular is* simplex.

$\sigma \in K$ is called a *simplex* of K, and its *dimension* is 1 less than its cardinality. The *vertices* of K have dimension 0. The dimension of K is the dimension of the largest simplex in K.

A geometric realization of σ is an embedding $\phi : \sigma \to \mathbb{R}^n$ such that the image of the vertices of σ are linearly independent. A geometric realization of K is an embedding of each simplex that preserves the adjacency relationships of K: That is, if σ, τ, ρ are simplices and $\sigma \cap \tau = \rho$, then $\phi(\sigma) \cap \phi(\tau) = \phi(\rho)$. Hence, an embedding is an isomorphism onto its image.

Next we define the orientation of a simplex:

Definition 50.3 *Let σ be a p-simplex. Define two orderings of its vertices of be equivalent if they differ by an even permutation. This defines two equivalence classes of vertex orderings for $p > 0$. (Vertices have only one class, hence one orientation.) Each of these classes is called an orientation of σ. An oriented simplex is a simplex σ together with an orientation ± 1 of σ.*

---

[1] See [32] for a nice review of this history.

Let $v_0, \ldots, v_p$ be linearly (affinely) independent points. We denote by $[v_0, \ldots, v_p]$ the oriented simplex consisting of the simplex with vertices $v_0, \ldots, v_p$ and the equivalence class of the particular ordering $v_0, \ldots, v_p$.

50.4 Homology Type Is Effectively Computable

One topological invariant is the *homology type*, by which we mean all the homology groups of a triangulation. Henceforth, all homology and chain groups will be taken to have integer coefficients. In this chapter, let Z denote the ring of integers. A finite triangulation's homology groups are all finitely generated Abelian groups. These groups have a structure theorem, which tells us that each can be expressed in normal form as the direct sum of a free Abelian group, and zero or more torsion groups (quotients of Z),

$$Z^b \oplus Z/d_1 \oplus \cdots \oplus Z/d_l, \tag{50.1}$$

where the each d_i divides the next, that is, $d_1|d_2|\cdots|d_l$. Furthermore, classical algebraic topology tells us that these homology groups are effectively computable.[2] Homology computation has been considered in a complexity-theoretic setting (for regular CW-complexes and for semi-algebraic sets) by Schwartz and Sharir [37], among others.

The homology computation for a simplicial complex K is effected as follows:

First, we construct the "chain groups" $\{C_p\}$, which are free Abelian groups generated by the (oriented) simplices of each dimension. A boundary homomorphism ∂_p maps from each C_p to C_{p-1}. The pth homology group is defined by

$$H_p(K) = (\ker \partial_p)/(\operatorname{im} \partial_{p+1}). \tag{50.2}$$

The *kernel* $\ker \partial_p$ of a map ∂_p is the set of chains it sends to zero. The *image* $\operatorname{im} \partial_{p+1}$ of a map $\partial_{p+1} : C_{p+1} \to C_p$ is the set of chains $\{ \partial_{p+1}(x) \mid x \in C_{p+1} \}$. See definition 50.6 and equation (50.7) below.

We calculate the matrix of the boundary homomorphism in "simplex basis coordinates." This matrix is then "diagonalized" over the ring of integers; the diagonal matrix is its "(Smith) normal form." The diagonal entries are

$$\overbrace{1, \ldots, 1}^{r\ 1's}, d_1, d_2, \ldots, d_l, \quad (r \geq 0).$$

From this matrix, we may immediately calculate the *Betti number b* and *torsion coefficients* $d_1, \ldots, d_l$ that give the cannonical form of the group (50.1).

---

[2] For a finite triangulation or regular cell complex. See textbooks such as [33, 35] for a good modern treatment.

50.4.1 Complexity

Suppose the input triangulation K has n_p simplices in dimension p, and dim $K = D$ for some dimension bound D. Let $n = \max_p n_p$. The matrix of each boundary homomorphism has size $O(n^2)$. The computational bottleneck in determining simplicial homology is the conversion of these matrices to (Smith) normal form.

We describe the classical algorithm for computing all the homology groups of a triangulation.

The rank n_p of C_p is the number of p-simplices in K. Hence, ∂_p may be represented by an $n_{p-1} \times n_p$ matrix A_p. Now, each p-simplex has a boundary which is a $(p-1)$ chain. For example, consider a triangular patch as an oriented 2-simplex $[v_0, v_1, v_2]$. Now, in general, for a p-simplex σ we have

$$\partial_p \sigma = \partial_p [v_0, \ldots, v_p] = \sum_{i=0}^{p} (-1)^i [v_0, \ldots, \hat{v}_i, \ldots, v_p], \tag{50.3}$$

where the "hat" $\hat{v}_i$ means that the vertex v_i is to be deleted to obtain a $(p-1)$-simplex. So, for our triangular patch we have

$$\partial_2 [v_0, v_1, v_2] = [v_1, v_2] - [v_0, v_2] + [v_0, v_1]. \tag{50.4}$$

Hence, "encoding" equation (50.4) into the matrix A_p requires $p + 1$ entries (which are ± 1's) into one column of A_p (here, $p = 2$). These are the only nonzero entries in that column.

Exercises

1. Compute the boundary $\partial_1([v_0, v_1])$ of a line segment $[v_0, v_1]$.

2. Compute the boundary $\partial_3([v_0, v_1, v_2, v_3])$ of a tetrahedron $[v_0, v_1, v_2, v_3]$.

3. Compute $\partial_0(\partial_1([v_0, v_1]))$ of a line segment $[v_0, v_1]$.

4. Compute $\partial_2(\partial_3([v_0, v_1, v_2, v_3]))$ of a tetrahedron $[v_0, v_1, v_2, v_3]$.

50.4.2 Applications

Mathematicians are interested in the efficient computation of the higher homotopy groups of simply-connected spaces (such as n-spheres). That computation uses the homology group computation as a subroutine.

However, there are many other applications as well. In structural biology and macromolecular design, we could take as input two triangulations representing macromolecules and compute whether they have the same homology groups. While isomorphism of all the homology groups does not imply homeomorphism, it does mean that their "topologies" are "similar." For example, the triangulations could represent solid finite-element meshes in macromolecular structural biology, or curved surfaces (e.g., Bézier systems). We know that surfaces with torsion homology are generally not embeddable in $\mathbb{R}^3$. For example, for the Klein bottle $\mathcal{K}$ and the projective plane

P^2, we have $H_1(\mathcal{K}) = Z \oplus Z/2$ and $H_1(P^2) = Z/2$, respectively. Hence the presence of torsion implies the physical unrealizability of a design. Thus, for example, we could compute whether a series of modifications to a design preserves its embeddability, connectivity, number of holes, and so on. There is a wealth of information in the homology groups of a topological space K. In geometric design, the Euler number $\chi(K)$ has been suggested as a topological invariant. The *Euler number* of K is computable from the Betti numbers in equation (50.1). Let β_p be the Betti number in dimension p. Then $\chi(K) = \sum_p (-1)^p \beta_p$. Hence, it is clear how the homology type provides strictly more information than the Euler characteristic. Because the homology groups carry so much information about the space, we could indeed imagine a protein design system where the homology type of the design was part of the design specification, or where designs were classified by homology type.

50.4.3 Foundations

The computability of the homology groups of a simplicial complex is well known basic mathematics; see textbooks on elementary algebraic topology (e.g., [33, 35]). Worst-case deterministic bounds have been given in several studies [34, 23, 8, 20, 21]; see section 50.5.3. For work on computing cohomology groups, see references [17, 19, 25]. For other work on computational algebraic topology, see references [5, 1, 37, 31, 14, 9, 2, 24, 6, 7], and also [42]. Other studies [40, 39] discuss related problems of interest.

50.5 Computing Homology Groups

50.5.1 Simplicial Homology

See section 50.3 for a review of simplicial complexes. We first define the *chain groups* for a simplicial complex K.

Definition 50.4   *A p-chain on a simplicial complex K is a function c from the oriented p-simplices of K to the integers that vanishes on all but finitely many p-simplices, such that*

$$c(\sigma) = -c(\sigma')$$

if σ and σ' are opposite orientations of the same simplex.

Definition 50.5   *We add p-chains by adding their values. This yields a group $C_p(K)$, called the chain group of dimension p. For $p < 0$ or $p > \dim K$, $C_p(K)$ is the trivial group.*

For every oriented simplex σ, we define the *elementary chain* c_σ as follows: $c_\sigma(\sigma) = 1$, $c_\sigma(\sigma') = -1$ (if σ' is the opposite orientation of σ), and c_σ vanishes on all other oriented simplices. Now, it is convenient to write σ to denote not only an oriented simplex σ, but also to denote the elementary p-chain c_σ. Hence, we write $\sigma' = -\sigma$.

We note the following fact: $C_p(K)$ is the direct sum of subgroups isomorphic to Z, one for each p-simplex of K. Hence, $C_p(K)$ is a free Abelian group generated by the oriented p-simplices of

K. To "compute" $C_p(K)$, we simply count the number n_p of p-simplices, and "construct" $C_p(K)$ as Z^{n_p}. This means we view the oriented p-simplices as a basis for $C_p(K)$, and we can write any chain in the group as a finite linear combination $\sum k_i \sigma_i$. One example of such a chain is the RHS of equation (50.4).

Next, we define the boundary homomorphism $\partial_p : C_p(K) \to C_{p-1}(K)$ using equation (50.3). An example of computing ∂_2 is shown in equation (50.4). This constructs a *free chain complex* (a differential graded group of degree -1, in which each C_p is free Abelian):

$$C_p(K) \xrightarrow{\partial_p} C_{p-1}(K) \xrightarrow{\partial_{p-1}} \cdots \xrightarrow{\partial_2} C_1(K) \xrightarrow{\partial_1} C_0(K). \tag{50.5}$$

It is straightforward to show that for all p

$$\partial_{p-1} \circ \partial_p = 0. \tag{50.6}$$

Finally, we can define the homology groups, formalizing (50.2):

Definition 50.6 *The kernel of $\partial_p : C_p(K) \to C_{p-1}(K)$ is called the group of p-cycles and is denoted $Z_p(K)$. The image of $\partial_{p+1} : C_{p+1}(K) \to C_p(K)$ is called the group of p-boundaries, and is denoted $B_p(K)$. By equation (50.6), each boundary of a $(p+1)$-chain is a p-cycle, so $B_p(K) \subset Z_p(K)$. We define*

$$H_p(K) = Z_p(K)/B_p(K) \tag{50.7}$$

to be the p-th homology group of K.

50.5.2   Computing the Homology Groups

Now, let G and H be free Abelian groups with bases $\sigma_1, \ldots, \sigma_n$ and $\rho_1, \ldots, \rho_m$, respectively. For a homomorphism $f : G \to H$, then

$$f(\sigma_j) = \sum_{i=1}^{j} a_{ij} \rho_i, \tag{50.8}$$

for unique integers a_{ij}. The matrix $A = (a_{ij})$ is called the *matrix* of f relative to the chosen bases for G and H.

We recall the following basic theorem from algebra:

Theorem 50.7 *Let $A = (a_{ij})$ be an $n \times m$ matrix over a principal ideal domain R. Then A can be "diagonalized" in the sense that we can obtain a diagonal matrix*

$$PAQ, \tag{50.9}$$

where $P \in GL_m(R)$ and $Q \in GL_n(R)$. If the diagonal elements of (50.9) are $d_1, d_2, \ldots, d_l$, then $d_1 | d_2 | \cdots | d_l$. Furthermore, if R is a Euclidean domain, then (50.9) may be obtained by elementary row and column operations on A.

This "diagonal" form is called the "normal form" of the matrix; the algorithm to compute it is called the "reduction algorithm." Z, of course, is a Euclidean domain. We further recall that the elementary row operations are as follows:

1. Add an integer multiple of row i to row j $(i \neq j)$.

2. Interchange rows i and j.

3. Multiply row i by a unit (in our case, ± 1).

Each of these corresponds to a change of basis in H. There are three similar "column" operations that correspond to a change of basis in G. Theorem 50.7 therefore states that we can always apply a sequence of these six operations on A to reduce it to the desired "normal form." By viewing the presentation matrices for Abelian groups as Z-modules, we obtain the structure theorem for finitely generated Abelian groups (equation 50.1). The reduction algorithm is central to computing the normal form of the boundary homomorphism, and hence, to computing the homology groups. We review the reduction algorithm in section 50.5.3. We proceed now to describe the following classical construction from algebraic topology (see, e.g., [33]).

1. Let $Z_p = \ker \partial_p$, and $B_p = \operatorname{im} \partial_{p+1}$.

2. Let W_p be all elements σ_p of C_p such that some nonzero multiple of σ_p belongs to B_p. W_p is a subgroup of C_p, called the *group of weak p-boundaries*. We have

$$B_p \subset W_p \subset Z_p \subset C_p.$$

3. Using the reduction algorithm, we diagonalize the matrix for ∂_p, that is, we choose bases $\sigma_1, \ldots, \sigma_n$ for C_p and $\rho_1, \ldots, \rho_m$ for C_{p-1}, relative to which the matrix of ∂_p has the normal form

$$
\begin{array}{c}
\\
\rho_1 \\
\vdots \\
\rho_l \\
\rho_{l+1} \\
\vdots \\
\rho_m
\end{array}
\left[
\begin{array}{ccc|ccc}
d_1 & & 0 & & & \\
 & \ddots & & & 0 & \\
0 & & d_l & & & \\
\hline
 & & & & & \\
 & 0 & & & 0 & \\
 & & & & &
\end{array}
\right]
\qquad (50.10)
$$

where all the d_i are positive, and $d_1 | d_2 | \cdots | d_l$. It is not hard to show that

(a) $\sigma_{l+1}, \ldots, \sigma_n$ is a basis for Z_p.

(b) $\rho_1, \ldots, \rho_l$ is a basis for W_{p-1}.

(c) $d_1 \rho_1, \ldots, d_l \rho_l$ is a basis for B_{p-1}.

A straightforward computation shows that

$$H_p(K) \cong (Z_p/W_p) \oplus (W_p/B_p). \tag{50.11}$$

Now, the group Z_p/W_p is free, and W_p/B_p is a torsion group. Hence, we can calculate $H_p(K)$ by computing these two groups.

50.5.3 The Algorithm for Homology Group Computation

We proceed as follows. First, we choose bases for the chain groups by arbitrarily orienting the simplices of K. These orientations remain fixed for the computation. Then we construct the matrix of the boundary homomorphism relative to these bases. Its entries will all be 0, 1, or -1. We next reduce this matrix to normal form using the reduction algorithm, obtaining a matrix like (50.10). Now,

1. The rank of Z_p is the number of zero columns in matrix (50.10).

2. The rank of W_{p-1} is the number of nonzero rows in (50.10).

3. There is an isomorphism

$$W_{p-1}/B_{p-1} \cong Z/d_1 \oplus \cdots \oplus Z/d_l.$$

Hence, from the normal form of the boundary homomorphism (50.10) we can "read off" the torsion coefficients of K in dimension $p-1$; they are simply the entries of the matrix that are greater than 1. We can also "read off" the rank of Z_p and W_{p-1} (see above). Finally, the normal form for ∂_{p+1} gives us the rank of W_p. The Betti number of K in dimension p is simply $b = \text{rank}(Z_p/W_p) = \text{rank}(Z_p) - \text{rank}(W_p)$.

The Reduction Algorithm We now prove theorem 50.7 by giving the classical reduction algorithm. In the process, we keep track of the complexity. We observe first that all elementary row (resp. column) operations take time $O(m)$ (resp. $O(n)$). The proof uses the division algorithm (i.e., the fact that Z is a Euclidean domain).

Proof of Theorem (50.7) Let $A = (a_{ij})$ be an $m \times n$ matrix over Z.

Step 1 (preprocessing) We find the smallest nonzero element, ℓ_1 of A. (Here *smallest* means smallest "size" in the ring—i.e., smallest absolute value). Permute the rows and columns to move the smallest element to the upper lefthand corner, a_{11}. Multiply the first row by ± 1 to make a_{11} positive.

Step 2 Suppose a_{11} doesn't divide some entry in the first column, say, $a_{11} \nmid a_{i1}$:

$$\begin{pmatrix} a_{11} & \cdot & \cdot & \cdot \\ \cdot & \cdot & \cdot & \cdot \\ \cdot & \cdot & \cdot & \cdot \\ \cdot & \cdot & \cdot & \cdot \\ a_{i1} & \cdot & \cdot & \cdot \\ \cdot & \cdot & \cdot & \cdot \\ \cdot & \cdot & \cdot & \cdot \end{pmatrix} \tag{50.12}$$

Use the division algorithm to obtain

$$a_{i1} = qa_{11} + r, \tag{50.13}$$

where $q, r \in Z$ and $0 < r < a_{11}$. Now we add $(-q)$ times the first row to the ith row. This replaces a_{i1} by r, while increasing the "size" of the numbers in row i by a factor of q. Since $r < a_{11}$, a_{11} is no longer the minimal element of A. The new minimal element is somewhere in row i (it may not be r). Repermute the matrix so that the new minimal element is in the upper left corner.

Step 3   Repeat step 2 until a_{11} divides every a_{i1} in the first column.

Step 4   Similarly, we may use column operations to reduce to the case where a_{11} divides every a_{1j} in the first row.

We observe that step 1 takes time $O(mn)$, step 2 takes time $O(m+n)$. We observe that steps 3 and 4 terminate because at each step, a_{11} is decremented by at least 1. Hence, we reach step 5 after $O(mn + \ell_1(n+m))$ time.

Step 5   Subtract off multiples of the first row and column to make all the elements a_{i1} in the first column and a_{1j} in the first row zero, for $i, j \neq 1$. Now the matrix looks like this:

$$\begin{pmatrix} a_{11} & 0 & \cdot & \cdot & \cdot & 0 \\ 0 & & & & & \\ \cdot & & & & & \\ \cdot & & & B & & \\ \cdot & & & & & \\ 0 & & & & & \end{pmatrix} \tag{50.14}$$

Now, we wish to ensure that a_{11} divides every element of the submatrix B. We number B's rows and columns starting at 2.

Step 6   Suppose some entry b_{ij} of B is not divisible by a_{11}. We add row i to row 1 to obtain:

$$\begin{pmatrix} a_{11} & b_{i2} & \cdots & b_{ij} & \cdots & b_{in} \\ 0 & & & & & \\ \cdot & & & & & \\ \cdot & & & B & & \\ \cdot & & & & & \\ 0 & & & & & \end{pmatrix}. \tag{50.15}$$

Next, we apply the division algorithm to obtain

$$b_{ij} = qa_{11} + r \tag{50.16}$$

with $q, r \in Z, 0 < r < a_{11}$. We subtract q times column 1 from column j to obtain:

$$
\begin{pmatrix}
a_{11} & b_{i2} & \cdots & r & \cdots & b_{in} \\
0 & & & & & \\
\cdot & & & & & \\
\cdot & & & \mathbf{B} & & \\
\cdot & & & & & \\
\cdot & & & & & \\
0 & & & & &
\end{pmatrix}. \tag{50.17}
$$

Step 7 Now, at this point, a_{11} is no longer the minimal element of A, since $r < a_{11}$. We go back to step 1, moving r to a_{11}. Since the corner element is decreased by at least 1 each time, the (entire) process terminates after at most ℓ_1 iterations. Finally, a_{11} will divide every entry in B, and the first row and column will be zero, except for a_{11}. Hence, the matrix has the form (50.14), and a_{11} divides each element of B.

Step 8 *(Recursive Step)* We note that if a is an integer that divides each entry of A, and that if A' is obtained from A by any row or column operation, then a still divides each entry of A'. Hence, we can recursively invoke the algorithm on submatrix B. □

All of the steps in the algorithm given in section 50.5.3 are essentially linear-time or quadratic-time operations except the reduction algorithm. The time $R(n)$ required by the reduction algorithm (section 50.5.3), in fact, dominates the computation of homology type, which can clearly be done in time $O(D(n^2 + R(n)))$. The worst-case analysis of the algorithm indicates that it could run in doubly exponential time, due to the potential growth of the integer coefficients. While worst-case bounds for the reduction algorithm are poor, the algorithm is straightforward to implement and in practice it often runs very efficiently for a geometric object (such as a protein) that is embedded in three-dimensional Euclidean space.

In 1861, Smith [34] gave what is now considered the classical "reduction" algorithm showing that an integer matrix can be diagonalized using elementary row and column operations (section 50.5.3). It was noted in the 1970s that neither the classical algorithm nor several well-known subsequent algorithms for reduction were known to be polynomial (see [23] for a review of these observations). Kannan and Bachem then gave a polynomial-time algorithm for reduction to Smith normal form. This bound was subsequently improved: Chou and Collins [8] provided an $O(s^{11})$ algorithm, where for an $n \times m$ matrix with maximum coefficient $\wp$, we define $s = m + n + \log |\wp|$. Two papers [20, 21] give fast worst-case bounds for computing Smith normal form, and hence the canonical structure of Abelian groups. The worst-case bounds given by Iliopoulos are roughly $O(s^5)$. The algorithms of Chou and Collins and Iliopoulos [20, 21, 8] are different from the classical algorithm. Either of these results could be used to compute the homology type in (deterministic) times $O(Dn^{11})$ or $O(Dn^5)$.

Exercise Compare and contrast the concepts of mathematical versus biological homology. How well does mathematical homology (e.g., integral homology groups) capture the notion of protein structural homology? Explain your answer, and give concrete examples from the PDB.

50.6 Alpha Shapes (α-Shapes) and Applications to Protein Structure

Herbert Edelsbrunner and colleagues [12] developed an efficient and useful version of homology group computation and applied it to structural biology. In this section, we will use the term *face* to describe both 2-simplices and 1-simplices. First, they restricted their attention to three-dimensional simplicial complexes, and to the part of the homology groups that do not have torsion (the so-called *Betti numbers,* discussed in section 50.4). They developed an efficient algorithm to compute Betti numbers for a triangulation. Then they imagined using an eraser to eliminate faces in the triangulation. The eraser would have a particular width, and so there would be some vertices that would be too close together for the eraser to fit between them, or to erase the face between them. Given an eraser with a particular width, they would thus remove some faces and then recompute the Betti numbers. Using a sophisticated computational geometry algorithm, they could then amortize the computation and sweep the eraser widths to vary from large to small (0), thereby calculating a measure of the topology as parameterized by eraser size (α) and face removal. Hence, the Betti numbers are parameterized by a continuous scalar value α, and where the faces of the triangulation generating the homology groups are removed according to this width (size) parameter. The topology of the triangulation, and the Betti numbers, only change at a finite set of α-values, which are called the *critical values.* The essence of the algorithm is to compute these critical values, and the topological changes (in Betti numbers) that passing through them triggers. This provides a nuanced representation of the topological structure of the triangulation. The approach, called α-*shapes,* has been a useful tool in analyzing protein structure.

The size and shape of macromolecules such as proteins and nucleic acids play an important role in their functions. In molecular modeling, these properties are usually quantified based on various discretization or tessellation procedures involving analytical or numerical computations. Liang et al. [26] presents an analytically exact method for computing the metric properties of macromolecules based on the alpha (α) shape theory. This method uses the duality between the alpha complex and the weighted Voronoi decomposition of a molecule. The authors applied the method to compute areas and volumes of a number of protein systems.

Although the structures of most proteins are well-packed, they contain numerous cavities that play key roles in accommodating small molecules, or enabling conformational changes. From high-resolution structures, it is possible to identify these cavities. Liang et al. [27] developed a precise algorithm based on alpha shapes for measuring space-filling–based molecular models (such as van der Waals, solvent accessible, and molecular surface descriptions). They applied this method to compute the surface area and volume of cavities in several proteins. In addition, all of the atoms/residues lining the cavities were identified. They used this method to study the structure and the stability of several proteins, as well as to locate cavities that could contain structural water molecules in the proton transport pathway in the membrane protein bacteriorhodopsin.

Identification and size characterization of surface pockets and occluded cavities are often initial steps in protein structure–based ligand design. An algorithm called CAST, for automatically locating and measuring protein pockets and cavities, was developed based on alpha shapes and discrete flow theory [28]. CAST identifies and measures the atoms lining pockets, pocket openings, and buried cavities; the volume and area of pockets and cavities; and the area and circumference of mouth openings. CAST analysis of over 100 proteins was carried out; these proteins included a set of 51 monomeric enzyme-ligand structures, several elastase-inhibitor complexes, the FK506 binding protein, 30 HIV-1 protease-inhibitor complexes, and a number of small and large protein inhibitors.

Medium-sized globular proteins typically were found to have 10–20 pockets/cavities. Most often, binding sites were pockets with 1–2 mouth openings; much less frequently, they were cavities. Ligand-binding pockets varied widely in size, most within the range 10^2–10^3 Å^3. Statistical analysis revealed that the number of pockets and cavities was correlated with protein size, but there was no correlation between the size of the protein and the size of binding sites. Most frequently, the largest pocket/cavity was the active site (but there are a number of interesting exceptions [28]).

Ligand volume and binding site volume were somewhat correlated when binding site volume is ≤ 700 Å^3, but the ligand seldom occupies the entire site [28]. Auxiliary pockets near the active site have been suggested as additional ("allosteric") binding surfaces for designed ligands.[3] Analysis of elastase-inhibitor complexes suggests that CAST can identify ancillary pockets suitable for recruitment in ligand design strategies. Analysis of the FK506 binding protein, and of compounds developed in SAR by NMR,[4] indicated that CAST pocket computation may provide a priori identification of target proteins for linked-fragment design. CAST analysis of 30 HIV-1 protease-inhibitor complexes showed that the flexible active site pocket can vary over a range of roughly 850 to 1,500 Å^3, and that there are two pockets near or adjoining the active site that may be recruited for ligand design.

50.7 Conclusions and Future Work

We have described an algorithm for computing all the homology groups of a triangulation. The algorithm can be generalized to compute cohomology groups. Cohomology group computation is one of the essential steps in computing the higher homotopy groups of a simply connected triangulation. A great deal of work remains to be done. As is well known, the cohomology groups of a space are completely determined by its homology. However, they come equipped with a natural "cohomology operation" that gives these groups a ring structure. Spaces with

[3] For example, see [30] and [18].

[4] See chapter 15 and [38].

identical homology and cohomology groups can have different ring structure, hence, the ring structure distinguishes "more finely" between spaces. Deciding ring isomorphism is a central question here. Fast algorithmic solutions to these problems will prove a challenge to researchers in computational algebraic topology.

This chapter provided an introduction to some aspects of computational topology and protein structure. Other work abounds. For example, by using topological techniques from knot theory and line weavings, Michael Erdmann has analyzed protein shape similarity in terms of structure-preserving transformations [13]. His paper defines similarity based on atomic motions that preserve local backbone topology without incurring significant distance errors. Such geometric algorithms for computational topology represent a parallel branch of development to those exploiting integral homology groups. Moreover, Erdmann's approach anticipates, as it were, protein design algorithms that consider bounded backbone flexibility (either continuous voxels [15] or discrete "backrubs" [16]) while still maintaining guarantees to compute the GMEC and enumerate a gap-free list of conformations in order of energy (see chapter 12 and [15, 16]). Such algorithms can be viewed as searching a conformation/mutation space subject to a topological constraint.

References

[1] D. Anick. Computing rational homotopy groups is #$\mathcal{P}$-hard, in *Computers in Geometry and Topology*, Lecture Notes in Pure and Applied Mathematics. M. Tangora (ed.). New York: Marcel Dekker, 1989.

[2] D. Arnon, G. Collins, and S. McCallum. *Cylindrical Algebraic Decomposition I: The Basic Algorithm,* CDS TR-427, Department Computer Science Purdue University, 1982.

[3] M. Ben-Or, D. Kozen, and J. Reif, The complexity of elementary algebra and geometry, *Journal of Computer and System Sciences* 32 (1986): 251–264.

[4] W. Boone, W. Haken, and V. Poenaru. On recursively unsolvable problems in topology and their classification, in *Contributions to Mathematical Logic*, H. Schmidt et al. (eds.). North Holland, 1968, pp. 37–74.

[5] E. H. Brown. Finite computability of Postnikov complexes., *Annals of Mathematics*, 65 (1957): 1–20.

[6] J. F. Canny. A New algebraic method for robot motion planning and real geometry, *FOCS* (1987). Foundations of Computer Science (FOCS) 1987

[7] J. F. Canny, and B. R. Donald. Simplified Voronoi diagrams *Discrete and Computational Geometry* 3;3 (1990): 219–236.

[8] T. J. Chou, and G. E. Collins. Algorithms for the solution of linear Diophantine equations, *SIAM Journal on Computing* 11 (1982): 687–708.

[9] G. E. Collins. Quantifier Elimination for Real Closed Fields by Cylindrical Algebraic Decomposition. *Lecture Notes in Computer Science*, No. 33. New York: Springer-Verlag, 1975, pp. 135–183.

[10] H. Crapo. Applications of geometric homology in *Geometry and Robotics*. Lecture Notes in Computer Science 391, I.D. Boissonnat and I.P. Laumond (eds.). New York: Springer-Verlag 1989.

[11] P. D. Domich. *Residual Methods for Computing Hermite and Smith Normal Forms* PhD thesis, Cornell University, 1985.

[12] H. Edelsbrunner and E. P. Mucke. Three-dimensional alpha shapes., *ACM Transactions on Graphics* 13 (1994): 43–72.

[13] M. A. Erdmann. Protein similarity from knot theory: Geometric convolution and line weavings. *Journal of Computational Biology* 12; 6 (2005): 609–637.

[14] H. Hironaka. Triangulations of algebraic sets. *Proceedings of the Symposia in Pure Mathematics American Mathematical Society* 29 (1975): 165–185.

[15] I. Georgiev and B. R. Donald. Dead-end elimination with backbone flexibility. *Bioinformatics* 23; 13 (2007). Special issue on papers from the International Conference on Intelligent Systems for Molecular Biology (ISMB 2007), Vienna, Austria: July 21–25, 2007.

[16] I. Georgiev, D. Keedy, J. Richardson, D. Richardson, and B. R. Donald. Algorithm for backrub motions in protein design. *Bioinformatics* 4;13 (2008):i196–i204. Special issue on papers from International Conference on Intelligent Systems for Molecular Biology (ISMB 2008) Toronto,: July 2008.

[17] N. M. Glazunov. Algorithms for calculations of cohomology groups of finite groups, in Computations in Algebra and Combinatorial Analyis, Kiev: *Akadem. Nauka Ukrainsk. SSR Instituta Kibernetika*, (Math. Reviews 82f 20080), 1978.

[18] M. J. Gorczynski, J. Grembecka, Y. Zhou, Y. Kong, L. Roudaia, M.G. Douvas, M. Newman, et al. Allosteric inhibition of the protein-protein interaction between the leukemia-associated proteins Runx1 and CBFb. *Chemistry & Biology* 14; 10 (2007):1186–1197.

[19] D. F. Holt. The mechanical computation of first and second cohomology groups, *Journal of Symbolic Computation* 1, (1985):351–361.

[20] C. S. Iliopoulos. Worst-case complexity bounds on algorithms for computing the canonical structure of finite Abelian groups and Hermite and Smith normal form of an integer matrix, *SIAM Journal on Computing*, 18;4, (1989):658–669.

[21] C. S. Iliopoulos. Worst-case complexity bounds on algorithms for computing the canonical structure of infinite Abelian groups and solving systems of linear diophantine equations, *SIAM Journal on Computing* 18;4, (1989):670–678.

[22] N. Jacobson. *Basic Algebra, I.* New York Freeman, 1985.

[23] R. Kannan and A. Bachem. Polynomial algorithms for computing the Smith and Hermite normal forms of an integer matrix. *SIAM Journal on Computing.* 8 (1979): 499–507.

[24] D. Kozen and C. K. Yap. Algebraic cell decomposition in NC. *Proceedings of the 26th Annual Symposium on Foundations of Computer Science* 515–521. (1985 Oct.).

[25] L. Lambe. *Algorithms for Computing the Cohomology of Nilpotetnt Groups,* in *Computers in Geometry and Topology.* Lecture Notes in Pure and Applied Mathematics. M. Tangora (ed.). New York: Marcel Dekker, 1989.

[26] J. Liang, H. Edelsbrunner, P. Fu, P. V. Sudhakar, and S. Subramaniam. *Analytical shape computation of macro-molecules: I. Molecular area and volume through alpha shape, Proteins* 33;1: (1998 oct. 1) 1–17.

[27] J. Liang, H. Edelsbrunner, P. Fu, P. V. Sudhakar, and S. Subramaniam. Analytical shape computation of macro-molecules: II. Inaccessible cavities in proteins, *Proteins* 33;1:18–29. (1998 Oct 1).

[28] J. Liang, H. Edelsbrunner, and C. Woodward. Anatomy of protein pockets and cavities: Measurement of binding site geometry and implications for ligand design. *Protein Science* 7;9: (1998 sep.) 1884–1897.

[29] A. A. Markov. Nerazreshimost' Problemy Gomeomorfiy, *Proceedings of the International Congress of Mathematicians*. J.A. Todd (ed.). Cambridge University Press, 1958, pp. 300–306.

[30] C. Mattos, B. Rasmussen, X. Ding, G.A. Petsko, and D. Ringe. Analogous inhibitors of elastase do not always bind analogously. *Nature Structural Biology* 1 (1994):55–58.

[31] S. McCallum. *Constructive Triangulation of Real Curves and Surfaces.* Sidney: University of Sidney, 1979.

[32] W. S. Massey. *Algebraic Topology: An Introduction*, New York: Springer-Verlag, 1967.

[33] Munkres. *Elements of Algebraic Topology*, Menlo-Park, CA: Addison-Wesley; 1984.

[34] H.J.S. Smith. On systems of indeterminate equations and congruences. *Philosophical Transactions* 151 (1861): 293–326.

[35] E. H. Spanier. *Algebraic Topology.* New York: Springer-Verlag, 1966.

[36] H. Seifert, and W. Threlfall. *Lehrbuch der Topologie.* New York: Chelsea, 1947, chap. 7.

[37] J. Schwartz, and M. Sharir. On the 'piano movers' problem, II. General techniques for computing topological properties of real algebraic manifolds, in *Planning, Geometry and Complexity of Robot Motion.* J. Schwartz, J. Hopcroft, and M. Sharir, (eds.). Norwood, NJ: Ablex. 1987, chap. 5, pp. 154–186.

[38] S. B. Shuker, P. J. Hajduk, R. P. Meadows, and S. W. Fesik. Discovering high-affinity ligands for proteins: SAR by NMR. *Science* 274;5292(1996 Nov 29):1531–1534.

[39] J. Snoeyink. A trivial knot whose spanning disks have exponential size, *Proceedings of ACM Symposium on Computational Geometry*, Berkeley, CA (1990).

[40] G. Vegter and C. K. Yap. Computational complexity of combinatorial surfaces, *Proceedings of ACM Symposium on Computational Geometry*, Berkeley, CA (1990).

[41] J. Warren. *The Effect of Base Points on Rational Bezier Surfaces*, Department of Computer Science, Rice University (1990).

[42] H. Whitney. Elementary structure of real algebraic varieties, *Annals of Mathematics*. 66;3 (1957): 545–556.

[43] C. Yap. *Algorithmic Motion Planning Advances in Robotics:* Vol. 1 J. Schwartz and C. Yap (eds.) Lawrence Erlbaum Associates, 1986.

Index

●●● Computational Molecular Biology
Sorin Istrail, Pavel Pevzner, and Michael Waterman, editors

Computational molecular biology is a new discipline, bringing together computational, statistical, experimental, and technological methods, which is energizing and dramatically accelerating the discovery of new technologies and tools for molecular biology. The MIT Press Series on Computational Molecular Biology is intended to provide a unique and effective venue for the rapid publication of monographs, textbooks, edited collections, reference works, and lecture notes of the highest quality.

Computational Molecular Biology: An Algorithmic Approach
Pavel A. Pevzner, 2000

Computational Methods for Modeling Biochemical Networks
James M. Bower and Hamid Bolouri, editors, 2001

Current Topics in Computational Molecular Biology
Tao Jiang, Ying Xu, and Michael Q. Zhang, editors, 2002

Gene Regulation and Metabolism: Postgenomic Computation Approaches
Julio Collado-Vides, editor, 2002

Microarrays for an Integrative Genomics
Isaac S. Kohane, Alvin Kho, and Atul J. Butte, 2002

Kernel Methods in Computational Biology
Bernhard Schölkopf, Koji Tsuda, and Jean-Philippe Vert, editors, 2004

An Introduction to Bioinformatics Algorithms
Neil C. Jones and Pavel A. Pevzner, 2004

Immunological Bioinformatics
Ole Lund, Morten Nielsen, Claus Lundegaard, Can Keşmir, and Søren Brunak, 2005

Ontologies for Bioinformatics
Kenneth Baclawski and Tianhua Niu, 2005

Printed in the United States
by Baker & Taylor Publisher Services